A C S   S Y M P O S I U M   S E R I E S **562**

Isopentenoids and Other Natural Products

Evolution and Function

W. David Nes, EDITOR

Texas Tech University

Developed from a symposium sponsored
by the Division of Agricultural and Food Chemistry
at the 205th National Meeting
of the American Chemical Society,
Denver, Colorado,
March 28–April 2, 1993

American Chemical Society, Washington, DC 1994

Library of Congress Cataloging-in-Publication Data

Isopentenoids and other natural products: evolution and function /
W. David Nes, editor

p.   cm.—(ACS symposium series, ISSN 0097–6156; 562)

Includes bibliographical references and indexes.

ISBN 0–8412–2934–1

1. Isopentenoids—Evolution. 2. Natural products—Evolution.

I. Nes, W. David. II. Series.

QP752.T47I86   1994
574.19′243—dc20   94–25742
                                                    CIP

1994 Advisory Board

Foreword

THE ACS SYMPOSIUM SERIES was first published in 1974 to provide a mechanism for publishing symposia quickly in book form. The purpose of this series is to publish comprehensive books developed from symposia, which are usually "snapshots in time" of the current research being done on a topic, plus some review material on the topic. For this reason, it is necessary that the papers be published as quickly as possible.

Before a symposium-based book is put under contract, the proposed table of contents is reviewed for appropriateness to the topic and for comprehensiveness of the collection. Some papers are excluded at this point, and others are added to round out the scope of the volume. In addition, a draft of each paper is peer-reviewed prior to final acceptance or rejection. This anonymous review process is supervised by the organizer(s) of the symposium, who become the editor(s) of the book. The authors then revise their papers according to the recommendations of both the reviewers and the editors, prepare camera-ready copy, and submit the final papers to the editors, who check that all necessary revisions have been made.

As a rule, only original research papers and original review papers are included in the volumes. Verbatim reproductions of previously published papers are not accepted.

M. Joan Comstock
Series Editor

Contents

Preface

Evolution, SIMPLY PUT, IS ADAPTIVE CHANGE through time. The idea that different organisms can change morphologically one into another is not new. Philosophers have noted these occurrences in the natural world and speculated on the process of change for thousands of years. Much debate has occurred, particularly in this century, on which forces promote change; whether species are unchanging, special creations; and whether life evolved ultimately from nonlife, that is, biochemical evolution from abiotic sources.

Charles Darwin and A. I. Oparin were the first to address the problem of evolution in book form: In 1859 Darwin wrote the book *On the Origin of Species by Means of Natural Selection*, and in 1924 Oparin wrote the book *The Origin of Life*. Many scientists today accept the Darwin–Oparin paradigm as a working hypothesis for the origin and further evolution of life. However, because of the revolution that has taken place in the past 10 years in structural chemistry and molecular biology, "facts" regarding lines of descent and other aspects of the Darwin–Oparin model of evolution have been reexamined. Interestingly, the modern concept of evolution proposed by biologists integrates the contributions of many fields of knowledge, including chemistry and biochemistry; this concept is often referred to as the "synthetic theory of organic evolution".

The study of biochemical evolution by the examination of natural product chemistry is a relatively recent development. Classically, the natural product chemist isolated compounds from plants and other sources, determined their structures, then compared their distribution in nature. Phylogenetic trees showed the differential operation of biochemical pathways based on the similarities and differences in the occurrence of compounds—this is chemotaxonomy. The modern approach to studying evolution of natural products is to consider multiple aspects of the molecules' chemistry, biosynthesis, and biology, notwithstanding chemotaxonomy, and also to establish the function of the end products, to determine their distribution in tissues (during ontogenetic changes) or in sediments and rocks (molecular fossils), to characterize the structure and molecular biology of enzymes and their respective genes, and to establish the transformation process in mechanistic detail.

This book will be useful for specialists involved in isopentenoid and related natural product biochemistry. Particular attention has been given to isopentenoids and sterols because of their importance in biology. Furthermore, possibly several aspects of evolution of a biochemical pathway can be examined in one forum, for example, from gene to enzyme to enzyme product, including its structure, distribution, and function. Other chapters are concerned with lipids, lignans, and membranes.

Acknowledgments

I thank the authors of the chapters of this book for their commitment to meeting the necessary deadlines and for their efforts to address a subject matter (evolution) that is peripheral to their major research focus. Thanks also to Rhonda Bitterli, Meg Marshall, and the rest of the ACS Books staff for their assistance during the editing and production of this volume.

W. DAVID NES
Department of Chemistry and Biochemistry
Texas Tech University
Lubbock, TX 79409–1061

May 2, 1994

Evolution of Isopentenoids

Chapter 1

Isopentenoids and Geochemistry

Simon C. Brassell

Biogeochemical Laboratories, Department of Geological Sciences, Indiana
University, Bloomington, IN 47405–5101

Natural products survive beyond the demise of their source organism,
although their fate depends on biological and geological factors, and
their chemistry. Sediments accumulate organic debris from numerous
sources, including diagnostic compounds whose structural specificity
permits recognition and decipherment of their origins and degradation
pathways. The complex compositions of sediments include components
that either:
   (i)   originate directly from organisms,
   (ii)   are transformed by microbial activity, or
   (iii) are altered by geological processes, especially thermal effects
          accompanying burial.
Compounds without biochemical precedent may derive from hitherto
unknown or defunct biosynthetic pathways, or be geological
transformation products. The geological record of isopentenoids reflects
biochemical evolution. The diverse range, including steroids and
hopenoids, in sediments billions of years old demonstrates the antiquity
of their biosynthetic pathways and other occurrences coincide with the
appearance of their putative source organisms.

The evolution of Life on Earth can be explored in several distinct, yet inter-related ways.
Historically, evolutionary changes in biota were primarily deduced from recognition of
progressive changes in fossil morphology. Lineages inferred from such paleontological
investigations could be further complemented by studies of the physiology of extant
organisms viewed as descendants of fossilized ancestors. In combination these lines of
evidence enabled reconstruction of detailed taxonomic relationships (1, 2). In addition,
biochemical evidence has been included in assessment of phylogeny, and, in general,
has supported relationships based on morphology. Most recently such chemical studies
have addressed the genetic affinities of organisms indicated by their RNA or DNA
sequences (3, 4), an approach that has also tended to substantiate established
phylogenies. Attempts have now been made to extend the application of these
techniques into the geological past by examination of ancient fossilized tissues (5, 6),
but such efforts are fraught with complications and remain controversial. Exploration of
the evolution of natural products requires knowledge of the occurrence of components
within organisms and the evolutionary sequence of biological development.

0097–6156/94/0562–0002$09.62/0

The principal approach adopted seeks to derive phylogenetic relationships by combining the genetic composition of extant organisms with paleontological and geological evidence of their evolution (7). An alternative approach involves investigation of the occurrences of natural products in the geological record. Specifically, it aims to identify the components present in rocks and petroleums of different ages and attempts to relate their occurrences with evolutionary changes in the organisms that constitute the biological sources of organic matter (8). It probably represents the most direct practical means to assess evolutionary changes in natural products over geological time, although it is beset with concerns centered on the survival of biosynthetic compounds. These issues form the focus of this paper.

Geochemical Evidence Contained in Sedimentary Organic Matter

Evaluation of the nature of geochemical evidence for the evolution of natural products first requires assessment of the various factors that determine limitations on the chemical nature and information content of sedimentary organic matter (9). These issues bear on a broader range of concerns, including:

(i)   How extensive and comprehensive is the rock record?
(ii) What are the sources and sinks of organic matter in contemporary environments?
(iii) What conditions favor and influence the preservation of organic matter in the sedimentary record?
(iv) To what extent is sedimentary organic matter representative of the range of its biological sources?
(v)   How does the onslaught of bacterial processes in surface sediments influence the composition of sedimentary organic matter?
(vi) What structural and composition information survives and persists during sediment burial and thermal alteration?

Answers to the above questions provide the necessary background information to determine whether evidence of evolutionary changes is preserved in the geological record of isopentenoids.

Survival of the Rock Record. The preservation of the rock record and its constituent organic matter over geological time exerts a critical control on efforts to elucidate geochemical evidence for evolutionary changes in the biosynthesis of isopentenoids. Specifically, various geological processes, principally consisting of burial, metamorphism, uplift, weathering and erosion, effect the consolidation of sediments as sedimentary rocks, their subsequent transformation by heat and pressure, their deformation under stress by fracturing and folding, and their mechanical and chemical degradation and breakdown to particles that can be redeposited as sediments. All of these operations are part of the rock cycle which continuously shapes and changes the sediment record over geological time, encompassing both the formation of new deposits and the alteration, recycling and ultimate destruction of ancient sediments. As a direct result of the cumulative effects of these influences, the worldwide prevalence of sedimentary rocks within a given stratigraphic time unit tends to be inversely proportional to age. Thus, occurrences of the oldest rocks are more scant that those formed during more recent epochs and examination of the earliest record is hindered by the paucity of surviving strata from the most ancient eons.

Gaps in the Geological Record. A second factor affecting the completeness of geochemical evidence for evolutionary processes is the fragmentary nature of the geological record. Sedimentary sequences inevitably contain gaps termed hiatuses which include periods of non-deposition, or non-continuous deposition, and episodes of erosion. All of these represent time intervals for which the rock record provides no vestige of any events that took place during these specific periods. Such hiatuses are found in virtually all sedimentary deposits, though their prevalence varies. Despite such constraints, efforts to derive a comprehensive history of changes occurring through geological time are still feasible if constructed in a piecemeal fashion from the combination of records collated from series of fragmentary sedimentary sequences. One encouraging consideration is that the entire documentation of evolution based on paleontology relies on interpretation of fossil remains preserved in sediments, despite the incomplete nature of individual stratigraphic sections and other limitations, such as the scarcity of evidence for earliest life forms (10). In principle, however, it should be possible to explore chemical evidence of evolutionary changes in parallel with the morphological fossil record, although the conditions that favor preservation of organic materials differ from those that lead to the survival of morphological remains, primarily skeletal material and other hard body parts. Hence, an important aspect of the exploration of organic records is the need to identify depositional settings and associated environmental conditions, and lithologic variations that are conducive to the preservation of organic debris.

Mineral, Fossil and Chemical Composition of Sediments. Sediments are primarily composed of a mineral matrix variously constituted by combinations of clastic detritus (e.g. sand, silt and clay particles containing silicate minerals), chemical precipitates (e.g. carbonates, gypsum), biogenic skeletal debris that may be composed of calcareous (e.g. coccolithophorids, foraminifera), and siliceous (e.g. diatoms, silicoflagellates, radiolaria) materials, augmented by primary (e.g. phosphorites, barite) and secondary minerals (e.g. pyrite). Co-occurring with this material are biogenic particles that are principally composed of organic matter (e.g. dinoflagellate cysts, fecal pellets). Open marine sediments tend to accumulate detritus from pelagic sources that are typically a combination of biogenic and clastic components originating from the overlying water column coupled with advected material. Hence, it is to be expected that sediments should contain organic compounds in markedly lower abundance than living cells because of the dilution effects inherent in the process of sediment formation and the lability of organic matter, when compared to carbonate and silicate minerals, which arises from its biological degradation.

Major Biological Processes within the Geochemical Carbon Cycle. The principal process effecting carbon fixation is the photosynthetic uptake of CO_2 by phytoplankton and terrestrial vegetation. A diverse range and biological specific series of enzymatic reactions then serve to build, modify and transform the initial biosynthetic building blocks into the wide variety or organic compounds found within the biosphere. Subsequently, this transient reservoir of carbon in biota is extensively degraded by further biological, physical and chemical processes so that the vast majority of primary production is recycled and does not enter the sedimentary domain. Thus, the annual flow of carbon from the atmosphere or hydrosphere through the biosphere to the lithosphere (ca. 2×10^{14} gCa-1; 11) represents a comparatively minor proportion of that contained in living systems.

Sedimentary Accumulation of Carbon. The amount of carbon accumulated in sediments (ca. 6×10^{17} gC; 11) greatly exceeds the present day inventory of carbon in the biosphere (ca. 5×10^{21} gC; 11). Inevitably the comparative size of these carbon reservoirs reflects the longevity of geological time whereby even the small flux of carbon that escapes remineralization and recycling to become incorporated in sediments ultimately amounts to the largest accumulation of carbon on earth. The sequestering of carbon through geological time, however, has not been constant; there are specific periods in the past that are associated with enhanced organic carbon accumulation in petroleum source rocks or coals (12). Thus, the sedimentary reservoir of carbon has fluctuated over time driven, in part, by evolutionary developments. For example, the diversification of pteridophytes and gymnosperms in the Carboniferous (*ca.* 360-290 Ma) appears to precede a major era of coal formation (*12*). The precise causes of developments or perturbations in carbon production and burial are the subject of much active research, not least because of their direct links with global carbon climate. In particular, global carbon budgets are potentially induced by events, such as increases in the frequency or magnitude of volcanic eruptions or the steadfast growth in anthropogenic combustion of fossil fuels, that cause changes in atmospheric CO_2 levels and may lead to global warming.

Efforts to evaluate ancient records prompts an exploration of the factors that influence the sources of organic matter and their alteration during and after deposition (*9*). Toward this objective, studies of contemporary depositional settings can provide evidence of the principal influences on the production and preservation of organic matter in the rock record. Most importantly, a detailed perspective of contemporary processes permits consideration of the varied controls that determine the survival of both total organic matter and specific organic compounds like isopentenoids.

The Abundance and Nature of Sedimentary Organic Matter. The proportion of organic carbon in sediments tends to be small, typically totaling <1% of sediment dry weight. It can be significantly greater in regions of high productivity, where preservation is enhanced by anoxia, when clastic inputs are minor, or where deposition in the deep sea occurs below the lysocline (or calcite compensation depth, between 2,000 - 4,000m) leading to dissolution of carbonate minerals. Under exceptional circumstances the organic carbon contents of sediments can exceed 30%, and is even higher in most coals. Part of this organic matter can be extracted with solvents, but the majority is composed of solvent-insoluble material called kerogen. Until recently it was thought that kerogen was largely formed by the varied and complex chemical interactions of mixtures of amino acids, carbohydrates and lipids, principally involving condensation and polymerization reactions (*13*) akin to those leading to melanoidin formation (*14*). It now appears that these reactions are only of minor importance; the major process in kerogen formation involves the selective presentation of biopolymeric materials (*15, 16*). This reappraisal of the nature of kerogen has important implications for the preservation of molecular information given that insoluble organic matter can be presumed to contain intact fragments of cell membranes. Such material is therefore likely to include the biosynthetic components that comprise membranes, including isopentenoid lipids.

The organic matter in contemporary sediments is highly varied in its chemistry and its biological origin, deriving from a multitude of sources (Figure 1). In open marine environments organic matter is primarily autochthonous in character and is largely composed of pelagic and benthic material of algal and bacterial origin (*17*). In near shore, deltaic and lacustrine systems this is typically augmented, diluted or largely replaced by significant allochthonous contributions from vascular plants (*18*); eolian processes may also deposit materials derived from terrigenous sources (*19*).

Biological Sources of Sedimentary Organic Matter. Identification of the biological sources of aquatic components in the sedimentary record is facilitated by reference to results from examination of the composition of living phytoplankton, including coccolithophorids, diatoms, dinoflagellates (*20*). Similarly, investigation of the organic constituents of bacteria and terrigenous plants provides evidence of their chemical signals and signature compounds (*21, 22*). To benefit geochemical investigations such studies need ideally to be focused on the collection and analysis of natural populations, or laboratory cultures, of species that represent major contributions to sediments. This condition is rarely met, but cumulative results from examination of the chemical composition of contemporary organisms do provide sufficient evidence of

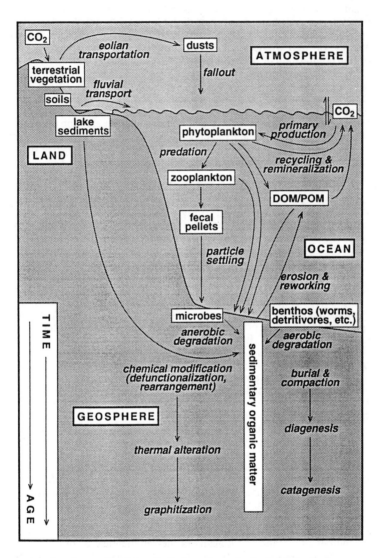

Figure 1. A schematic view depicting the principal biological sources and reservoirs of organic matter, the processes that govern its movement within the environment and the factors that influence its fate in the geosphere.

the occurrences and prevalence of individual compounds that numerous source assignments are possible. However, significant limitations inherent in this approach remain. For example, it is not remotely possible to examine even a tiny proportion of the multitudinous array of extant species, let alone their variations in chemical composition that may arise from fluctuations in temperature and salinity, and from changes in their nutrient and light supply. Furthermore, in this paper, source-specific compounds are applied in the interpretation of the evolutionary record, attempting to account for changes in the principal sources of organic matter that have occurred over geological time. Yet it is not possible to determine the potential contributions to ancient sediments that may have originated from organisms that are now extinct (23).

Temporal Changes. During the Lower Paleozoic (*ca.* 550-400 Ma), none of the classes of phytoplankton that tend to dominate the contemporary oceans, principally diatoms, coccolithophorids and dinoflagellates, had evolved. Thus, the characteristics of planktonic organic matter were inevitably different from those of the present day during earlier periods in the geological past, and can be expected to have left their mark in the sedimentary record. For example, *Gloeocapsomorpha prisca* is a prominent fossil constituent of unknown biological affinity, perhaps algal or bacterial, with a distinct chemical signature, that appears only in Ordovician (*ca.* 510-440 Ma) marine shales (24, 25). In general, Paleozoic black shales provide a testament of an oceanic regime wherein depositional conditions favored widespread burial of marine organic matter (26) markedly different in character from that which typifies the extensive deposits of Cretaceous black shales (*ca.* 145-65 Ma) formed during the so called oceanic anoxic events (27), and also different from that in contemporary settings. Similarly, contributions of organic matter to sediments from land plants were non-existent prior to their evolution in the Silurian (*ca.* 440-410 Ma), and subsequent diversification in the Carboniferous (28). Thus, the character of organic matter from terrestrial sources has undergone significant changes over geological time as, in turn, pteridophytes, gymnosperms and angiosperms have evolved and proliferated.

A further concern is the absence of direct evidence regarding the chemical uniformity of a specific plant group over time. There is no means to assess the composition of extinct species other than by comparison with their closest living relative. Similarly it is impossible to determine whether there have been changes in the composition or constituents of an organism prior to the evolution of its specific contemporary form. The presumption is made that evolutionary changes in the chemistry of a single species are minimal.

Other factors that may potentially influence the nature of the global history of carbon include climatic controls its production and burial. For example, do sedimentary records of organic matter indicate a reduction in land biota, especially at higher latitudes, during major glaciations? Such considerations serve to illustrate the absence of temporal and spatial consistency in the nature of the biological sources of sedimentary organic matter, and, thus, in its chemical composition through geological history. Inevitably, the conclusion that organic matter varies with time stems from the fact that it ultimately derives from biota that have changed dramatically as evolution proceeded.

Preservation of Sedimentary Organic Matter

Any search for evolutionary trends in natural products is constrained by the paucity and incompleteness of the ancient rock record, coupled with the fact that sediments contain a biased molecular record. All of the process affecting the preservation of organic matter serve to favor the survival of certain components more than others. Specifically, it is in the oxygen deficient environment of sediments that the best prospects for molecular preservation occur, and components derived from aquatic organisms are more likely to survive the process of settling to reach the sediments than those from terrestrial sources that require significant, and sometimes extensive, transportation. Similarly, environmental constraints including nutrient supply, biological populations, bathymetry and oxicity levels, all serve to regulate the production of individual compounds within a specific depositional setting, and determine their potential to survive. In essence, these factors affect the survival of individual molecules in the same way that they influence the amount of organic matter preserved in sediments determining both the quality and quantity of organic carbon that survives in sediments (9, 29). The major types of influences which merit detailed and separate consideration are therefore:

  (i)   factors internal to the organic matter, specifically the structural and chemical characteristics of individual molecules,
  (ii) variations in external physicochemical controls, notably environmental conditions that determine the survival of organic matter,
  (iii) the action of biological agents, notably microbial alteration,
  (iv) thermal effects encountered as sediments become increasingly deeply buried.

Molecular Controls. Undoubtedly, the survival of individual natural products within sedimentary environments is largely dependent on their structural characteristics which determine their propensity to biological modification and their resistance to physical or chemical transformation. The heterogeneous character of organic matter means that its constituents, monomeric, polymeric or macromolecular, differ in their stability to these degradative processes (15). The recognition that these are discrepancies in the geochemical behavior of the biosynthetic components of organisms is not surprising given that their principal constituent chemical parts, namely lipids, proteins and carbohydrates, are fundamentally different in their chemistry, are typically located in separate units within the cell, and play disparate functional roles within organisms.

Inevitably, it is the more refractory compounds that escape degradation whereas labile constituents succumb to alteration (15). For example, long-chain aliphatic hydrocarbons, especially in alkanes, derived from terrestrial plant waxes can survive extensive eolian transportation (19). By contrast, other components, notably polyunsaturated carboxylic acids (30) and highly functionalized carotenoids (31, 32) are less frequently encountered in sediments because they are partially or completely lost to oxidation and remineralization within the water column. Simple alkanes, alkenes and mono functional lipids (e.g. alkanols, alkanoic acids) are more robust molecules from the standpoint of geological stability than polyfunctional compounds, such as sugars, and tend therefore to be the analytical targets of most geochemical investigations (33). However, less stable functionalities can be preserved within macromolecular biopolymers (15).

Sulfur Incorporation. It is also possible for unsaturated and polyunsaturated lipids to survive in sediments by intra- or inter-molecular incorporation of sulfur (15, 34). The wide variety of organo-sulfur compounds (OSC) recognized in sediments in recent years (34, 35) demonstrates the viability and prevalence of such reactions. The structures of OSC and the inferred sites of sulfur incorporation demonstrate that this process preserves molecular characteristics that would otherwise be expected to be lost by degradation. For example, specific OSC derive from and retain the structural features of carotenoids and other polyunsaturated isopentenoids (34, 35).

Environmental Controls. The preservation of organic matter is affected by environmental conditions in both the water column and the underlying bottom sediment. In oxic settings, extensive reworking, recycling and remineralization of organic matter can occur, leading to a substantial quantitative loss within the water column (Figure 1). A markedly greater proportion of organic matter tends to survive in anoxic or dysaerobic systems given the diminished possibility for oxidative reworking (36). However, preservation of organic matter is also a function of productivity in the environment because its oxidation is dependent on the availability of sufficient oxygen (*37*). Seasonal or episodic productivity cycles and their associated phytoplankton blooms, or the change in flux of organic matter in nutrient-rich waters (e.g. upwelling areas) that descends through the water column can lead to oxygen depletion at intermediate water depths and, hence, enhanced quantitative preservation of organic matter in underlying sediments.

Biological Controls. Sedimentary environments are not sterile regimes. In particular, the faunal and microbial communities that populate both depositional environments and their overlying water columns can radically modify the molecular composition of sedimentary or sedimenting organic matter in several distinct ways (*9*). They may effect changes through extensive degradation and/or subtle modification of components already present in the organic detritus (*38*), and they may augment the molecular record by adding significant contributions of their own constituent organic compounds (*39*). Thus, the primary organic matter principally formed by photosynthesis is supplemented by predators, consumers and secondary producers. Initially, the feeding habits and digestive processes of predators, notably zooplankton, can lead to selective losses of constituents biosynthesized by primary producers (*40*). For example, sterols with multiple nuclear double bonds (e.g. $\Delta^{5,7}$-steradienols) are more readily assimilated by zooplankton predation than their less unsaturated analogues (e.g. Δ^5-sterol); whereas 4-methylstanols tend not to be assimilated (*41*). Subsequently, organic matter is subject to degradation by benthic fauna with organisms like filter feeders and detritivores active predominantly at the sediment/water interface and burrowers like worms effective at shallow sediment depths (*42*). This aerobic activity is accompanied and eventually succeeded by a hierarchical sequence of microbial degradation of organic matter governed by various redox reactions in the subsurface. Thus, oxidative degradation by fauna and microbes in surficial sediments is typically underlain by successive depth zones of nitrate reduction, sulfate reduction and methanogenesis (*43*). Within each zone sedimentary organic matter is sequentially altered by the different microbial processes so that its character continues to change with depth.

Exposure and Accessibility of Organic Matter to Microbes. The rate of sediment accumulation or burial can be critical to the survival of organic matter. In environments where deposition is rapid sedimented material can pass quickly through the various zones of benthic and microbial degradation consequently reducing its exposure to such alteration processes. The biological or sedimentary packaging of organic matter also influences its accessibility to microbial attack and therefore its survival. For example, compounds derived from phytoplankton that are incorporated into zooplankton fecal pellets may be protected from alteration during rapid settling through the water column and after deposition if the pellet remains intact (*44*). Similarly, organic constituents that are strongly absorbed on clay particles may not be accessible to microbial alteration. Hence, sediment lithology may also influence the survival of organic matter.

Thermal Controls. Sediment deposition and consolidation does not, however, lead to a cessation of chemical activity. The zone of microbial alteration of sedimentary organic matter is followed by series of systematic and sequential chemical transformations as thermal processes become the dominant controls on the molecular composition of sediments. Initially these processes consist of a diverse range of reactions that involve defunctionalization (e.g. decarboxylation, dehydration, dehydrogenation) structural rearrangements, stereochemical isomerizations, and both oxidations and reductions (*33, 45*). These transformations occur within the shallow, cooler zone of sediment burial given the collective term diagenesis. Subsequently, the thermal breakdown of organic matter, primarily consisting of cracking reactions that first involve carbon-sulfur and later carbon-carbon bond cleavages, come to dominate within the deeper catagenesis zone. It is these reactions that ultimately lead to a loss in the molecular information content of thermally altered sediments and severely limit their value or preclude their use as measures of ancient biota.

Cumulative Effects on Sedimentary Organic Matter. The sum total of sedimentary processes serves therefore to transform the organic matter of recently deposited sediments, wherein its chemical characteristics closely resemble those of living systems, into a more complex mixture in ancient mature sediments which possesses distinct molecular features, dominated by aliphatic and aromatic hydrocarbons and insoluble macromolecules. The primary task of the geochemist attempting to explore evolutionary records in such materials is therefore to first unravel the identities and quantities of individual components within the sedimentary organic matter, and seek chemical evidence that can help determine its original, biological form.

Analytical Approaches in Isopentenoid Geochemistry

The principal task is to decipher the complex mixture of molecules that comprise sedimentary organic matter. In general, this aim focuses on the extractable components, but a variety of procedures including pyrolysis and chemical degradation permit assessment of compounds released from the insoluble kerogen (*15*).

GC-MS. One method is critical to analysis of the molecular constituents of sediments and petroleums, namely gas chromatography-mass spectrometry (GC-MS; *46*, *47*). The capability of such instruments to characterize components within complex mixtures represents a necessary prerequisite for examining sedimentary organic matter and has led to GC-MS becoming the principal tool for molecular geochemical determinations. The use of GC-MS circumvents the major problem that target compounds are rarely present in sufficient quantities for fully diagnostic spectroscopic methods to be employed. In addition, the chemical similarity of co-occurring sedimentary components present in varied abundance greatly complicates efforts at separation and purification of individual compounds. Instead, sediment extracts tend to be fractionated to yield mixtures of components which possess similar chemical functionality (e.g. alkanes, or carboxylic acids, or ketones; *46*).

Compound Identification. Assignments of components within these mixtures by GC-MS are typically based on a combination of mass spectra, GC retention times and/or mass chromatography of diagnostic molecular or fragment ions. The difficulty of distinguishing the mass spectral characteristics and retention times of closely allied components, such as stereoisomers, creates the need to perform GC-MS co-injections with authentic standards. The cumulative efforts of years of synthesis of such standards has helped create an extensive repertoire of confirmed assignments and characteristics for numerous compounds, especially isopentenoids, which now permit identification of many components from GC-MS data to be made with confidence. These efforts have been important for the geochemical community because reference libraries contain comparatively few standard mass spectra for components that are important constituents of sediments. More recently, the use of GC-MS/MS has further improved the ability to determine sedimentary distributions of organic components because of the enhanced selectivity of this technique achieved by monitoring series of parent to daughter ion relationships (*46*, *48*). All of these analytical approaches also permit compound quantification based on comparison of the intensity of a diagnostic mass spectral response with that of internal or external standards. Thus, these instrumental methods enable recognition of the identify of geologically occurring compounds and determination of their abundance, which, in turn, permit assessment of their biological source(s) and serve as the principal information basis for elucidation of evolutionary trends.

Determination of Carbon Isotope Compositions. A further recent development in analytical instrumentation, namely gas chromatography-isotope ratio mass spectrometry (GC-IRMS or CSIA — compound specific isotope analysis) enables determination of the carbon isotopic signatures of individual molecules within sedimentary complex mixtures (49, 50). Previously, only the $\delta^{13}C$ value of the total organic matter of a sediment could be measured. This new GC-IRMS technique has proven uniquely valuable in assessment of the biological origins of specific compounds and whether they derive from single or multiple sources. The basis for this approach lies in the discrepancies among the $\delta^{13}C$ values of components synthesized by organisms that utilize different sources of carbon (e.g. CO_2 or CH_4) and employ discrete pathways for carbon fixation. For example, hopanoids (Figure 2) derived from methylotrophs are highly depleted in ^{13}C (49), wheras aryl isoprenoids (Figure 2) originating from photosynthetic sulfur bacteria are enriched in ^{13}C (51).

Sedimentary Records of Isopentenoids

Occurrences and Abundances in Geological Materials. In general, isopentenoids are present in comparatively low abundances in geological materials, though they are among the most prevalent molecular forms within sedimentary organic matter. The concentrations of individual compounds within extractable organic matter typically lie in the ppb range. For some components their amounts in this bitumen are augmented by further contributions liberated from macromolecular compounds within the kerogen during the thermal alteration of sediments. In some instances it appears that the insoluble organic matter of kerogens represent the major source of specific components. For example, treatment of kerogens with boron halides selectively cleaves ether bonds and has been found to yield biphytane (52; Figure 2), a ether-linked constituent of archaebacterial membranes. However, the absence of other compounds, notably 28,30-dinorhopane (Figure 3), in pyrolysates indicates that they are not present in kerogen (53). There is therefore no established or uniform relationship between the abundance of sedimentary compounds in extractable and inextractable form. This observation is not unexpected given the diverse influences of chemical variability, biological function and other factors that control their presence within these two operationally defined carbon reservoirs.

Most major classes of isopentenoids have been reported in sediments (54). They exist with various functionalities, occurring as saturated, unsaturated and aromatic hydrocarbons, carboxylic acids, alcohols, ketones, aldehydes and as polyfunctional components. Many are components directly biosynthesized by organisms, whereas others are formed by microbial or geological processes in the sediment subsurface.

Source-Specific Isopentenoids. A significant proportion of the individual isopentenoid components found in geological materials can be attributed to contributions from discrete biological sources. Such assignments are typically limited to those components that have a restricted biological occurrence, which may be at any taxonomic level — species, genus, family, order, class, phylum or kingdom. For example, sterols appear restricted to the eukaryote kingdom (Figure 2; 55). They have been reported in prokaryotes, but none has been proven to synthesize sterols *de novo*, except the methylotroph *Methylococcus capsulatus* which contains 4-methyl and 4,4-dimethyl, but not 4-desmethyl, sterols (56). Other structural groups found in geological materials can also be related to contributions from specific biota (Figure 2) and, in turn, both reflect

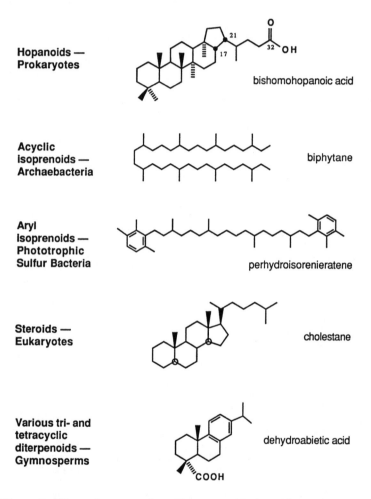

**Hopanoids —
Prokaryotes**

bishomohopanoic acid

**Acyclic
isoprenoids —
Archaebacteria**

biphytane

**Aryl
Isoprenoids —
Phototrophic
Sulfur Bacteria**

perhydroisorenieratene

**Steroids —
Eukaryotes**

cholestane

**Various tri- and
tetracyclic
diterpenoids —
Gymnosperms**

dehydroabietic acid

Figure 2. Illustrative examples of isopentenoids found in sediments that can be directly related to specific biological sources.

the sources of organic matter and enable their assessment. Indeed, the origins of diagnostic components are sufficiently diverse that sediment inputs from several of the major groups of algae (e.g. diatoms, coccolithophorids, dinoflagellates), various bacteria (e.g. archaebacteria, purple sulfur bacteria, cyanobacteria), and major divisions of vascular plants (e.g. gymnosperms, angiosperms) can all be represented (*54*). Such source-specific compounds therefore provide an indication of contributions of organic matter from particular biota and when quantified, also offer a measure of the comparative importance of their various biological sources (*57-59*). In addition, determination of the carbon isotopic composition of individual molecules can aid assessment of their biological origin (*49, 50*) and help recognize when a given component derives from multiple sources (*60*). This approach relies on the dependence of carbon isotopic signatures on carbon source and biosynthetic pathways. Thus, the isopentenoids synthesized by organisms that utilize different sources of carbon, for example CO_2 fixation by photoautotrophs versus CH_4 uptake by methylotrophs (*49, 50*; Table I), and those that use the Krebs pathway rather than the Calvin cycle (Table I) can be distinguished. Comparison of the carbon isotopic composition of individual molecules with the $\delta^{13}C$ of bulk organic matter can also provide a quantitative measure of the importance of different biological sources of carbon.

Inevitably the molecular and isotopic signals preserved in sedimentary organic matter are biased. They tend to emphasize or overestimate contributions from those organisms that biosynthesize refractory compounds which survive better in the geological record. Thus, assessment of the relative importance of different biota based on quantification of their diagnostic constituents (*58*) in sediments may not accurately

Table I. Selected geolipids, their carbon isotope values and inferred biological sources

Compound Structure and Name	$\delta^{13}C$ (‰)	Biological Source	Ref.
phytane	-31.8	photosynthetic algae	50
perhydroisorenieratene	-10.7	photosynthetic sulfur bacteria	61
biphytane	-25.5	archaebacteria	62
28,30-dinorhopane	-32.3	unknown	62
17β(H),21β(H)-30-norhopane	-65.3	methylotrophic bacteria	49,50

reflect their comparative importance as contributors of total biomass. A more comprehensive approach requires a broader data set. For example, comparison of molecular indicators of biological contributions with other geochemical measures, such as bulk carbon isotopes. In contemporary settings it is appropriate to seek comparison with direct biological measurements, such as cell counts, whereas for ancient sediments paleontological evidence, especially fossil assemblages, provides an independent measure of the distributions and populations of organisms within the environment. It is also important, however, to recognize that the records of biological populations separately based on geochemical and paleontological evidence should be regarded as complementary, not identical. Indeed, it may be possible to interpret apparent discrepancies between such records as measures of biological activity at different trophic levels. Specifically, molecular indicators can provide evidence of the legacy of bacterial activity that is not readily assessed by paleontology (63).

Biologically Unprecedented Isopentenoids.   Many organic constituents of sediments have not been reported in living organisms. Such components reflect both the incompleteness of the chemical inventory of organic compounds in biota and the diverse range of sedimentary transformation pathways. Efforts to understand and interpret the origins of these components demand an appreciation of biosynthetic pathways and the range of potential reactions that could occur in the sediment subsurface generating unprecedented compounds from known biosynthetic precursors. Consideration of these factors can help distinguish components that are formed as products of geological transformations from those that are deemed to be produced biosynthetically either by organisms that have not been studied, or by pathways that are, as yet, unknown. The concept of biologically unprecedented (or ophan) components in the geological record is best illustrated by specific examples (Figure 3; 9, 63, 64).

Unprecedented Steroids Formed by Diagenesis?   Various biosynthetic modifications of steroidal structures are well documented, but the geological occurrences of steryl ethers (63, 65, 66), A-norsteroids (67), and 3β-methylsteroids (68) cannot be directly related to known biological sources, nor can they be readily attributed to specific diagenetic transformation pathways. Cholesteryl ethers with a C_{16} alkyl group have been reported in bovine cardiac muscle (69), but this biological source provides no viable explanation for the occurrence of compounds with shorter alkyl chains in contemporary sediments. Furthermore, the diagenetic formation of steryl ethers from sterol precursors seems somewhat improbable because the carbon number range of their side chains (C_8-C_{10}; 65, 66) is incompatible with the major alkyl groups found in sediments. The most probable origin for steryl ethers in sediments is that they derive from an as yet unidentified biological source. The apparent absence of A-norsteroids (Figure 3) in contemporary environments, coupled with their occurrence in older, diagenetically altered sediments (67) suggests that they are probably transformation products formed from sterol precursors. The precise reaction pathways that lead to their formation and their specific precursors, however, cannot be determined from the rather limited current information on their sedimentary occurrences.

Isopentenoid Ophans.   The principal biological sources of sedimentary hopanoids are prokaryotes (70) and it seems probable that 28,30-dinorhopane derives from this kingdom, although no obvious precursor has been identified in living biota which can adequately account for its geological occurrences (62, 71, 72) and its absence in kerogen pyrolysates (53). The carbon isotope compositions of 28,30-dinorhopane isomers are significantly depleted in ^{13}C (62) indicating that they, or their precursor(s), are synthesized by organism(s) utilizing a ^{13}C-depleted carbon source. The $\delta^{13}C$ values of 28,30-dinorhopanes also differ from those of co-occurring hopanes (62) proving that not all hopanoids originate from the same biological source, as previously shown by the

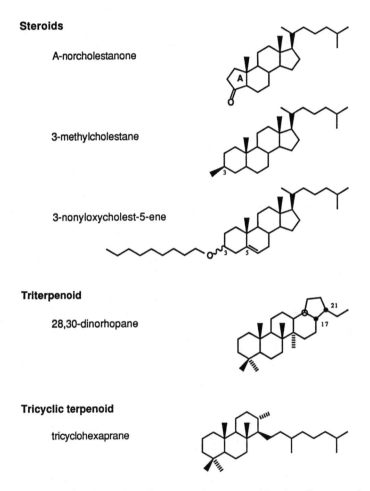

Steroids

   A-norcholestanone

   3-methylcholestane

   3-nonyloxycholest-5-ene

Triterpenoid

   28,30-dinorhopane

Tricyclic terpenoid

   tricyclohexaprane

Figure 3. A selection of sedimentary isopentenoids that have no known biological precurors, although their structures attest to an origin from biolipids.

range in their $\delta^{13}C$ values in other sediments (*49, 50*). The precise origin of 28,30-dinorhopane remains enigmatic and awaits further evidence of its biological source, or diagenetic precursors and mode of formation. The presence of tricyclohexapane in sediments (*54*) provides supportive evidence for the existence of tricyclohexaprenol in organisms (*73*). This postulated isopentenoid alcohol is held to be the precursor of the tricyclic terpanes that occur widely in ancient sediments and petroleums (*74, 75*), and is thought to act as a surrogate for sterols in prokaryotic membranes (*73*).

Biological or Geological Origins? The existence of many organic compounds in sediments whose precise origins cannot be assessed solely on the basis of their occurrences in sediments and/or petroleums serves to emphasize the necessity of furthering our understanding of potential biological or geological sources of these components. There is a need to integrate and compare data accumulated from evaluation of the biosynthetic products of organisms that are environmentally significant with results obtained from elucidation of transformation reactions operative in the sediment subsurface and the conditions that favor them. A coupling of these objectives offers promise in further clarifying the origin of molecular constituents in sediments and petroleums, as it has in the past.

Biolipids and Geolipids.

The term "geolipid" is used to denote sedimentary compounds that are the products of geological transformations of biolipids. The basis for this terminology is the implication that geolipids are derived from biolipid precursors (*33, 55*), a presumption typically based on structural similarities between the two components.

Botryococcene and Botryococcane. Perhaps the best example of unequivocal structural evidence for an inferred link between a highly specific biolipid and a geolipid counterpart is that of botryococcene and botryococcane (Figure 4). The peculiar linkage of isopentenoid units in botryococcene and its unusual sequential pattern of methylation provides uniquely diagnostic structures (*76, 77*) that can be readily related to the saturated equivalent hydrocarbons found in sediments and petroleums (*78-80*). Remarkably, botryococcane is a prominent aliphatic hydrocarbon in both of the giant oil fields of Sumatra (*78*; Duri and Minas; 2 and 7 BBO, i.e. billion barrels of oil, respectively) where its total accumulation amounts to *ca.* 10^{13} g of this single molecule derived from *Botryococcus braunii*.

β-Carotene and β-Carotane. A second clear link between biolipid and geolipid based on structural comparability is that of β-carotene and β-carotane (Figure 4; *33*). The presence of β-carotane among the products from desulfurization of sedimentary OSC (*34, 35*) suggests that the hydrogenation of β-carotene may proceed via sulfur incorporation (*12*). β-carotane is another geolipid which occurs in vast geological accumulations. It is a prominent aliphatic hydrocarbon component of the Green River oil shale formation (*81*) that in places is over 600 m thick and extends over 65,000 km^2 in Colorado, Utah and Wyoming (*82*). Furthermore, β-carotane is the single most abundant alkane in reservoired petroleums from the Karamayi oil field (2-3 BBO) within the Junggar Basin in northwest China (*83, 84*). The combined amount of β-carotane in these two locations probably exceeds *ca.* 10^{13} g, and such vast deposits may be attributed to contributions from organisms like *Dunaliella salina* that produce high concentrations of β-carotene in saline environments populated by restricted biological assemblages.

Confirmation of Biolipid to Geolipid Transformations. The examples of botryococcane and β-carotane serve to illustrate the value of preserved structures as

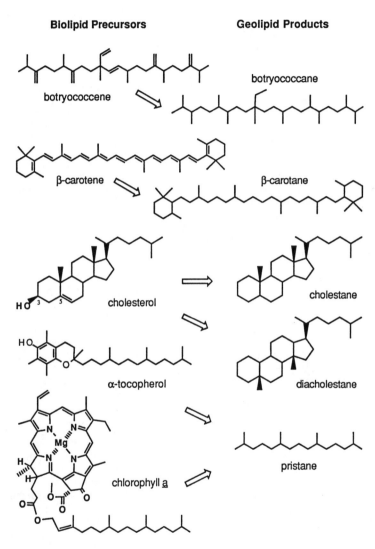

Figure 4. Inferred relationships between a number of biolipid precurors and their geolipid products.

clues to the biological origins of sedimentary constituents. As described above these inferred relationships between biolipids and geolipids are proposed principally on the basis of their structural comparability, but such assessments can also be supported by either:

(i)   the geological occurrences of precursor and product (and potential intermediates), and/or

(ii) laboratory simulation experiments that can help determine biolipid transformation pathways.

Geological Evidence. The best evidence to support inferred precursor/product relationships is obtained from investigation of compound distributions through a sequence of samples from a uniform stratigraphic section. In such instances it is expected that the downcore abundance of the biolipid should gradually decrease while that of the geolipid increases, consistent with the idea of a gradual conversion of the former to the latter. However, when the comparative amounts of the components are expressed as a ratio of product to precursor, as is often the case, it is also possible that the apparent transformation actually may reflect preferential degradation of the precursor rather than chemical interconversion (46, 85). Hence, further conformation, or rejection, of such proposed transformation reactions can be provided by absolute quantitation of compound abundances (46, 85).

Experimental Evidence. The second approach entails demonstration of the possibility of chemical interconversion of biolipid to geolipid in the laboratory (46). Preferably, the reagents (e.g. chemicals and mineral substrates) and the time/temperature of the simulation experiment should attempt to emulate conditions encountered during sediment compaction and diagenesis to ensure geological realism. No laboratory simulation, however, can replicate the enormity of geological time. Thus, the experimental temperature typically must be increased to compensate for a shorter time scale preferably without changing the nature of the reaction by favoring transformations with kinetics that are less viable geologically.

Laboratory simulations with cholesterol (55, 86, 87) have demonstrated that it can be transformed by a number of different, geologically feasible, reaction pathways, giving rise to a variety of geolipid products that can co-occur. Thus, cholestane and diacholestane (Figure 4), which exist in numerous isomeric forms in sediments (46, 55), can both be interpreted as geolipid products derived from the biolipid precursor cholesterol. Other biolipids are held, similarly, to be precursors of multiple products with different structures, and these geolipids may also exist as a broad range of stereochemical isomers. Hence, a diversity exists in geolipids, especially in terms of their stereochemical configurations, that exceeds the variation found in biolipids. The stability or ease of formation of some geolipids, however, is sufficient that they are products that can be derived from multiple precursors. One example is the isoprenoid hydrocarbon pristane which is formed primarily from chlorophylls and/or tocopherols (Figure 4). The difference in the chemical character of these two biolipid precursors for pristane, when coupled with evidence from pyrolysis studies (88, 89), suggests that their alteration to yield pristane may occur at different stages of diagenesis. Specifically, pristane formation from chlorophylls begins during the earliest phases of sedimentation (90), whereas the breakdown of tocopherols represents a later-stage, thermally driven process (89). A given geolipid may therefore not only derive from multiple precursors, but the proportional contributions from each of its separate origins may be dependent on the extent of diagenetic alteration and change as diagenesis proceeds.

Geochemical Interconversion of Structural Types. Interpretation of the biological origins of sedimentary lipid constituents and assessment of their transformation pathways in the subsurface presumes that individual isopentenoid structures are retained during the process of diagenesis and catagenesis even though their functionalities may be lost or modified. In those instances where structural rearrangement is known to occur (e.g. conversion of sterenes or spirosterenes to diasterenes; *45, 55, 87, 91*) the products of the reactions are still directly and uniquely related to their precursors and could not otherwise have been formed.

Taraxerene to Oleanene Conversion. Recent studies have shown that one component undergoes a rearrangement during diagenesis that transforms it into a different structural class. Specifically, the alkene, Δ^{14}-taraxerene, which is thought to derive from the triterpenoid taraxerol, is converted into olean-12-ene (*92*), which can itself be formed diagenetically from olean-12-en-3β-ol (β-amyrin; Figure 5) and is an intermediate in oleanane formation. Thus, the inability to differentiate between the two potential precursors of olean-12-ene also applies to its alkane equivalent, which is widely employed as a marker for higher plant, specifically angiosperm, contributions to sediments (*54, 63, 92*). The presence of oleanane in an ancient sediment or petroleum cannot therefore now be presumed to indicate that its original environment received contributions of oleanane triterpenoids, such as β-amyrin. Fortunately, the reported biological occurrences of taraxerol appear generally similar to those of oleanane precursors, and taraxerol and β-amyrin commonly co-occur (*92, 93*). The potential impact of the discovery of structural interconversion on the use of oleanane as a source indicator is therefore perhaps limited, although this presumption is not fully tested.

Structural Interconversions: General Phenomena? Recognition of the taraxerene to oleanene interconversion (*92*) illustrates the possibility of such reactions, but it seems unlikely that they represent widespread phenomena in the diagenetic modifications of biolipids, primarily because of the particular chemical requirements necessary for such transformation reactions. Specifically, the taraxerene/oleanene interconversion entails a 1,2-methyl shift, a reaction that may be favored geologically by a difference in the comparative stability of the two compounds, and probably mediated by clay catalysis. There are several other comparable diagenetic reactions, including clay-catalyzed formation of diasterenes and spirosterenes (*55, 87, 91*), but perhaps the closest parallel is the transformation of hop-17(21)-ene into neohop-13(18)-ene (*94*), which also involves a single 1,2-methyl shift. It is therefore necessary to recognize that structural interconversion may not be restricted to taraxerene/oleanene and perhaps invoke this process as a possible explanation for other geological occurrences. For example, the apparent disappearance with increasing diagenesis of certain triterpenoids that are abundant in shallow sediments (e.g. fernenes; *95*) prompts concern that they may be similarly transformed into another lipid class. However, it is presumptive and speculative to infer any other significant structural interconversion without further supportive evidence.

Evolutionary Trends in the Geological Record

The exploration for evolutionary changes in the molecular constituents of sediments rests on the principal assumption that evolutionary developments in life on Earth should be matched by biochemical changes that can be expected to survive in the geological record. This presumption is predicated on a number of other assumptions which need to be considered, and undoubtedly complicate the assessment and interpretation of the molecular records contained in sediments. However, it can be forcefully argued that such potential constraints do not invalidate the approach.

Potential Constraints on Geological Records of Biochemical Evolution. First, it is necessary to assume that the process of structural interconversion is not a general phenomena, but is restricted to a narrow range of related triterpenoids. Thus,

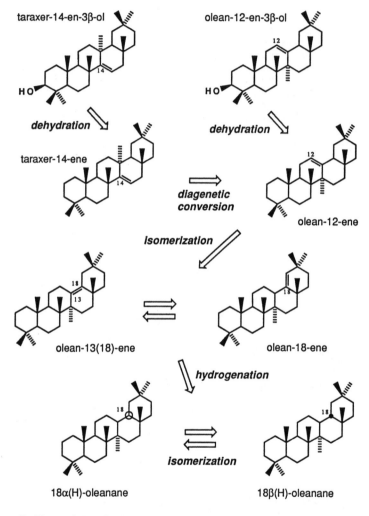

taraxer-14-en-3β-ol

olean-12-en-3β-ol

dehydration

dehydration

taraxer-14-ene

*diagenetic
conversion*

olean-12-ene

isomerization

olean-13(18)-ene

olean-18-ene

hydrogenation

isomerization

18α(H)-oleanane

18β(H)-oleanane

Figure 5. The pathway for interconversion of taraxerene to oleanene (*92*) and its subsequent diagenetic conversion to oleananes.

the integrity of structural preservation implicit in assessment of biolipid/geolipid relationships is not compromised. Second, it is assumed that the first appearance of a given geolipid structure can be related to the evolution of the biosynthetic pathway that produces its biolipid precursor. Inevitably, the evolutionary advent of a given compound may be revised by subsequent discoveries that extend its lineage back in time. Third, the disappearance and persistent absence of a compound can be presumed to correspond to the extinction of its biological source, excluding the absence of diagenetically formed geolipids in contemporary and immature sediments because they have not yet been generated from biolipid precursors. At present, however, no well documented compounds meet these criteria, an observation that points to a conservatism in biosynthetic pathways. Fourth, it is presumed that there is no biochemical re-invention. Thus, the recognition of substantial gaps in the geological record is held to be unimportant, reflecting the lack of analyses of suitable samples. Specifically, hiatuses in the sediment record may preclude assessment of the particular time period, or the known geological samples may not be representative of the depositional and diagenetic conditions under which the specific compound occurs, or is formed. The only major temporal gap that falls into this category is the intermittant sedimentary occurrences of dinosterane stereoisomers (4,23,24-trimethylcholestanes; *96, 97*). These compounds have been found in Proterozoic sediments (*64*), but not in any other strata before the Late Permian (*ca*. 250 Ma; *97*). Their presence in post-Permian sediments is generally attributed to inputs from dinoflagellates, whereas the earlier occurrences may perhaps reflect contributions from acritachs, thought to be possible evolutionary ancestors of dinoflagellates. In this instance, as with smaller gaps that span a less dramatic time interval, the apparent absence of specific marker compounds may be attribable to a number of factors, such as problems inherent in sample selection, in the survival of components during diagenesis, and in the uniformity and prevalence of their occurrence.

Controls on Molecular Records. The principal controls that restrict the comprehensive character of the sedimentary record for a given compound, include:
  (i)   the paucity of the sediment record where some time intervals are better represented than others,
  (ii) a lack of analyses of samples deposited under appropriate environmental conditions,
  (iii) changes or limitations in the environmental niche of their precursor caused by competitive evolutionary pressures,
  (iv) modification of their degradation pathways perhaps arising from evolutionary developments in predators, and
  (v)   the abundance of the compound within its source biota and the importance of those organisms as contributors of sedimentary biomass.
These types of controls on compound occurrences are entirely plausible given the the biological specificity of many isopentenoids, the limited biological assemblages found in many environmental settings, and the low tolerance to environmental variations (e.g. salinity, oxicity) of organisms within certain ecological niches. In addition, the evolution of more advanced or competitive species has ousted some organisms from their original depositional settings where they were widespread, restricting them to a narrower range of environments. The molecular record attributable to biota affected by such changes would be expected to show a reduction in spatial distribution, and therefore in their global abundance, which, in turn, makes their recognition more difficult.

Geological History of Isopentenoids. A compilation of some of the earliest occurrences of individual isopentenoids (Figure 6) illustrates a number of features that merit specific comment. It is evident that major classes of isopentenoids evolved

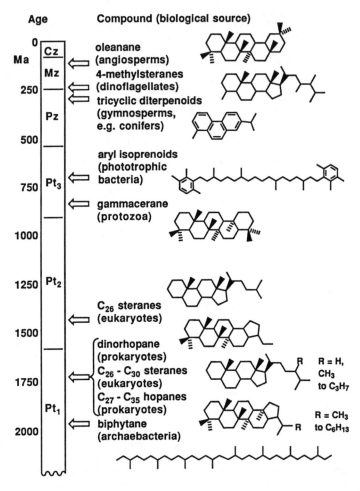

Figure 6. A compilation of the first reported occurrences of selected biolipids in sediments. Abbreviations of geological eras are as follows: Cz = Cenozoic, Mz = Mesozoic, Pz = Paleozoic, Pt$_1$, Pt$_2$, and Pt$_3$, = Proterozoic.

comparatively early in the history of life on Earth. Specifically, hopanoids and steroids, together with acyclic isoprenoid hydrocarbons like pristane and phytane, occur in the oldest organic-rich sediments that contain significant amounts of extractable compounds from the McArthur Basin in Australia (*64, 98*). Much of the structural diversity in acyclic isoprenoids, steroids and hopanoids found throughout the geological record is apparent in these Early Proterozoic (Pt$_1$) sediments (*ca.* 1.7 Ga). All older organic-rich rocks are either graphitic, or preserve no extractable compounds that are unequivocally indigenous. However, selective chemical treatments of kerogen from one Early Proterozoic rock provided an assessment of molecular structures retained among the insoluble organic matter. Specifically, treatment with ether-cleavage reagents (boron trihalides) and subsequent hydrogenation (LiAlH$_4$) released biphytane, a component of archaebacterial lipids, from the kerogen of the Gunflint shale (*ca.* 1.9 Ga; *99*).

The Unsubstantiated Archean Record. None of the accounts of compounds in the most ancient strata has withstood critical examination. Literature reports, principally publications from the 1960s (*100, 101*), of extractable components in Archean sedimentary rocks (4 -2.5 Ga) cannot now be substantiated. The compounds alleged to be derived from the earliest life forms are now viewed as mobile hydrocarbons that migrated from younger strata long after the original sediment was consolidated (*102*), or are attributed to extraneous contaminants added accidentally during sample handling and work-up. The latter explanation is mindful of the fact that the reports of components in Archean rocks predate the introduction of vigorous analytical standards of cleanliness (e.g. in solvent purity) that have been deemed necessary in subsequent investigations to preclude sample contamination by pervasive hydrocarbons. Regretfully, the lack of viable samples from the Archean limits extension of biochemical characteristics to this earliest eon of geological time. However, the carbon isotopic of sedimentary organic matter within the Archean does record signals that attest to the advent of an oxygenated atmosphere (*103*).

Links between Isopentenoid Occurrences and Biological Evolution. Examination of the appearances of individual compounds reveals several connections between their molecular occurrences and the evolution of their presumed biological sources (*104*). The occurrence of acyclic isoprenoids, hopanoids and steroids in the early Proterozoic is consistent with the antiquity of their biological sources, namely Archaebacteria, prokaryotes and eukaryotes, respectively. Perhaps the earliest occurrence of a diagnostic compound that can be confidently linked to a specific group of organisms is gammacerane. This component first appears in Late Proterozoic (Pt$_3$) sediments that are thought to contain some of the earliest examples of fossil protozoa, the inferred biological source of this hydrocarbon (*105*). Aryl isoprenoids, such as isorenieratene (Figure 6), have been reported in Paleozoic sediments and Late Proterozoic oils (*51, 64*). These occurrences are consistent with the interpretation that they derive from phototrophic sulfur bacteria, although the phylogenetic lineage of these ancient biota significantly predates their first geochemical occurrence of aryl isoprenoids. Tricyclic (and tetracyclic) diterpenoids such as retene (Figure 6), formed from cyclization of phytenyl precursors, are marker compounds for sediment contributions from gymnosperms (*54, 106*). Hence, it is not unexpected to find that they are present in Permian sediments, significantly pre-dating the appearance of pentacyclic triterpenoids, such as oleanane, that derive from angiosperms. Oleanane first occurs as a prominent constituent of sediments in the Cretaceous (Figure 6), seemingly coincident with the major diversification of angiosperms from which it derives (*54, 104*). Similarly, the presence of 4,23,24-trimethylsteranes as major components in Permian sediments (*96, 97*) corresponds with a period of diversification in their presumed source organisms, dinoflagellates, which may have been accompanied by an enhancement in their importance as contributors of sedimentary biomass.

However, as described above, these compounds have also been reported in Proterozoic sediments (*64*), where they are attributed to a different biological source, possibly acritachs.

Overall, this small selection of compounds whose occurrences can be well defined temporally provides an illustration of the extent to which molecular analyses of sediments and petroleums can provide evidence of evolutionary developments in isopentenoid biosynthesis.

Summary

The capability of molecular indicators to provide evidence of the evolution of biosynthetic processes can be reviewed in terms of several challenges that need to be undertaken to further develop and improve the future prospects of this approach.

Biological Challenges. A major task is better documentation of the distributions and abundances of isopentenoids in biota from all trophic levels (e.g. primary producers, predators, consumers, and secondary producers) to further evaluate the breadth of the potential biological sources for given compounds. There is a particular need to continue and expand the study of diagnostic components in the most critical biological groups, namely those organisms that are thought to represent the principal contributors of sedimentary biomass. In addition, it is necessary to determine the importance of external controls on the production and abundance of these components under various environmental conditions, including changes in nutrient supply, temperature, light levels, CO_2 concentrations, and other influential variables. Sustained activity in these areas continues to lead to an accumulation of information on the compositions of living organisms which, in turn, aids recognition of the potential biological sources for isopentenoids in the geosphere. Periodically, new analyses lead to a revision in the source assignments of sedimentary molecules, as in the case of fernenes (*95*).

Chemical Challenges. The need for accurate identification and verification of compound occurrences in geological materials is paramount. However, it is only through synthesis of target geolipid compounds that most of the necessary standards are obtained that permit confirmation, substantiation, or rejection of tentative assignments solely based on spectrometric and chromatographic data. The synthesis of standards to verify the structural and stereochemical characteristics of geolipids provides a viable alternative for recognition of the constituents of sediments and/or petroleums that are present in insufficient quantities to enable isolation and rigorous spectroscopic assessment. Also, the process of compound elucidation represents a continuous endeavor; as the inventory of geolipids expands, so too does the need for synthesis of authentic standards.

Transformation Reactions. Another requirement is recognition of sedimentary transformation processes, especially those that influence the character of geolipid products formed from biolipid precursors. Further assessment and confirmation of precursor/product relationship requires examination of compound trends in sedimentary sequences and laboratory simulation experiments best coupled with molecular mechanics calculations that assess the comparative stability of the products (*107*). It is also important to explore the relationships between soluble (bitumen) and insoluble (kerogen) organic matter as the transformation pathways of organic compounds in sediments often involves the transfer of components between these two operationally defined parts of sedimentary organic matter. Thus, evaluation of degradation reactions, whether physical, chemical or thermal, and whether individual or sequential, provides additional evidence that describes possible locations for geolipid molecules (and/or their functionalities) within kerogen and the nature of their bonding. In recent years the physical technique of thermal breakdown using pyrolysis methods has been complemented by the development and application of selective and specific techniques for chemical degradations. The success of these latter approaches, such as ether-cleavage, hydrogenolysis, and Raney nickel desulfurization, has helped to provide the impetus for these research efforts.

Geological Challenges. Apart from the continued quest for sedimentary sequences that offer the chance for progress in the recognition of evolutionary trends for isopentenoids, three other principal targets for future focussed research can be proposed.

Temporal Records of Isopentenoid Occurrences. First, it seems both appropriate and worthwhile to continue the search for evidence that differences in molecular compositions reflect evolutionary changes over time intervals on a geological scale. In part, this reflects the recognition that the geochemical information is unique in itself, although it is also important that it is used to complement evidence from other approaches. It also bears on the comparative lack of detailed studies across major stratigraphic intervals, such as the Permian/Triassic (245 Ma) or Cretaceous/Tertiary (65 Ma) extinction boundaries, where dramatic changes in life occurred. Unfortunately, past efforts to tackle such critical time intervals have been largely thwarted by the lack of well preserved sediment sections that are sufficiently rich in organic matter and adequately span the boundary without a major change in lithology or depositional environment.

Geological Records of Isopentenoid Abundances. Second, it is timely to exploit not only the presence or absence of components as indicators of evolutionary developments, but also to determine whether their relative (or even absolute) abundances have the potential to provide evidence of evolutionary change. This raises the question whether it is possible to interpret a gradual stratigraphic decrease or increase in the abundance of a source-specific compound as a measure of either a reduction, or a growth, respectively, in the prevalence of its source organism(s). The promise of this approach has been demonstrated by one recent study which showed that the comparative abundances of 24-propyl and 24-isopropyl steranes varies through geological time (*108*). The markedly higher proportion of the latter steranes during the Late Proterozoic and Cambrian eras (*ca.* 900 - 510 Ma) may perhaps reflect changes in the dominance of porifera (sponges) as contributors of sedimentary organic matter (*108*). This example shows that the relative proportions of individual geolipids can indeed possess the potential to denote specific period of geological time. Depletion of the $\delta^{13}C$ values for organic carbon from Cretaceous sediments (*109*) further demonstrates that temporal signals can be represented by the carbon isotopes of bulk organic matter. Therefore, it seems probable that enriched or depleted $\delta^{13}C$ values for a given compound may be indicative of particular stratigraphic intervals. In addition, time-dependent differences in the $\delta^{13}C$ of components of marine organisms can also reflect variations in oceanic pCO_2 given that the uptake of carbon isotopes during carbon fixation is influenced by CO_2 concentration (*50, 110*). All of these possibilities illustrate the potential value of any systematic stratigraphic variation in molecular or carbon isotopic characteristics, especially if such measures are found to be diagnostic to particular time periods. By comparison, the presence of oleanane can be used to indicate that the organic matter of a sediment or derived petroleum is of Late Cretaceous to Tertiary age, but cannot provide further discrimination within this substantial time span (>90 Ma).

Records of Climatic Change and Isopentenoid Distributions. Third, a final consideration pertinent to examination, evaluation and elucidation of the geological history of isopentenoids is the assessment of controls other than evolutionary factors that may influence their sedimentary distributions and abundances. It is apparent, for example, that fluctuations in isopentenoid compositions can reflect climatic trends. Major variations in the concentrations of phytoplankton-specific isopentenoids (e.g. dinosterol, a sterol marker for dinoflagellates) in Pleistocene sedimentary sequences from both the Atlantic and Pacific Oceans delineate temporal differences in phytoplankton productivity associated with glacial/interglacial cycles (57, 59). Similar fluctuations in molecular abundances in recent sediments from the Santa Barbara basin correspond to El Niño events (111). Thus, both long- and short-term changes in climate can be recorded in the molecular composition of sedimentary organic matter. There are also variations in the spatial distributions of molecular signals that accompany these temporal changes. For example, wind strength and direction is influenced by climate and can be assessed from the occurrence of compounds derived from eolian fallout (112). Similarly, in the oceanic realm plankton productivity tends to be driven by the supply of nutrients, which may change with time and space, especially in regions of upwelling. Hence, assessment of spatial variations in the abundances and distributions of isopentenoids within sediment sequences may provide documentation of local, regional and global scale changes in climate.

Acknowledgments.

I am grateful for research support from a Fellowship in Science and Engineering awarded by the David and Lucile Packard Foundation, from the National Science Foundation (OCE 88-22583; OCE 93-03672), and from the Petroleum Research Fund administered by the American Chemical Society (23056-AC8).

Literature Cited.

1. Tappan, H. *The Paleobiology of Plant Protists*; Freeman: San Francisco, Calif., 1980; 1028pp.
2. Aubry, M. -P. In *Nannofossils and their Applications*; Crux, J. A.; van Heck, S., Eds.; E. Horwood: Chichester, 1989; pp. 21-40.
3. Woese, C. R.; Kandler, O. Wheelis, M. L. *Proc. Natl. Acad. Sci.* 1990, *87*, 4576-4579.
4. Lake, J. A. *Nature* 1988, *331*, 184-186.
5. Golenberg, E. M. *Phil. Trans. R. Soc. Lond. B* 1991, *333*, 419-427.
6. Sidow, A.; Wilson, A. C.; Pääbo, S. *Phil. Trans. R. Soc. Lond. B* 1991, *333*, 429-433.
7. Runnegar, B.; Schopf, J. W. *Molecular Evolution and the Fossil Record*; Broadhead, T. W., Ed.; Paleontological Society: Knoxville, TN, 1988; 167pp.
8. Grantham, P. J.; Wakefield, L. L *Org. Geochem.* 1988, *12*, 61-73.
9. Summons, R. E. In *Organic Geochemistry*; Engel, M. H.; Macko, S. A., Eds.; Plenum Press: New York, 1993; pp. 3-21.
10. Schopf, J. W.; Walter, M. R. In *Earth's Earliest Biosphere; Its Origin and Evolution*; Schopf, J. W., Ed. Princeton University Press: Princeton, 1983; pp. 187-213.
11. Holmén, K. In *Global Biogeochemical Cycles;* Butcher, S. S.; Charlson, R. J.; Orians G. H.; Wolfe, G. V., Eds.; Harcourt Brace Jovanovich: London, 1992; pp. 239-262.
12. Tissot, B. *Nature* 1979, *277*, 463-465.
13. Tissot, B. P.; Welte, D. H. *Petroleum Formation and Occurrence*; Springer-Verlag: Berlin, 1984; 699p.
14. Hedges, J. I. *Geochim. Cosmochim. Acta* 1978, *42*, 69-76.

15. de Leeuw, J. W.; Largeau, C. In *Organic Geochemistry: Principles and Applications*; Engel, M. H.; Macko, S. A., Eds.; Plenum: New York, N.Y., 1993; pp. 23-72.
16. Tegelaar, E. W.; Derenne, S.; Largeau, C.; de Leeuw, J. W. *Geochim. Cosmochim. Acta* 1989, *3*, 3103-3107.
17. Brassell, S. C.; Eglinton, G. In *Coastal Upwelling: Its Sediment Record, Part A*; Suess, E.; Thiede, J., Eds.; Plenum Press: New York, 1983; p. 545-571.
18. Simoneit, B. R. T. In *Chemical Oceanography, Volume 7*; Riley, J. P.; Chester, R., Eds.; Academic Press: London, 1978; pp. 233-311.
19. Simoneit, B. R. T.; Chester, R.; Eglinton, G. *Nature* 1977, *267*, 682-685.
20. Volkman, J. K. *Org. Geochem.* 1986, *9*, 83-99.
21. Harwood, J. L.; Russell, N. J. *Lipids in Plants and Microbes*; Allen and Unwin: London, 1984; 162p.
22. Hedges, J. I.; Mann, D. C. *Geochim. Cosmochim. Acta* 1979, *43*, 1803-1807.
23. Marlowe, I. T.; Brassell, S. C.; Eglinton, G.; Green, J. C. *Chem. Geol.* 1990, *88*, 349-375.
24. Reed, J. D.; Illich, H. A.; Horsfield, B. In *Advances in Organic Geochemistry 1985*; Leythaeuser, D. and Rullköter, J., Eds.; Pergamon: Oxford, 1986; pp. 347-358.
25. Hoffmann, C. F.; Foster, B.; Powell, T. G.; Summons, R. E. *Geochim. Cosmochim. Acta* 1987, *51*, 2681-2697.
26. Schopf, T. J. M. In *Coastal Upwelling: Its Sediment Record, Part B*; Suess, E.; Thiede, J., Eds.; Plenum Press: New York, 1983; p. 579-596.
27. Jenkyns, H. C. *J. Geol. Soc., Lond.* 1980, *137*, 171-188.
28. Stewart, W. N. *Paleobotany and the Evolution of Plants*; Cambridge University Press: Cambridge, 1983; 405p.
29. Eglinton, G.; Logan, G. A. In *Molecules Through Time: Fossil Molecules and Biochemical Systematics*; Eglinton, G. and Curry, G. B., Eds.; Phil. Trans. R. Soc. Lond. B 333: London, 1991; pp. 315-328.
30. Smith, D. J.; Eglinton, G. *Nature* 1983, *304*, 259-262.
31. Repeta, D. J.; Gagosian, R. B. *Geochim. Cosmochim. Acta* 1984, *48*, 1265-1277.
32. Klok, J.; Baas, M.; Cox, H. C.; de Leeuw, J. W.; Schenck, P. A. *Tetrahedron Lett.* 1984, *25*, 5577-5580.
33. Brassell, S. C. In *Productivity, Accumulation and Preservation of Organic Matter in Recent and Ancient Sediments*; Whelan, J. K.; Farrington, J. W., Eds.; Columbia University: New York, N. Y., 1992; pp. 339-367.
34. Sinninghe Damsté, J. S.; de Leeuw, J. W. *Org. Geochem.* 1990, *16*, 1077-1101.
35. Sinninghe Damsté, J. S.; de Leeuw, J. W.; Kock-van Dalen, A. C.; de Zeeuw, M. A.; de Lange, F.; Rijpstra, W. I. C.; Schenck, P. A. *Geochim. Cosmochim. Acta* 1987, *51*, 2369-2391.
36. Demaison, G. J.; Moore, G. T. *Am. Assc. Pet. Geol. Bull.* 1980, *64*, 1179-1209.
37. Calvert, S. E.; Pederson, T. F. In *Organic Matter: Productivity, Accumulation, and Preservation in Recent and Ancient Sediments*; Whelan, J. and Farrington, J. W., Eds.; Columbia University: New York, N.Y., 1992; pp. 231-263.
38. Deming, J. W.; Baross, J. A. In *Organic Geochemistry*; Engel, M. H.; Macko, S. A., Eds.; Plenum Press: New York, 1993; pp. 119-144.
39. Volkman, J. K.; Johns, R. B. *Nature* 1977, *267*, 693-694.
40. Prahl, F. G.; Eglinton, G.; Corner, E. D. S.; O'Hara, S. C. M.; Forsberg, T. E. V. *J. Mar. Biol. Assoc. U.K.* 1984, *64*, 317-334.
41. Harvey, H. R.; Eglinton, G.; O'Hara, S. C. M.; Corner, E. D. S. *Geochim. Cosmochim. Acta* 1987, *51*, 3030-3041.
42. Bradshaw, S. A.; Eglinton, G. In *Organic Geochemistry*; Engel, M. H.; Macko, S. A., Eds.; Plenum Press: New York, 1993; pp. 225-235.

43. Jørgensen, B. B. In *The Major Biogeochemical Cycles and their Interactions*; Bolin, B.; Cook, R. B., Eds.; Wiley: Chichester, 1983; pp. 477-515.
44. Corner, E. D. S.; O'Hara, S. C. M.; Neal, A. C.; Eglinton, G. In *The Biological Chemistry of Marine Copepods*; Corner, E. D. S.; O'Hara, S. C. M., Eds.; Oxford University Press: Oxford, 1986; pp. 261-321.
45. Brassell, S. C. *Phil. Trans. R. Soc. Lond. A* 1985, *315*, 57-75.
46. Brassell, S. C. In *Geochemistry of Organic Matter in Sediments and Sedimentary Rocks*; Pratt, L. M.; Comer, J. B.; Brassell, S. C., Eds., SEPM Short Course 27; Society for Economic Paleontologists and Mineralogists: Tulsa, OK, 1992; pp. 29-72.
47. Brassell, S. C.; Gowar, A. P.; Eglinton, G. In *Advances in Organic Geochemistry 1979*; Douglas, A. G.; Maxwell, J. R., Eds.; Pergamon: Oxford, 1980; pp. 421-426.
48. Philp, R. P.; Oung, J. N.; Lewis, C. A. *J. Chromatog.* 1988, *446*, 3-16.
49. Freeman, K. H.; Hayes, J. M.; Trendel, J. -M.; Albrecht P. *Nature* 1990, *343*, 254-256.
50. Hayes, J. M.; Freeman, K. H.; Popp, B. N.; Hoham, C. H. *Org. Geochem.* 1990, *16*, 1115-1128.
51. Summons, R. E. and Powell, T. G. *Geochim. Cosmochim. Acta* 1987, *51*, 557-566.
52. Chappe, B.; Michaelis, W.; Albrecht, P. In *Advances in Organic Geochemistry 1979*; Douglas, A. G.; Maxwell, J. R., Eds.; Pergamon Press: Oxford, 1980; pp. 265-274.
53. Moldowan, J. M.; Seifert, W. K.; Arnold, E.; Clardy, J. *Geochim. Cosmochim. Acta* 1984, *48*, 1651-1661.
54. Brassell, S. C.; Eglinton, G.; Maxwell, J. R. *Biochem. Soc. Trans.* 1983, *11*, 575.
55. Mackenzie, A. S.; Brassell, S. C.; Eglinton, G.; Maxwell, J. R. *Science* 1982, *217*, 491-504.
56. Bird, C. W.; Lynch, J. M.; Pirt, S. J.; Reid, W. W.; Brooks, C. J. W.; Middleditch, B. S. *Nature* 1971, *230*, 473-474.
57. Brassell, S. C.; Brereton, R. G.; Eglinton, G. Grimalt, J.; Leibezeit, G.; Marlowe, I. T.; Pflaumann, U.; Sarnthein, M. *Org. Geochem.* 1986, *10*, 649-660.
58. ten Haven, H. L.; Baas, M.; Kroot, M.; de Leeuw, J. W.; Schenck, P. A.; Ebbing, J. *Geochim. Cosmochim. Acta* 1987, *51*, 803-810.
59. Prahl, F. G.; Muehlhausen, L. A.; Lyle, M.; *Paleoceanogr.* 1989, *4*, 495-510.
60. Collister, J. W.; Summons, R. E.; Lichtfouse, E.; Hayes, J. M. *Org. Geochem.* 1992, *19*, 649-660.265-276.
61. Kohnen, M. E. L.; Schouten, S.; Sinninghe Damsté, J. S.; de Leeuw, J. W.; Merritt, D. A.; Hayes, J. M. *Science* 1992, *321*, 832-838.
62. Schoell, M.; McCaffrey, M. A.; Fago, F. J.; Moldowan, M. J.; *Geochim. Cosmochim. Acta* 1992, *56*, 1391-1399.
63. Brassell, S. C.; Eglinton, G. In *Organic Marine Geochemistry*; Sohn, M., Ed.; American Chemical Society Symposium Ser. 305; ACS: Washington, D.C., 1986; pp.10-32.
64. Summons, R. E.; Walter, M. R. *Am. Jour. Sci.* 1990, *290*, 212-244.
65. Boon, J. J.; de Leeuw, J. W. *Mar. Chem.* 1979, *7*, 117.
66. Brassell, S. C.; Comet, P. A.; Eglinton, G.; Isaacson, P. J.; McEvoy, J.; Maxwell, J. R.; Thomson, I. D.; Tibbetts, P. J. C.; Volkman, J. K. In *Initial Reports of the Deep Sea Drilling Project;* Scientific Party, Eds.; U.S. Government Printing Office: Washington D.C., 1980; Vol. LVI, LVII Pt. 2; pp. 1367-1390.
67. McEvoy, J.; Maxwell, J. R. *Org. Geochem.* 1986, *9*, 101-104.
68. Summons, R. E.; Capon, R. J. *Geochim. Cosmochim. Acta* 1988, *42*, 473-484.
69. Funasaki, J.; Gilbertson, J. R. *J. Lipid Research* 1968, *9*, 766-768.

70. Ourisson, G.; Albrecht, P.; Rohmer, M. *Pure Appl. Chem.* 1979, *51*, 709.
71. Seifert, W. K. *Geochim. Cosmochim. Acta* 1978, *42*, 473-484.
72. Katz, B. J.; Elrod, L. W. *Geochim. Cosmochim. Acta* 1983, *47*, 389.
73. Ourisson, G.; Rohmer, M.; Poralla, K. *Ann. Rev. Microbiol.* 1987, *41*, 301-333.
74. Rohmer, M.; Bouvier, P.; Ourisson, G. *Proc. Natl. Acad. Sc. USA*, 1979, *765*, 847-851.
75. Aquino Neto, F. R.; Restle, A.; Connan, J.; Albrecht, P.; Ourisson, G. *Tetrahedron Lett.* 1982, *23*, 2027-2030.
76. Maxwell, J. R.; Douglas, A. G.; Eglinton, G.; McCormick, A. *Phytochem.* 1968, *7*, 2157-2171.
77. Metzger, P.; Casadevall, E. *Tetrahedron. Lett.* 1983, *24*, 4013-4016.
78. Seifert, W. K.; Moldowan, J. M. *Geochim. Cosmochim. Acta* 1981, *45*, 783-794.
79. Brassell, S. C.; Eglinton, G.; Fu, J. *Org. Geochem.* 1986, *10*, 927-941.
80. Curiale, J. *Org. Geochem.* 1987, *11*, 233-244.
81. Murphy, M. T. J.; McCormick, A.; Eglinton, G. *Science* 1967, *157*, 1040-1042.
82. Bradley, W. H. *U.S. Geol. Survey Prof. Paper* 1929, *158*, 87-110.
83. Jiang, Z. S.; Fowler, M. G. *Org. Geochem.* 1986, *10*, 831-839.
84. Carroll, A. R.; Brassell, S. C.; Graham, S. A. *Am. Assc. Petrol. Geol. Bull.* 1992, *76*, 1874-1902.
85. Requejo, A. G. In *Biological Markers in Sediments and Petroleum*; Moldowan, J. M.; Albrecht, P.; Philp, R. P., Eds. Prentice-Hall: Englewood Cliffs, N.J., 1992; pp. 222-240.
86. Rubinstein, I.; Sieskind, O.; Albrecht, P. *J. Chem. Soc. Perkin Trans. I* 1975, 1833-1836.
87. Peakman, T. M.; Maxwell, J. R. *Org. Geochem.* 1988, *13*, 583-592.
88. van Graas, G.; de Leeuw, J. W.; Schenck, P. A. *Geochim. Cosmochim. Acta* 1981, *45*, 2465-2474.
89. Goossens, H.; de Leeuw, J. W.; Schenck, P. A.; Brassell, S. C. *Nature,* 1984, *312*, 440.
90. Didyk, B. M.; Simoneit, B. R. T.; Brassell, S. C.; Eglinton, G. *Nature,* 1978, *272*, 216.
91. Brassell, S. C.; McEvoy, J.; Hoffmann, C. F.; Lamb, N. A.; Peakman, T. M.; Maxwell, J. R. *Org. Geochem.* 1984, *6*, 11-23.
92. ten Haven, H. L.; Rullkötter, J. *Geochim. Cosmochim. Acta* 1988, *52*, 2543-2548.
93. Brassell, S. C.; Eglinton, G.; Howell, V. J. In *Marine Petroleum Source Rock;* Brooks, J.; Fleet, A. J., Eds.; Blackwell: Oxford, 1987; pp. 79-98.
94. Brassell, S. C.; Comet, P. A.; Eglinton, G.; Isaacson, P. J.; McEvoy, J.; Maxwell, J. R.; Thomson, I. D.; Tibbetts, P. J. C.; Volkman, J. K. In *Advances in Organic Geochemistry 1979*; Douglas, A. G.; Maxwell, J. R., Eds.; Pergamon: Oxford, 1980; pp. 375-391.
95. Brassell, S. C.; Eglinton, G. In *Advances in Organic Geochemistry 1981*; Bjorøy, M. et al., Ed.; Wiley: Chichester, 1983; pp. 684-697.
96. Summons, R. E.; Volkman, J. K.; Boreham, C. J. *Geochim. Cosmochim. Acta* 1987, *51*, 3075-3082.
97. Summons, R. E.; Thomas, J.; Maxwell, J. R.; Boreham, C. J. *Geochim. Cosmochim. Acta* 1992, *56*, 2437-2444.
98. Summons, R. E.; Volkman, J.; Boreham, C. J. *Geochim. Cosmochim. Acta* 1988, *52*, 1747-1763.
99. Chappe, B.; Albrecht, P.; Michaelis, W. *Terra Cognita* 1983, *3*, 216.
100. Eglinton, G.; Scott, P. M.; Belsky, T.; Burlingame, A. L.; Calvin, M. *Science* 1964, *145*, 263-264.
101. McKirdy, D. M. *Precambrian Res.*, 1974, *1*, 75-137.

102. Hayes, J. M.; Kaplan, I. R.; Wedeking, K. W. In *Earth's Earliest Biosphere; Its Origin and Evolution*; Schopf, J. W., Ed. Princeton University Press: Princeton, 1983; pp. 93-134.
103. Hayes, J. M. In *Earth's Earliest Biosphere; Its Origin and Evolution*; Schopf, J. W., Ed. Princeton University Press: Princeton, 1983; pp. 291-301.
104. Peters, K. E.; Moldowan, J. M. *The Biomarker Guide: Interpreting Molecular Fossils in Petroleum and Ancient Sediments*; Prentice Hall: Englewood Cliffs, N. J., 1993; 363p.
105. Summons, R. E.; Brassell, S. C.; Eglinton, G.; Evans, E. J.; Horodyski, R. J.; Robinson, N.; Ward, D. M. *Geochim. Cosmochim. Acta* 1988, *52*, 2625-2673.
106. Alexander, R.; Larcher, A. V.; Kagi, R. I.; Price, P. L. In *Biological Markers in Sediments and Petroleum*; Moldowan, J. M.; Albrecht, P.; Philp, R. P., Eds.; Prentice-Hall: Englewood Cliffs, N.J., 1992; pp. 201-221.
107. Van Graas, G.; Baas, J. M. A.; de Graaf, V.; de Leeuw, J. W. *Geochim. Cosmochim. Acta* 1982, *46*, 2399-2402.
108. McCaffrey, M. A.; Moldowan, J. M.; Lipton, P. A.; Summons, R. E.; Peters, K. E.; Jeganathan, A.; Watt, D. S. *Geochim. Cosmochim. Acta* 1994, *58*, 529-532.
109. Dean, W. E.; Arthur, M. A.; Claypool, G. E. *Mar. Geol.* 1986, *70*, 119-157.
110. Popp, B. N.; Takigiku, R.; Hayes, J. M.; Louda, J. W.; Baker, E. W. *Am. Jour. Sci.* 1989 *289*, 436-454.
111. Kennedy, J. A.; Brassell, S. C. *Org. Geochem.* 1992, *19*, 235-244.
112. Poynter, J. G.; Farrimond, P.; Robinson, N.; Eglinton, G. In *Paleoclimatology and Paleometerology: Modern and Past Patterns of Global Atmospheric Transport*; Leinen, M.; Sarnthein, Eds.; Kluwer: Dordrecht, 1989; pp. 435-462.

RECEIVED May 2, 1994

Chapter 2

Hopanoid and Other Polyterpenoid Biosynthesis in Eubacteria

Phylogenetic Significance

Michel Rohmer and Philippe Bisseret

Laboratoire de Chimie Microbienne, Associé au Centre National de la Recherche Scientifique, Ecole Nationale Supérieure de Chimie de Mulhouse, 3 rue Alfred Warner, 68093 Mulhouse, France

Triterpenoids of the hopane series are frequently found as membrane stabilizers in Eubacteria. Owing to their usually high intracellular concentration and their stability, hopanoids are better suited for the elucidation of the isoprenoid biosynthetic routes in bacteria than the more wide-spread polyprenol derivatives (bactoprenol, ubiquinones and menaquinones) or carotenoids. Incorporations of ^{13}C labelled precursors (acetate, glucose, pyruvate and D-erythrose) into hopanoids and bacterial ubiquinones allowed the discovery and partial elucidation of a novel biosynthetic pathway for the early steps of isoprenoid biosynthesis leading to IPP in bacteria. The resulting hypothetical biogenetic scheme excludes HMG CoA and mevalonate as precursors and includes direct formation of the C_5 isoprenic skeleton i) by condensation of a C_2 subunit derived from pyruvate decarboxylation on the C-2 carbonyl group of a triose phosphate derivative and ii) a transposition step. This peculiar feature combined with other specific aspects linked to the formation of bacterial hopanoids (lack of specificity of squalene cyclase, formation of the amphiphilic C_{35} bacteriohopane skeleton by a carbon/carbon linkage between a D-pentose and a triterpene, unusual methylations) points out the diversity of the routes employed by living organisms to obtain their membrane constituents.

Abbreviations. Ac CoA: acetyl Coenzyme A; DMAPP: dimethylallyl alcohol diphosphate; HMGCoA: 3-hydroxy-3-methyl Coenzyme A; IPP: isopentenyl diphosphate.

Although wide-spread and inordinately diverse in Eukaryotes, isoprenoids are in comparison rather scarcely present in Eubacteria. Those of general occurrence, the bactoprenols acting as carbohydrate carriers for the biosynthesis of the cell-wall and the ubiquinones and menaquinones involved in the electron transport chains, are derived from regular C_{40} to C_{55} all-*trans* polyprenols and are present at low concentrations only. Those found at much higher levels such as the carotenoids, the triterpenoids of the hopane series or some lower terpenoids from Actinomycetes, although fairly widely distributed, are found scattered in numerous taxa. Worst, the hopanoids (Fig. 1) which are probably the polyterpenoids found in highest concentrations (up to 30mg/g, dry weight, in *Zymomonas mobilis*) remained undisclosed until the early 70's, owing to the difficulties linked to their isolation and purification (1,2). For all these reasons isoprenoid biosynthesis has been only scarcely studied in Eubacteria (3). Investigations were furthermore restricted to very few selected strains and to the polyprenol derived isoprenoids, using in most cases radiolabelled precursors usually without identification of the labelled positions. When this has been done, discrepancies with the classical pathway involving Ac CoA and mevalonate appeared immediately in at least two bacteria, *Escherichia coli* (4,5) and a *Streptomyces* species (6). In fact hopanoids are admirably suited for investigations in the isoprenoid metabolic pathway. They are present in a broad spectrum of bacteria. Their intracellular concentration is usually sufficiently high, allowing incorporation of ^{13}C labelled precursors and yielding isotopic enrichments readily measured by ^{13}C-NMR. They are quite stable, and their isolation is in fact easy once the proper methodology has been set up.

Hopanoids from Eubacteria

Structural Variety. All bacterial hopanoids are 3-deoxytriterpenes directly derived from squalene cyclization. The most simple ones, diploptene **1** and diplopterol **2**, could be detected in nearly all hopanoid producing bacteria (7) (Fig. 1). They are not distinctive for these prokaryotes as these C_{30} triterpenes or more oxidized derivatives are found in several eukaryotic phyla (*e.g.* fungi, lichens mosses, ferns or protozoa) (8). The C_{35} bacteriohopanepolyols are in fact characteristic for Eubacteria (Fig. 1). They have never been reported from the few investigated Archaebacteria or from Eukaryotes. The traces amounts we could detect in some higher plants (*e.g. Trifolium repens* or *Viola odorata*) reflect merely the presence of bacterial epibionts living on the leaf surfaces (Knani and Rohmer, unpublished results). This bacteriohopane skeleton presents the unusual feature of an additional C_5 n-alkyl polyfunctionalized side-chain linked by a carbon/carbon bond to the hopane isopropyl group. The side-chains are usually all derived from bacteriohopanetetrol **3** or from aminobacteriohopanetriol **4**, although pentols, aminotetrols and aminopentols are

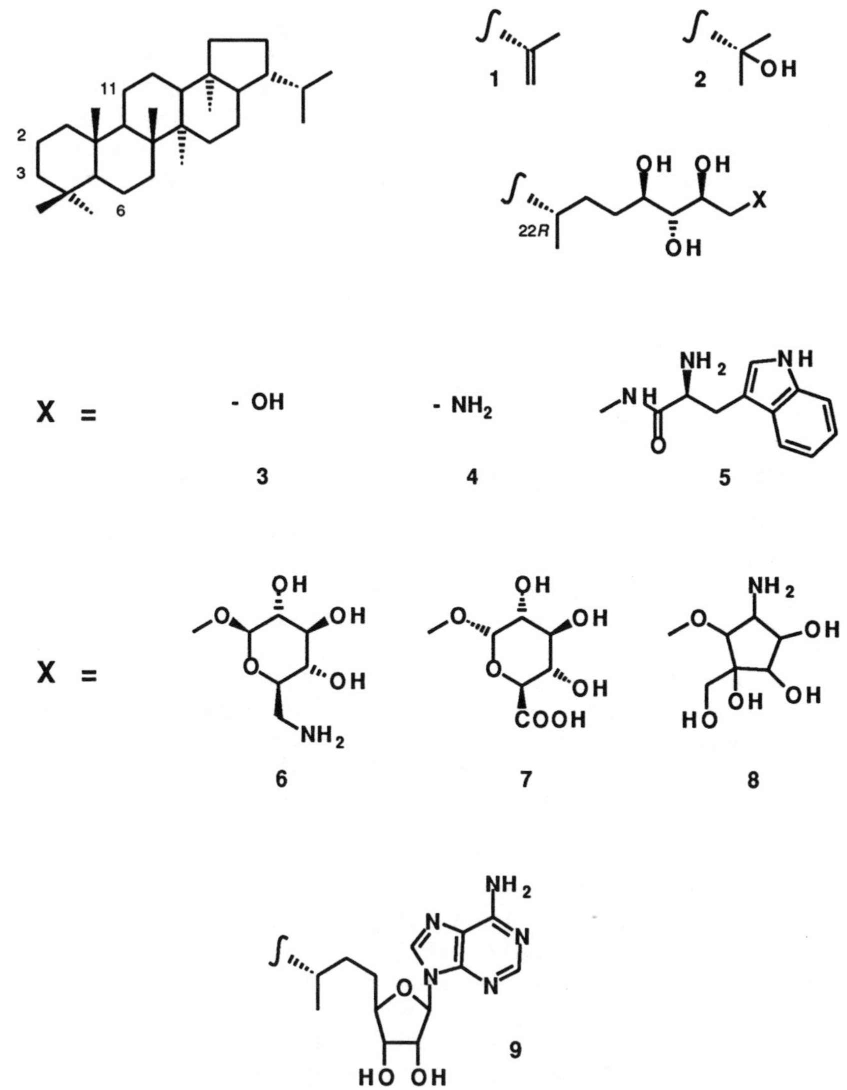

Fig. 1. Selected bacterial hopanoids.

known. With the exception of the *N*-aminoacyl derivatives (*e.g.* 5) from *Rhodomicrobium vannielii* (9), aminopolyols are always found as free compounds. The polyols however are in most cases linked to other polar moieties via the terminal C-35 hydroxy group (1). The most common derivatives are tetrol glycosides involving either aminohexoses (*e.g.* 6) or uronic acids (*e.g.* 7) or tetrol ethers with a peculiar carbapseudofuranopentose (*e.g.* 8) resembling the framework found in some enzyme inhibitors or nucleoside analogues from Actinomycetes and fungi (1, 10-12). Modifications may also be introduced on the pentacyclic hopane skeleton, probably at a later stage of the biosynthesis, and are probably of minor physiological importance. They are found until now rather unfrequently and might be characteristic for some taxonomic groups: double-bonds in rings B or C in the hopanoids from the acetic acid bacteria, additional methyl groups either at C-2 in cyanobacteria, the prochlorophyte *Prochlorothrix hollandica* (Simonin, Jürgens and Rohmer, unpublished results) and in the *Methylobacterium* species or at C-3 in acetic bacteria and obligate methylotrophs (7, 13, 14). Until now 29 different kinds of side-chains have been identified. Including the structural variations found on the hopane skeleton, this leads to a huge diversity for bacterial hopanoids. The list of the bacterial hopanoids is by far not closed as several new structures are always under investigation in this laboratory. For a recent review, see for instance reference (1).

Detection, Isolation and Distribution. Until now there is no general method allowing the detection of free bacterial C_{35} hopanoids. Because of their poor solubility in most organic solvents and their high polarity, these compounds are lost (or the recovery yields are very poor) by all attempts of direct isolation using chromatographic methods. Derivatization, directly performed on the crude $CHCl_3/CH_3OH$ extract by acetylation of the hydroxy and amino groups and methylation of the carboxylic groups, is essential for the isolation of intact hopanoids (1, 9, 11, 12)

Because of the structural diversity of bacterial hopanoids and the variability of the hopanoid composition of all bacteria examined so far, the separation methodology has to be adjusted in each case. This has been done for instance for the quantitative determination of the hopanoids from *Zymomonas mobilis* (15). Furthermore, there is no specific technique for the detection of hopanoids. After a chromatographic separation, all fractions have to be checked by ^1H-NMR spectroscopy for the possible presence of the characteristic methyl pattern of the hopane framework. In order to get rid of all these problems which are essentially linked to the structural variability of the side-chain, a method of rather general application has been developped: oxidation of the polyhydroxylated side-chains by periodic acid followed by reduction by sodium borohydride of the resulting aldehydes into primary alcohols which can be readily isolated by thin layer chromatography and analyzed by gas chromatography and gas chromatography/ mass spectrometry after acetylation (7, 16) or by high performance liquid chromatography after naphtoylation (17). This is to date the most rapid and expedient method for the detection of hopanoids in bacteria. One should just keep in mind some limitations. i) The presence of two free vicinal hydroxy groups next to the hopane moiety is required: hopanoids such as adenosylhopane 9 are not detected (18). ii) For unknown reasons, the hopanoid concentration may be in some cases (*e.g. Zymomonas mobilis*) underestimated. iii) Finally, we are aware that at least in one bacterium (an *Acetobacter aceti* ssp. *xylinum* strain) some hopanoids can not be extracted by organic solvents, escaping thus completely to the usual detection procedures (Herrmann and Rohmer, unpublished results).

This side-chain degradation method is however the most suited one for a fast screening for hopanoids in bacteria. We have analyzed more than 250 species until now. Although some trends concerning the distribution of this triterpenic series may already appear, no definitive and clear-cut conclusions can be drawn at the moment (2,7). Hopanoid distribution in Eubacteria will probably appear much clearer when on the one hand the relationship between taxonomy and phylogeny is better understood in bacteria and when on the other hand a much broader range of microorganisms is investigated. Indeed, the current knowledge on hopanoid distribution reflects rather the availability of the strains in collections and the easiness of their isolation and cultivation rather than their representativity and significance in the environment. Marine bacteria are for instance poorly represented in our lists.

Intracellular Localization and Rôle. Hopanoids are essential metabolites for the bacteria synthesizing them. Indeed, *in vivo* inhibition of hopanoid biosynthesis by addition of a squalene cyclase inhibitor to the culture medium leads selectively to no growth and death of the tested hopanoid producers, whereas hopanoid non-producers were not affected by these compounds (19). Intracellular localization has been determined in Gram-negative bacteria, firstly in *Zymomonas mobilis* were they have been found in the cytoplasmic membrane as well as in the outer membrane of the cell wall (20), and more recently in photosynthesizing bacteria such as the cyanobacterium *Synechocystis* sp. PCC 6714 and the prochlorophyte *Prochlorothrix hollandica*, where they were localized in the cell wall and in the thylakoids of the photosynthetic apparatus, plasma membranes from these prokaryotes being not available (21, Simonin, Jürgens and Rohmer, unpublished results). Furthermore, in the symbiotic nitrogen-fixing Gram-positive bacterium *Frankia alni*, isolated vesicle envelopes consisted mainly of hopanoids which were supposed to protect the nitrogenase from oxygen (22).

Hopanoid concentrations in bacterial cells are of the same order of magnitude as the sterol content in eukaryotic cells, *i.e.* usually in the 0.5 to 5 mg/g (dry weight) range (7). All these data suggest a structural rôle for the hopanoids in bacterial membranes. Comparison of the common structural features of sterols and hopanoids (*i.e.* amphiphilic molecules with a rigid, planar and polycyclic ring system, similar dimensions) led to the assumption that structural similarity may reflect functional similarity, hopanoids acting as bacterial membrane stabilizers much like sterols do in eukaryotic membranes (23,24). This assertion has been verified for hopanoids on artificial membrane models and on several biological systems (2,25). They modulate, like sterols, membrane fluidity and permeability, at least qualitatively. Finally, the structural diversity of bacterial hopanoids is not in contradiction with this rôle, as most of them meet all features required for membrane stabilizers. This diversity has to be compared to the huge number of sterols found in Eukaryotes, especially in marine invertebrates, and may just reflect a primitive feature with only minimal adequation between structure and rôle.

Hopanoid Biosynthesis

Early Steps of Hopanoid Biosynthesis in Eubacteria: a Non-Mevalonate Route towards IPP and DMAPP. The rather high concentration of hopanoids in bacterial cells (with records up to 30 mg/g for *Zymomonas mobilis*) compared to the tiny amounts found for the other bacterial isoprenoids of general occurrence (*i.e.* the bactoprenol derivatives and the isoprenoid chain bearing quinones) makes them admirably suited for incorporation experiments with precursors labelled with stable isotopes. Such experiments using essentially ^{13}C labelled acetate, glucose and pyruvate were primarily designed in order to determine the origin of the additional C_5 side-chain of the bacteriohopanepolyols (see next paragraph). They threw in fact light on an unexpected and yet undisclosed novel route for the formation of isoprenoids. Indeed, whereas IPP and DMAPP were always successfully incorporated into isoprenoids by bacterial systems, indicating that these two C_5 alcohols are universal precursors for isoprenoids, the detection of the enzymic activities corresponding to the early steps revealed in many cases problematic (3,26). The results of our labelling experiments can be summarized as follows (3,27).

i) The observed labelling patterns did not correspond to the expected ones according to the Bloch-Cornforth route via acetyl CoA, acetoacetyl CoA, HMG CoA and mevalonate. Exogenous acetate was never directly incorporated into isoprenoids. The labelling pattern could eventually be explained assuming that three different non-interconvertible acetate pools were involved in the formation of an isoprenic unit. Such a compartmentation is however incompatible with the regulation of metabolic pathways in prokaryotic cells.

ii) In all incorporation experiments no scrambling occurred. The labelling patterns (labelled positions as well as their isotopic abundances) could be in all cases unambiguously interpreted, taking into account for the formation of the postulated intermediates only classical, well known metabolic pathways.

iii) Incorporation of doubly labelled (4,5-$^{13}C_2$)glucose showed clearly that the carbon atoms derived from C-4 and C-5 of glucose were **simultaneously** incorporated and corresponded respectively to the C-4 and C-3 carbon atoms of IPP: a transposition step is therefore required for the formation of IPP and DMAPP, and the acetyl CoA route could be definitively excluded (Fig. 2).

iv) The origin of all five carbon atoms of an isoprenic unit could be determined for each labelling experiment (Fig. 2), allowing to propose an hypothetical biogenetic scheme. Formation of the C_5 carbon atom skeleton of an isoprenic units results most probably from the condensation of the C_2 thiamine adduct resulting from pyruvate decarboxylation on carbon C-2 of a triose phosphate derivative and a transposition step. These two reactions should be followed by phosphorylation, water eliminations and reductions occuring in an unknown reaction sequence (Fig. 3). This pathway resembles the L-valine biosynthesis. This amino-acid, as well as its C_5 precursors, could however be excluded as precursors, as its labelling pattern in *Zymomonas mobilis* after feeding of ^{13}C labelled glucose corresponded as expected to a synthesis from two pyruvate subunits and was different from that of isoprenic units.

v) This novel pathway is probably of general occurrence. It was found in phylogenetically unrelated bacteria, Gram-negative as well as in the Gram-positive. It concerns not only the hopanoids from the *Methylobacterium* and *Rhodopseudomonas* species, *Zymomonas mobilis* or *Alicyclobacillus acidoterrestris*. It has been also detected for the formation of the ubiquinones of *Methylobacterium fujisawaense* and *Escherichia coli* which does not produce any detectable hopanoid, and most probably in a *Streptomyces* species for the formation of sesquiterpenoids of the pentalenolactone series (6).

Of course very little is known on the distribution of this novel pathway in Eubacteria. It is also quite clear that the classical mevalonate pathway is also operating in some of these microorganisms. This could be for instance shown by the normal incorporation of ^{13}C labelled acetate or mevalonate into bacterial isoprenoids or by the careful identification of all enzymatic steps between Ac CoA and IPP (3,26). However there is no indication that both pathways are simultaneously present in one single bacterium.

Formation of the Pentacyclic C_{35} Bacteriohopane Skeleton. Formation of the C_{35} bacteriohopane skeleton involves at least three kinds of enzymic reactions presenting unusual features:
i) squalene cyclization leading to the hopane pentacyclic ring system, ii) coupling via a carbon/carbon bond between a triterpene and a D-pentose and iii) finally modifications of the bacteriohopane skeleton by methylation reactions.

Squalene Cyclization. A bacterial squalene cyclase has been firstly characterized in a cell-free system from an *Acetobacter* species (28). This enzymic activity has been reported later from several other bacteria (29). More recently the enzyme has been isolated from *Alicyclobacillus acidocaldarius* (30), *Rhodopseudomonas palustris* (31) and *Zymomonas mobi-*

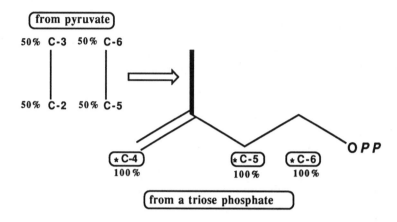

Fig. 2. Origin of the carbon atoms in the hopanoids from *Zymomonas mobilis* after feeding of ^{13}C labelled glucose. For the sake of clarity, only one isoprenic materialized by IPP has been represented.

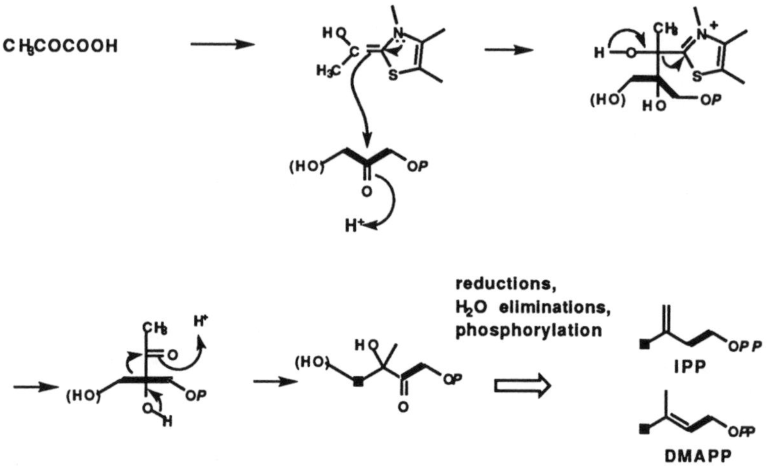

CH₃COCOOH

reductions,
H₂O eliminations,
phosphorylation

IPP

DMAPP

Fig. 3. Novel hypothetical pathway for the formation of isoprenic units in Eubacteria from triose phosphate derivatives.

lis (*32*). In the case of *A. acidocaldarius*, the amino-acid sequence has even been determined (*33*). The major feature of the bacterial squalene cyclases compared to the eukaryotic squalene oxide cyclases is their lack of specificity towards the substrate. Whereas the eukaryotic cyclases convert only the 3S-enantiomer of squalene epoxide into 3β-hydroxy triterpenes, the bacterial enzymes cyclize next to squalene, their normal substrate, both enantiomers of squalene epoxide, yielding 3β-hydroxy- or 3α-hydroxyhopanoids respectively from the 3S- or 3R-enantiomer. They were even shown to be able to convert 2,3-dihydrosqualene into tetracyclic triterpenes with a five membered D-ring of the dammarane or the transposed euphane series. Enzymes linked to the formation of pentacyclic ring systems seem therefore capable of performing reactions similar to those involved involved in the formation of the sterol precursors (*34*). Furthermore, a closely related enzyme, the squalene cyclase from the ciliate *Tetrahymena pyriformis*, cyclizing normally squalene into diplopterol and the quasi-hopanoid tetrahymanol, transforms regular polyprenol derivatives into polycyclic compounds although several methyl groups occupy positions different from those of the normal substrate: the C_{25} derivative leads to te tetracyclic scalarane skeleton, and the C_{30} one to a mixture of di-, tri-, tetra- and pentacyclic cyclization products (*35*). All these aspects have been extensively described in a recent review and will not be discussed in details here (*29*). It has just to be pointed out that a similar lack of specificity was also found for the geranylgeraniol diphosphate cyclase from higher plants involved in the formation of kaurene, the precursor of gibberellins. The latter enzyme cyclizes geranylgeraniol diphosphate into a 3-deoxyditerpene as well as its two enantiomeric analogues bearing an oxirane ring in place of the terminal double-bond into the corresponding 3β- and 3α-hydroxy diterpenes (*36*).

Formation of the C_{35} Bacteriohopane Skeleton. The presence of an additional C_5 polyhydroxylated *n*-alkyl chain linked to a triterpene is unique in the natural product chemistry. Incorporation experiments using ^{13}C labelled acetate and glucose were performed in order to elucidate the origin of this moiety. this side-chain is indeed a D-pentose derivative arising either from non-oxidative pentose phosphate pathway in the *Methylobacterium* and *Rhodopseudomonas* species or in *Zymomonas mobilis* or from oxidative pentose phosphate pathway in *Alicyclobacillus acidocaldarius* (*27, 37-40*) This carbohydrate is linked via its C-5 carbon atom to the hopane isopropyl group, and according to the most common stereochemistry of bacteriohopanetetrol or aminobacteriohopanetriol, the precursor arises most probably from D-ribose (*41-43*), or eventually in the *Acetobacter* species also from D-arabinose (*44*). Nothing is known until now about the precursors of the hopane or the carbohydrate moieties, and the nature of the enzymic reaction allowing this unusual linkage. Anyway, this reaction yields an amphiphilic molecule acting as membrane stabilizers with a stable non-hydrolysable carbon/carbon bond between both moieties and with a polar head which is not derived from dioxygen like the eukaryotic sterol hydroxy group.

Methylation Reactions of the Bacteriohopane Skeleton. Additional methyl groups are often present in bacterial hopanoids. They are most probably introduced after squalene cyclization and/or after the formation of the bacteriohopane skeleton. They can be located usually either at C-2β (*e.g.* in *Methylobacterium* and cyanobacteria species) (*14*) or C-3β (*e.g.* in *Acetobacter* species or in obligate methylotrophs) (*45*), rarely at C-2α (*Methylobacterium organophilum*) (*46*) or at C-31 (*Acetobacter europaeus*) (*40*). Such methylation patterns are nearly never found in eukaryotic isoprenoids with the exception of iris triterpenoids (*47, 48*), the precursors of the irones, or in chalmicrin, a fungal sesquiterpenoid, which resemble the 3β-methylhopanoid (*49*). Successful incorporation of doubly-labelled [^{3}H,^{14}C]-methyl methionine and/or [^{2}H$_3$]-methyl methionine showed that the methyl donor is derived from this amino-acid and is most probably S-adenosylmethionine. All three protons of the transferred methyl group were retained in all methylated hopanoids much like in enzymic methylation reactions of aromatic rings and not like in most methylations of the phytosterol side-chains (*50*). This excludes peculiarly intermediates with methylene or cyclopropyl groups.

Conclusion

The bacterial hopanoids and the eukaryotic sterols are structurally and functionally similar, the members of both series acting as building blocks as well as stabilizers in biological membranes. Their respective biosyntheses show however considerable differences and point out how metabolites with identical functions can be obtained by different pathways. The assumed primitiveness of the hopanoid pathway was merely based upon an argumentation which will be probably impossible or at least very difficult to prove or disprove as the modern bacteria have already a long evolution behind them. Hopanoid biosynthesis is independent of dioxygen, and thus compatible with an anoxic primeval earth atmosphere. Squalene cyclization into the hopane skeleton occurs in the thermodynamically most stable all pre-chair conformation of the polyene whereas the formation of the sterol precursors, lanosterol or cycloartenol requires a less favored pre-chair/boat/chair/chair conformation of squalene oxide. Neither a transposition step, nor further demethylations of the polycyclic skeleton and double-bond migrations as they are found in the sterol formation are required in the hopanoid biosynthesis. Further, the lack of specificity of the squalene cyclases compared to the narrow specificity of the eukaryotic squalene oxide cyclases was also interpreted as a rather primitive characteristic (*24*). However the most simple hopanoids, diploptene and diplopterol directly derived from squalene cyclization, are unable to stabilize artificial phospholipid bilayers (*51*). Whether the crucial step required for the obtention of an amphiphilic membrane stabilizers, *i.e.* the formation of the bacteriohopanepolyols and the subsequent derivatization of the terminal functional group (leading for instance to the numerous composite hopanoids), is a primitive feature will be difficult to ascertain, and might be indeed a rather recent acquisition.

Finally the discovery of a novel pathway for the formation of isoprenic units points to another similar problem. Whereas this breakthrough was mainly made possible owing to the efforts we invested during the last twelve years in the bacterial hopanoid chemistry and biochemistry, it represents a more general problem as this pathway concerns triterpenoids of the hopane series as well as the ubiquinone of *Escherichia coli* or sesquiterpenoids of a *Streptomyces*. Nearly nothing is known about its distribution among Eubacteria and eventually some Eukaryotes such as higher plants. Indeed it is known for instance that the biosynthesis of the chloroplast isoprenoids is not affected by mevinolin, an HMGCoA reductase inhibitor, whereas sterol biosynthesis in the cytoplasm is efficiently blocked by this compound (*52, 53*). This result could be interpreted by the occurrence in plant of a non-mevalonate pathway for isoprenoid formation next to the classical Bloch-Cornforth pathway. This feature linked with the exotic characteristics of other plant enzymic reactions, such as the lack of specificity of the geranylgeraniol diphosphate cyclase involved in the formation of the plant hormones of the gibberelin series (*36*) and the methylation pattern of the iris triterpenoids (*37, 38*) resembling that of the 3-methylhopanoids might point out possible preserved relictual metabolic pathway inherited from prokaryotic ancestors.

Literature Cited

1   Rohmer, M. *Pure Appl. Chem.* **1993**, *65*, pp 1293-1298.
2   Sahm, H.; Rohmer, M.; Bringer-Meyer, S.; Sprenger, G.; Welle, R. *Adv. Micr. Physiol.* **1993**, *35*, pp 249-273.
3   Rohmer, M.; Knani, M.; Simonin, P.; Sutter, B.; Sahm, H. *Biochem. J.* **1993**, *295*, pp 517-524 and references cited therein.
4   Pandian, S.; Saengchjan, S.; Raman, T.S. *Biochem. J.* **1981**, *196*, pp 675-681.
5   Zhou, D.; White, R.H. *Biochem. J.* **1991**, *273*, pp 627-634.
6   Cane, D.E.; Rossi, T.; Tillman, A.M.; Pachlatko, J. *J. Am. Chem. Soc.* **1981**, *103*, pp 1838-1843.
7   Rohmer, M.; Bouvier-Navé, P.; Ourisson, G. *J. Gen. Microbiol.* **1984**, *130*, pp 1137-1150.
8   Ourisson, G.; Albrecht, P.; Rohmer, M. *Pure Appl. Chem.* **1979**, *51*, pp 709-729.

9   Neunlist, S.; Rohmer, M. *Eur. J. Biochem.* **1985**, *147*, pp 561-568.
10   Simonin, P.; Jürgens U.J.; Rohmer, M. *Tetrahedron Lett.* **1992**, *33*, pp 3629-3632.
11   Llopiz, P.; Neunlist, S.; Rohmer, M. *Biochem. J.* **1992**, *287*, pp 159-161.
12   Renoux, J.-M.; Rohmer, M. *Eur. J. Biochem.* **1985**, *151*, pp 405-410.
13   Rohmer, M.; Ourisson, G. *J. Chem. Res.* **1986** (S) pp 356-357, (M) pp 3037-3059.
14   Bisseret, P.; Zundel, M.; Rohmer, M. *Eur. J. Biochem.* **1985**, *150*, pp 29-34.
15   Schulenberg-Schell, H.; Neuss, B.; Sahm, H. *Anal. Biochem.* **1989**, *181*, pp 120-124.
16   Rohmer, M.; Ourisson, G. *Tetrahedron Lett.* **1976**, pp 3633-3636.
17   Barrow, K.D.; Chuck, J.-A. *Anal. Biochem.* **1990**, *184*, pp 395-399.
18   Neunlist, S.; Bisseret, P.; Rohmer, M. *Eur. J. Biochem.* **1988**, *171*, pp 245-252.
19   Flesch, G.; Rohmer, M. *Arch. Microbiol.* **1987**, *147*, pp 100-104.
20   Tahara, Y.; Yuhara, H.; Yamada, Y. *Agric. Biol. Chem.* **1988**, *52*, pp 607-609.
21   Jürgens, U.J.; Simonin, P.; Rohmer, M. *FEMS Microbiol. Lett.* **1992**, *92*, pp 285-288.
22   Berry, A.M.; Harriott, O.T.; Moreau, R.A.; Osman, S.F.; Benson, D.R.; Jones, A.D. *Proc. Natl Acad. Sci. US* **1993**, *90*, pp 6091-.
23   Nes, W.R. *Lipids* **1974**, *9*, pp 596-612.
24   Rohmer, M.; Bouvier, P.; Ourisson, G. *Proc. Natl. Acad. Sci. USA* **1979**, *76*, pp 847-851.
25   Ourisson, G.; Rohmer, M. *Acc. Chem. Res.* **1992**, *25*, pp 403-408.
26   Horbach, S.; Welle, R.; Sahm, H. *FEMS Microbiol. Lett.* **1993**, *115*, pp 135-140, and references cited therein.
27   Flesch, G.; Rohmer, M. *Eur. J. Biochem.* **1988**, *175*, pp 405-411.
28   Anding, C.; Rohmer, M.; Ourisson, G. *J. Am. Chem. Soc.* **1976**, *98*, pp 1274-1275.
29   Abe, I.; Rohmer, M.; Prestwich, G. *Chem. Rev.* **1993**, *93*, pp 2189-2206.
30   Ochs, D.; Tappe, C.H.; Gärtner, P.; Kellner, R.; Poralla, K. *Eur. J. Biochem.* **1990**, *194*, pp 75-80.
31   Kleemann, G. *Ph. D. Thesis*, Universität Tübingen, FRG, **1992**.
32   Tappe, C.H. *Ph. D. Thesis*, Universität Tübingen, FRG **1993**.
33   Ochs, D.; Kaletta, C.; Entian, K.-D.; Beck-Seckinger A.; Poralla, K. *J. Bacteriol.* **1992**, *174*, pp 298-302.
34   Abe, I.; Rohmer, M. *J. Chem. Soc. Perkin I* **1994**, in the press.
35   Renoux, J.-M.; Rohmer, M. *Eur. J. Biochem.* **1986**, *155*, pp 125-132.
36   Coates, R.M.; Conradi, R.A.; Ley, D.A.; Akeson, A.; Harada, J.; Lee, S.C.; West, C.A. *J. Am. Chem. Soc.* **1976**, *98*, pp 4659-4661.
37   Rohmer, M.; Sutter, B.; Sahm, H. *J. Chem. Soc. Chem. Commun.* **1989**, pp 1471-1472.

38   Sutter, B. *Ph. D. Thesis*, Université de Haute Alsace, Mulhouse, France, **1991**.

39   Knani, M. *Ph. D. Thesis*, Université de Haute Alsace, Mulhouse, France, **1992**.

40   Simonin, P. *Ph. D. Thesis*, Université de Haute Alsace, Mulhouse, France, **1993**.

41   Bisseret, P.; Rohmer, M. *J. Org. Chem.* **1989**, *54*, pp 2958-2964.

42   Neunlist, S.; Rohmer, M. *J. Chem. Soc. Chem. Commun.* **1988**, pp 830-832.

43   Zhou, P.; Berova, N.; Nakanishi, K.; Rohmer, M. *J. Chem. Soc. Chem. Commun.* **1991**, pp 256-258.

44   Peiseler, B.; Rohmer, M. *J. Chem. Res.* **1992**, (S) pp 298-299, (M) pp 2353-2369.

45   Zundel, M.; Rohmer, M. *Eur. J. Biochem.* **1985**, *150*, pp 23-27.

46   Stampf, P.; Bisseret, P.; Rohmer, M. *Tetrahedron* **1991**, *47*, pp 7081-7090.

47   Günthard, H.H.; Seidel, C.F.; Ruzicka, L. *Helv. Chim. Acta* **1952**, *35*, pp 1820-1826.

48   Marner, F.J.; Krick, W.; Gellrich, B.; Jaenicke, L.; Winter, W. *J. Org. Chem.* **1982**, *47*, pp 2531-2536.

49   Fex, T. *Phytochemistry*, **1982**, *21*, pp 367-369.

50   Zundel, M.; Rohmer, M. *Eur. J. Biochem.* **1985**, *150*, pp 35-39.

51   Bisseret, P.; Wolff, G.; Albrecht, A.-M.; Tanaka, T.; Nakatani, Y.; Ourisson, G. *Biochem. Biophys. Res. Commun.* **1983**, *110*, pp 320-324.

52   Schindler, S.; Bach, T.J.; Lichtenthaler, H. *Z. Naturforsch.* **1984**, *40c*, pp 308-214.

53   Döll, M.; Schindler, S.; Lichthenthaler, H.; Bach, T.J. in *Structure, Function and Metabolism of Plant Lipids*, Siegenthaler, P.-A., Eichenberger, W., Eds.; Elsevier Science Publishers B.V.: Amsterdam, The Netherlands, **1984**, pp 277-278.

RECEIVED May 2, 1994

Chapter 3

Evolution of Sterol and Triterpene Cyclases

Christopher J. Buntel and John H. Griffin

Department of Chemistry, Stanford University, Stanford, CA 94305-5080

Squalene- and oxidosqualene cyclase enzymes catalyze remarkable and complex cyclization/rearrangement reactions. Recently, two cyclase genes have been cloned and sequenced, making it possible to address questions of cyclase evolution and mechanistic enzymology. A comparison of the predicted amino acid sequences of the squalene-hopene cyclase from *Bacillus acidocaldarius* and the oxidosqualene-lanosterol cyclase from *Candida albicans* is presented. These enzymes have different substrate and product specificities but share four regions of substantial amino acid identity, consistent with a divergent evolutionary relationship. These cyclases also share a general richness in tryptophan and tyrosine residues. This feature, the cationic nature of the cyclization/rearrangement reactions, and literature precedents have led to formulation of the *aromatic hypothesis*--that electron-rich aromatic sidechains from tryptophan, tyrosine, and perhaps phenylalanine are essential features of cyclase active sites, where they direct the folding of substrate and stabilize positively charged transition states and/or high-energy intermediates during cyclization/rearrangement.

Squalene- and oxidosqualene cyclase enzymes constitute a family of biocatalysts which catalyze the transformation of (3*S*)-2,3-oxidosqualene to a diverse array of sterols and related triterpenes (*1*). Figure 1 presents reactions representing four subclasses of oxidosqualene cyclase enzymes, which are differentiated by the identities of cyclized intermediate cations that are presumed to be formed on the basis of product structures and mechanistic rationale. Product specificity is species-dependent: In animals, fungi, and the bacterium *Methylococcus capsulatus*, cyclization proceeds to form lanosterol. Higher plants and algae produce a related sterol, cycloartenol. Many plants also convert oxidosqualene to cyclic triterpenes such as euphol and β-amyrin. Lower plants as well as certain species of protozoa and bacteria generally synthesize cyclic triterpenes such as hopenes by cyclization of squalene rather than oxidosqualene (*2*), though cell-free systems from some of these organisms will also catalyze the cyclization of both enantiomers of the epoxide (*3,4*).

0097–6156/94/0562–0044$08.00/0

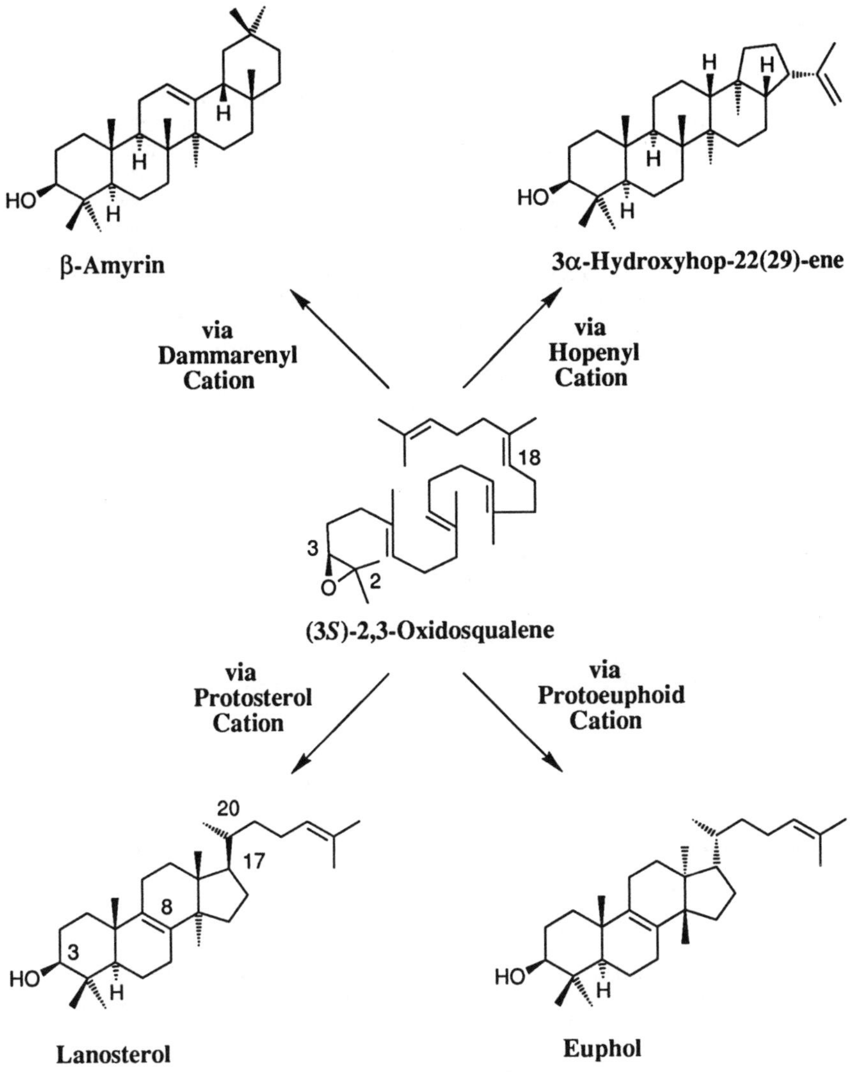

Figure 1. Diversity in oxidosqualene cyclization.

Complexity of Cyclization. Squalene- and oxidosqualene cyclizations are remarkable in that they convert relatively simple acyclic polyisoprenoid substrates to stereochemically rich polycyclic products in a single biosynthetic step. The details of these processes have intrigued scientists for decades. In 1953, Woodward and Bloch used radioisotope tracer experiments to provide evidence that cholesterol and lanosterol are derived via a mechanism involving both cyclization and rearragement of squalene (5), supplanting an earlier suggestion by Robinson involving cyclization alone (6). In 1955, a model for the stereochemical course of squalene cyclization and rearrangement was put forth independently by Eschenmoser, Ruzicka, Jeger, and Arigoni (7), and by Stork and Burgstahler (8). This model provided the basis for testable hypotheses which have been the subject of vigorous investigation from the 1960s to the present. In 1966 the Corey/Bloch and van Tamelen/Clayton groups reported that 2,3-oxidosqualene rather than squalene is the immediate biosynthetic precursor to sterols in mammalian systems (9,10). This finding was quickly extended to sterol biosynthesis in yeast (11) and triterpenoid biosynthesis in higher plants (12-14). Subsequent studies of the oxidosqualene-to-lanosterol transformation, culminating with recent work from the Corey laboratory (15-18), have provided a detailed understanding of the substrate structural requirements, specificity, and stereochemistry for this particular process. The covalent and stereochemical course of the oxidosqualene-to-lanosterol conversion is depicted in Figure 2. The transformation involves no less than ten discrete covalent changes and is compelling in its complexity. Protonation of the epoxide initiates a cascade of four cation-olefin cyclizations which lead to the formation of the protosterol cation. The trans-syn-trans geometry of the protosterol cation ABC ring system derives from a boat-like conformation of the developing B ring during the cyclization process. The protosterol cation then undergoes a series of four suprafacial hydride and methyl 1,2-shifts, followed by a terminating deprotonation at C-9 to afford lanosterol.

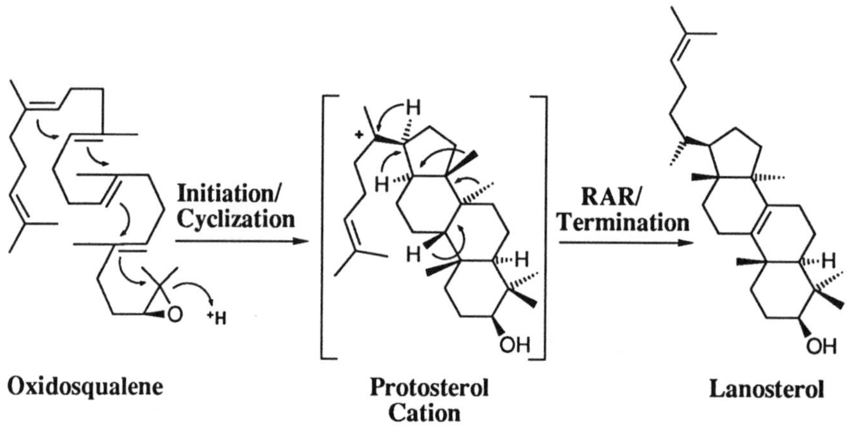

Oxidosqualene   Protosterol   Lanosterol
                    Cation

Figure 2. The complex transformation of oxidosqualene to lanosterol.

Questions of Evolution and Mechanistic Enzymology. The diversity and complexity of oxidosqualene cyclization reactions inspire a series of questions related to molecular evolution and mechanistic enzymology:

1) Have cyclase enzymes evolved by divergent and/or convergent forms of evolution?
2) What are the structures of cyclase enzymes?
3) What are the active site requirements for catalysis of complex cyclization/rearrangement reactions?
4) How is cyclase product specificity determined?

In this paper we will address or provide the basis for addressing these issues in the context of existing theoretical models and recent advances in cyclase biochemistry and molecular biology.

Models for Cyclase Evolution and Action. The detailed understanding of structures, stereochemistry, and dynamics involved in the cyclization/rearrangement of oxidosqualene to lanosterol described earlier stands in stark contrast to our rudimentary knowledge of the enzymes that catalyze this and related reactions. Difficulty with complete purification of membrane-associated cyclase enzymes have until recently blocked progress towards detailed mechanistic understanding of these important biosynthetic processes. In the face of these difficulties and based on an understanding of the stereochemistry and substrate specificity of cyclization reactions, the structures of effective oxidosqualene cyclase inhibitors (*19-22*), and the results of biomimetic acid-catalyzed polyene cyclization reactions (*23*), theoretical models for cyclase evolution and action have been advanced: Nes was the first to consider evolutionary aspects of sterol biosynthesis (*1,24*). With particular regards to polyene cyclization, Ourisson developed a model for evolution of cyclase enzymes (*25-27*) and Johnson suggested the means by which enzymes might promote cyclization of oxidosqualene to the protosterol cation (*28-30*). In the Ourisson model, extant cyclase enzymes evolved from a primitive cyclase through a process of gene duplication and divergence that maintained certain crucial elements within the active sites: an acidic site, a basic site, and nucleophilic sites. This model suggests that differences in product specificity among cyclase enzymes have arisen through mutations which change the shape of cyclizing active sites and/or the positions of the crucial functional groups. In the Johnson model (Figure 3, altered to reflect the recent stereochemical analysis of Corey and Virgil) (*16,18*), cyclization is initiated by proton donation from a specific amino acid residue which acts as a general acid catalyst. The ring connections and stereochemistry of the ensuing cation-olefin cyclization cascade are dictated by delivery of stabilizing, face-selective axial negative point charges to carbon centers which become cationic in transition states and/or high energy intermediates.

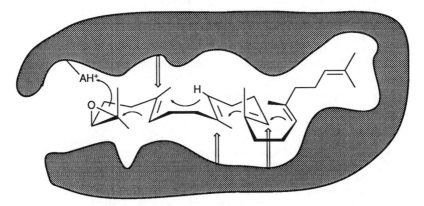

Figure 3. Johnson model for oxidosqualene cyclization. Shown are the proposed chair-boat-chair conformation of the substrate in the enzyme active site, the general acid initiating residue (AH+) and axial negative point charges (arrows).

Recent Advances in Cyclase Biochemistry and Molecular Biology. Two recent developments set the stage for testing of the existing theoretical models and herald a new era in cyclase enzymology.

   Purification. The first development is the purification, to apparent homogeneity, of several cyclase enzymes from natural sources. Abe, Ebizuka and coworkers worked out methods for the purification of oxidosqualene-β-amyrin and oxidosqualene-cycloartenol cyclases from *Pisum sativum* and *Rabdosia japonica*, and of oxidosqualene-lanosterol cyclase from rat liver (*31-34*). These purification protocols have been extended and simplified by Prestwich(*35*) and by Moore (*36*). Corey and Matsuda developed an affinity column bearing a cyclase inhibitor for the purification of the oxidosqualene-lanosterol cyclase from the yeast *Saccharomyces cerevisiae* (*37*). Relative molecular weights (Mr) for these cyclases, which have been purified from eukaryotic sources, are presented in Table I. Also reported in Table I are relative molecular weights for two squalene cyclases purified from prokaryotic sources by the Poralla and Poralla/Ourisson groups: the squalene-hopene cyclase from *Bacillus acidocaldarius* (*38*), and the squalene-tetrahymanol cyclase (*39*). As a group, the cyclases require detergents for their solubilization and activity. In two cases, cyclase enzymes with the same product specificity but from different sources are similar in size--the two oxidosqualene-β-amyrin cyclases isolated were found to be 28 and 35 kDa, while the two oxidosqualene-cycloartenol cyclases were found to be 54 and 55 kDa. Greater variation is observed among the reported relative molecular weights for oxidosqualene-lanosterol cyclases: 26 kDa for the yeast cyclase versus 65-78 kDa for the vertebrate cyclases. The two squalene cyclases are similar in size (72 and 75 kDa) but exhibit different product specificity.

Table I. Properties of Purified (Oxido)Squalene Cyclase Enzymes

Cyclase	Source	Mr (kDa)
Oxidosqualene-β-Amyrin	*P. sativum*	35
Oxidosqualene-β-Amyrin	*R. japonica*	28
Oxidosqualene-Cycloartenol	*P. sativum*	55
Oxidosqualene-Cycloartenol	*R. japonica*	54
Oxidosqualene-Lanosterol	Rat liver	65-78
Oxidosqualene-Lanosterol	Pig liver	75
Oxidosqualene-Lanosterol	*S. cerevisiae*	26
Squalene-Hopene	*B. acidocaldarius*	75
Squalene-Tetrahymanol	*T. thermophila*	72

Cloning and Sequencing. The second development is the cloning and DNA sequencing of two cyclase enzymes: the squalene-hopene cyclase from *B. acidocaldarius* and the oxidosqualene-lanosterol cyclase from *Candida albicans*.. The squalene-hopene cyclase was cloned by Ochs, Poralla, and coworkers using DNA probes derived from amino acid sequences obtained from the purified protein (*40*). DNA sequences containing the gene for the *C. albicans* oxidosqualene-lanosterol cyclase were isolated by Kelly et al. on the basis of their ability to complement cyclase-deficient (*erg7*) mutants of the yeast *Saccharomyces cerevisiae* (*41*). We further characterized these complementing DNA fragments by sequencing, and identified an open reading frame (ORF) encoding the cyclase enzyme (*42*). At the level of DNA sequence, the cyclase genes bear no significant similarity to one another or to sequences in the GenBank database.

Comparison of Squalene-Hopene and Oxidosqualene-Lanosterol Cyclase Sequences.

The predicted amino acid sequences for the squalene-hopene and oxidosqualene-lanosterol cyclases are presented in Figure 4. A comparison of these sequences has important implications for the evolution and mode of action of these enzymes. The squalene-hopene cyclase is predicted to consist of 627 amino acids with a molecular

B. acidocaldarius Squalene-Hopene Cyclase

```
  1  MAEQLVEAPAYARTLDRAVEYLLSCQKDEGYWWGPLLSNVTMEAEYVLLC
 51  HILDRVDRDRMEKIRRYLLHEQREDARGPCTRVGRRTSTRPSRRTSRSSI
101  SACRATRSRCRRRSGSFRARAGSSRRACSRDVAGAGGRISVGEGAHGPAG
151  DHVPRQAHAAQHLRVWLVGSGDRRGALDCDEPPAGVPAARAGARARAVRD
201  RRASAPAPPREGVGGSSTRSTGRCTGLRSCRCTRSAAAEIRALDWLLERQ
251  AGDGSWGGIQPPWFYALIGLKILDMTQHPAFIKGWEGLELYGVELDYGGW
301  MFQASISPVWDTGLAVLALPAGLPADHDRLVKAGEWLLDRQITVPGDWAV
351  KRPNLKPGGFAFQFDNVYYPDVDDTAVVVWALNTLRLPDEPQRDAMTKGF
401  RWIVGMQSSNGGWGAYDVDNTSDLPNHIPFCDFGEVTDPPSEDLTAHVLE
451  CFGSFGYDDAWKVIRRAVEYLKREQKPDGSWFGRWGVNYLYGTGAVVSAL
501  KAVGIDTREPYIQKALDWVEQHQNPDGGWGEDCRSYEDPAYAGKGASTPS
551  QTAWALMALIAGGRAESEAARRGVQYLVETQRPDGGWDEPYYTGTASPGD
601  FYLGYTMYRHVFPTLALGRYKQAIERR
```

C. albicans Oxidosqualene-Lanosterol Cyclase

```
  1  MYYSEEIGLPKTDISRWRLRSDALGRETWHYLSQSECESEPQSTFVQWLL
 51  ESPDFPSPPSSDIHTSGEAARKGADFLKLLQLDNGIFPCQYKGPMFMTIG
101  YVTANYYSKTEIPEPYRVEMIRYIVNTAHPVDGGWGLHSVDKSTCFGTTM
151  NYVCLRLLGMEKDHPVLVKARKTLHRLGGAIKNPHWGKAWLSILNLYEWE
201  GVNPAPPELWRLPYWLPIHPAKWWVHTRAIYLPLGYTSANRVQCELDPLL
251  KEIRNEIYVPSQLPYESIKFGNQRNNVCGVDLYYPHTKILDFANSILSKW
301  EAVRPKWLLNWVNKKVYDLIVKEYQNTEYLCIAPVSFAFNMVVTCHYEGS
351  ESENFKKLQNRMNDVLFHGPQGMTVMGTNGVQVWDAAFMVQYFFMTGLVD
401  DPKYHDMIRKSYLFLVRSQFTENCVDGSFRDRRKGAWPFSTKEQGYTVSD
451  CTAEAMKAIIMVRNHASFADIRDEIKDENLFDAVEVLLQIQNVGEWEYGS
501  FSTYEGIKAPLLLEKLNPAEVFNNIMVEYPYVECTDSSVLGLTYFAKYYP
551  DYKPELIQKTISSAIQYILDSQDNIDGSWYGCWGICYTYASMFALEALHT
601  VGLDYESSSAVKKGCDFLISKQLPDGGWSESMKGCETHSYVNGENSLVVQ
651  SAWALIGLILGNYPDEEPIKRGIQFLMKRQLPTGEWKYEDIEGVFNHSCA
701  IEYPSYRFLFPIKALGLYKNKYGDKVLV
```

Figure 4. Predicted amino acid sequence for the squalene-hopene cyclase from *B. acidocaldarius* (Top) and the oxidosqualene-lanosterol cyclase from *C. albicans* (Bottom). Regions of sequence identity are underlined. Hydrophobic regions of the *C. albicans* cyclase are italicized. Tryptophan residues (**W**) are bold.

weight of 69.5 kDa.   The oxidosqualene-lanosterol cyclase is predicted to be ~20% larger at 728 amino acids and 83.7 kDa.

Hydropathy.   A hydropathy analysis of the squalene-hopene and oxidosqualene-lanosterol cyclases reveals that they are moderately hydrophilic proteins.   As depicted in Figure 5, the squalene-hopene cyclase possesses no obviously hydrophobic sequence.   It has been proposed that this enzyme associates with membranes by virtue of its richness in arginine residues--its arginine content of 10.6% is equal to or exceeds that of 99% of bacterial proteins (*43*).   The oxidosqualene-lanosterol cyclase is not arginine-rich.   Arginine constitutes only 3.7% of the residues of this protein, which is lower than the median value (4.3%) found in proteins from *S. cerevisiae*.   The *C. albicans* cyclase does possess two notable hydrophobic regions which span amino acid residues 329-345 and 645-664.   We propose that one or both of these regions are involved in anchoring this enzyme to membranes.   Further, we suggest that addition of hydrophobic membrane-localizing elements to cyclase enzymes during the course of evolution may have removed selective pressures that maintained alternate mechanisms for membrane-localization, such as that provided by richness in arginine residues.

B. acidocaldarius Squalene-Hopene Cyclase

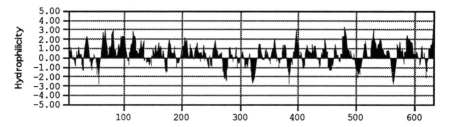

C. albicans Oxidosqualene-Lanosterol Cyclase

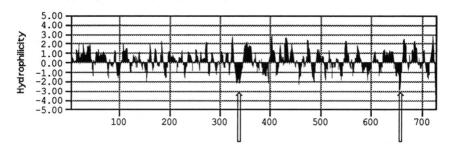

Figure 5.   Hydropathy plots for the squalene-hopene (top) and oxidosqualene-lanosterol (bottom) cyclases.   Arrows indicate notably hydrophobic regions of the oxidosqualene-lanosterol cyclase primary sequence.

Amino Acid Sequence Identity.   The predicted amino acid sequences for the squalene-hopene and oxidosqualene-lanosterol cyclases have no significant similarities to sequences in protein sequence databases PIR 31 and SWISS-PROT 21 (*40,42*).   However, when the predicted amino acid sequences of the cyclases are compared to one another, four regions of substantial similarity are observed.   As shown in Figure 6, these regions range from 30% identity over 81 residues to 46% identity over 37 residues. The presence of these identities suggests a divergent evolutionary relationship between the two enzymes, in accord with Ourisson's hypothesis (25-27). In the squalene-hopene cyclase, the identity regions are nearly contiguous and lie near the center to the C-terminus of the protein. In the oxidosqualene-lanosterol cyclase, the identity regions are split, with two located near the N-terminus and two in the central to C-terminal portion of the protein. This difference, and the fact that the regions of identity assigned by the FASTA program occur in different orders in the primary sequences (1-2-3-4 for the *C. albicans* enzyme versus 1-3-2-4 for the *B. acidocaldarius* enzyme) suggest that the ways in which the elements of secondary structure are connected in these two cyclase enzymes may have changed during the course of evolution.

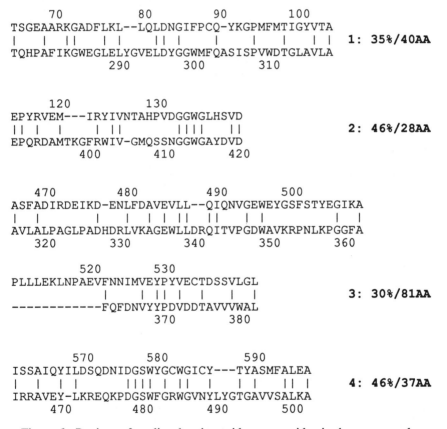

```
        70            80            90           100
TSGEAARKGADFLKL--LQLDNGIFPCQ-YKGPMFMTIGYVTA
 |   |  || |  | |   ||  |   |    |    |   |  | |       1: 35%/40AA
TQHPAFIKGWEGLELYGVELDYGGWMFQASISPVWDTGLAVLA
           290           300           310


        120           130
EPYRVEM---IRYIVNTAHPVDGGWGLHSVD
|| |   |     | ||       ||||    ||                     2: 46%/28AA
EPQRDAMTKGFRWIV-GMQSSNGGWGAYDVD
           400           410           420


      470           480           490           500
ASFADIRDEIKD-ENLFDAVEVLL--QIQNVGEWEYGSFSTYEGIKA
 |   |    |  | |     ||  |    |    |   | |       | |
AVLALPAGLPADHDRLVKAGEWLLDRQITVPGDWAVKRPNLKPGGFA
  320           330           340           350           360


           520           530
PLLLEKLNPAEVFNNIMVEYPYVECTDSSVLGL
            |   | ||| |  |    |   |             3: 30%/81AA
-----------FQFDNVYYPDVDDTAVVVWAL
                     370           380


           570           580           590
ISSAIQYILDSQDNIDGSWYGCWGICY---TYASMFALEA
 |    |   | |     |||| | ||  |    |   |    || |         4: 46%/37AA
IRRAVEY-LKREQKPDGSWFGRWGVNYLYGTGAVVSALKA
           470           480           490           500
```

Figure 6. Regions of predicted amino acid sequence identity between squalene-hopene and oxidosqualene-lanosterol cyclase enzymes.

In all, 66 identities over 186 residues are observed in the four regions presented in Figure 6. The identities include amino acids with basic, acidic, polar, and non-polar sidechains, which could play various roles in the initiation, propagation, control, and termination of cyclization and cyclization/rearrangement reactions. Of particular note are the four conserved tryptophan (W) residues, which in each case occur as part of the sequence GXWX. This motif is found a total of 10 times within the squalene-hopene cyclase sequence and 7 times within the oxidosqualene-lanosterol cyclase sequence. Poralla et al. have hypothesized that this sequence, within the context of a larger "QW Motif", plays a role in the evolution and function of cyclase enzymes. (Poralla, K.; Brünjes, J.; Prestwich, G. D.; Abe, I.; Reipen, I.; Sprenger, G. *FEBS Lett.*, in press.)

The Aromatic Hypothesis. In addition to the specifically conserved tryptophan residues, we had previously noted that the squalene-hopene cyclase and the oxidosqualene-lanosterol cyclase are generally rich in both tryptophan and tyrosine, amino acid residues with electron-rich aromatic sidechains (*42*). The squalene-hopene cyclase contains 3.2% tryptophan and 4.0% tyrosine. The median values for representation of these amino acids in *E. coli* proteins are 1.2% and 2.7%, respectively (*43*). The levels of tryptophan and tyrosine found in the *B. acidocaldarius* enzyme exceed those found in > 95% and ~ 85% of *E. coli* proteins, respectively. The oxidosqualene-lanosterol cyclase contains 3.0% tryptophan and 6.3% tyrosine. When compared to proteins from the yeast *S. cerevisiae* (for which the median values for representation of tryptophan and tyrosine are 0.9% and 3.3%, respectively), the *C. albicans* enzyme possesses these amino acids at levels greater than those found in 99% of all *S. cerevisiae* proteins.

These findings have led us to formulate the *aromatic hypothesis*--that electron-rich aromatic sidechains from tryptophan, tyrosine, and perhaps phenylalanine are essential features of cyclase active sites, where they direct the folding of substrate and stabilize positively charged transition states and/or high-energy intermediates during cyclization/rearrangement. This idea is schematized in the context of the Johnson model in Figure 7. Here, the π-systems of indole, phenyl, and phenol sidechains from tryptophan, phenylalanine, and tyrosine, respectively constitute the "negative point charges" postulated by Johnson. Relative to true anions such as those found in the sidechains of aspartic and glutamic acid, we expect that electron-rich π-systems would be less likely to be alkylated by cationic intermediates formed during cyclization/rearrangement.

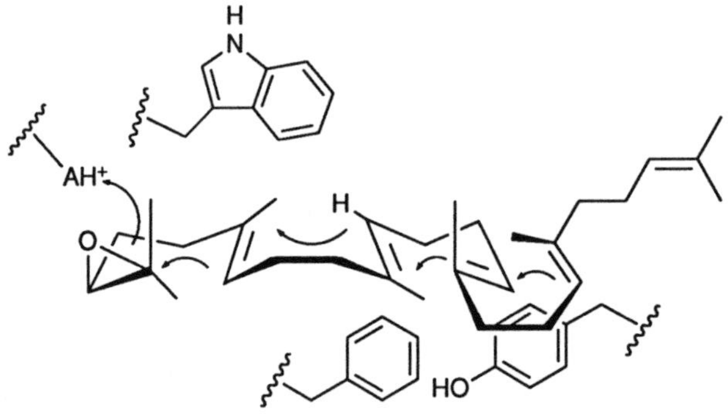

Figure 7. Hypothetical model for involvement of aromatic amino acid sidechains in the cyclization of oxidosqualene to the protosterol cation.

Support for the aromatic hypothesis comes from model studies and protein crystal structure analysis. In biomimetic model studies, Dougherty has shown that cation-π interactions can stabilize both ground and transition states (*44*), and has used these results to predict the involvement of aromatic sidechains in receptors and enzymes that operate on cationic species (*45,46*). In the structural realm, the widespread occurrence of amino-aromatic interactions have been noted (*47*). In addition, a cocrystal structure shows that the myeloma protein McPC603 interacts with the quaternary ammonium ion of its ligand phosphorylcholine through the sidechains of two tyrosine and one tryptophan residue (*48*). Finally, the recent x-ray analysis of acetylcholine esterase reveals that the cation-binding "anionic subsite" of this enzyme is actually an aromatic-rich trench (*45,49*).

We intend to test the aromatic hypothesis through detailed structure/function analyses of cyclase enzymes. We have developed a heterologous expression system for the *C. albicans* cyclase, and are beginning studies directed at determining the effects of both directed and random mutations on the activity and product specificity of this enzyme.

Acknowledgments. This work was supported by Stanford University, a starter grant from NSF (CHE-9018241), a Camille and Henry Dreyfus Foundation New Faculty Award, and a Shell Foundation Faculty Career Initiation Award.

Literature Cited.

(*1*) Nes, W.R.; McKean, M.L. *Biochemistry of Steroids and Other Isopentenoids.* University Park Press, Baltimore, MD: **1977**, pp. 229-270.

(*2*) Ourisson, G.; Rohmer, M.; Poralla, K. *Ann. Rev. Microbiol.* **1987**, *41*, 301-333.

(*3*) Rohmer, M.; Anding, C.; Ourisson, G. *Eur. J. Biochem.* **1980**, *112*, 541-547.

(*4*) Bouvier, P.; Berger, Y.; Rohmer, M.; Ourisson, G. *Eur. J. Biochem.* **1980**, *112*, 549-556.

(*5*) Woodward, R.B.; Block, K. *J. Am. Chem. Soc.* **1953**, *75*, 2023-2024.

(*6*) Robinson, R. *J. Soc. Chem. Ind.* **1934**, *53*, 1062-1063.

(*7*) Eschenmoser, A.; Ruzicka, L.; Jeger, O.; Arigoni, D. *Helv. Chim. Acta* **1955**, *38*, 1891-1904.

(*8*) Stork, G.; Burgstahler, A.W. *J. Am. Chem. Soc.* **1955**, *77*, 5068-5077.

(*9*) Corey, E.J.; Russey, W.E.; Ortiz de Montellano, P.R. *J. Amer. Chem. Soc.* **1966**, *88*, 4750-4751.

(*10*) van Tamelen, E.E.; Willett, J.D.; Clayton, R.B.; Lord, K.E. *J. Amer. Chem. Soc.* **1966**, *88*, 4752-4754.

(*11*) Barton, D.H.R.; Gosden, A.F.; Mellows, G.; Widdowson, D.A. *Chem. Comm.* **1968**, 1067-1068.

(*12*) Rees, H.H.; Goad, L.J.; Goodwin, T.W. *Tet. Lett.* **1968**, 723-725.

(*13*) Corey, E.J.; Ortiz de Montellano, P.R. *J. Am. Chem. Soc.* **1967**, *89*, 3362-3363.

(*14*) Gotfredsen, W.O.; Lorck, H.; van Tamelen, E.E.; Willett, J.D.; Clayton, R.B. *J. Am. Chem. Soc.* **1968**, *90*, 208-209.

(*15*) Corey, E.J. Virgil, S.C. *J. Am. Chem. Soc.* **1990**, *112*, 6429-6431.

(*16*) Corey, E.J.; Virgil, S.C. *J. Am. Chem. Soc.* **1991**, *113*, 4025-4026.

(*17*) Corey, E.J.; Virgil, S.C.; Liu, D.R.; Sarshar, S. *J. Am. Chem. Soc.* **1992**, *114*, 1524-1525.

(*18*) Corey, E.J.; Virgil, S.C.; Sarshar, S. *J. Am. Chem. Soc.*, **1991**, *113*, 8171-8172.

(*19*) Corey, E.J. ; Ortiz de Montellano, P.R.; Lin, K.; Dean, P.D.G. *J. Am. Chem. Soc.* **1967**, *89*, 2797-2798.

(*20*) Duriatti, A.; Bouvier-Nave, P.; Benveniste, P.; Schuber, F.; Delprino, L. Balliano, G.; Cattel, L. *Biochem. Pharm.* **1985**, *34*, 2765-2777.

(*21*) Taton, M.; Benveniste, P.; Rahier, A. *Pure Appl. Chem.* **1987**, *59*, 287-294.

(*22*) Taton, M.; Benveniste, P.; Rahier, A.; Johnson, W.S.; Liu, H.; Sudhakar, A.R. *Biochemistry* **1992**, *31*, 7892-7898.

(*23*) Johnson, W.S. *Bioorg. Chem.* **1976**, *5*, 51-98.

(*24*) Nes, W.R. *Lipids* **1974**, *9*, 596-612.

(*25*) Rohmer, M.; Bouvier, P.; Ourisson, G. *Proc. Natl. Acad. Sci. U.S.A.* **1979**, *76*, 847-851.

(*26*) Ourisson, G.; Rohmer, M.; Poralla, K. *Ann. Rev. Microbiol.* **1987**, *41*, 301-333.

(*27*) Ourisson, G. *Pure Appl. Chem.* **1989**, *61*, 345-348.

(*28*) Johnson, W.S.; Lindell, S.D.; Steele, J. *J. Am. Chem. Soc.* **1987**, *109*, 5852-5853.

(29) Johnson, W.S. *Tetrahedron* **1991**, *47*, xi-1.
(30) Johnson, W.S.; Buchanan, R.A.; Bartlett, W.R.; Tham, F.S.; Kullnig, R.K. *J. Am. Chem. Soc.* **1993**, *115*, 504-515.
(31) Abe, I.; Ebizuka, Y.; Sankawa, U. *Chem. Pharm. Bull.* **1988**, *36*, 5031-5034.
(32) Abe, I.; Sankawa, U.; Ebizuka, Y. *Chem. Pharm. Bull.* **1989**, *37*, 536-538.
(33) Abe, I.; Ebizuka, Y.; Seo, S.; Sankawa, U. *FEBS Lett.* **1989**, *249*, 100-104.
(34) Kusano, M.; Abe, I.; Sankawa, U.; Ebizuka, Y. *Chem. Pharm. Bull.* **1991**, *39*, 239-241.
(35) Abe, I.; Bai, M.; Xiao, X.-y.; Prestwich, G.D. *Biochem. Biophys. Res. Comm.* **1992**, *187*, 32-38.
(36) Moore, W.R.; Schatzman, G.L. *J. Biol. Chem.* **1992**, *267*, 22003-22006.
(37) Corey, E.J.; Matsuda, S.P.T. *J. Am. Chem. Soc.* **1991**, *113*, 8172-8174.
(38) Ochs, D.; Tappe, C.H.; Gärtner, P.; Kellner, R.; Poralla, K. *Eur. J. Biochem.* **1990**, *194*, 75-80.
(39) Saar, J.; Kader, J.-C.; Poralla, K.; Ourisson, G. *Biochim. Biophys. Acta* **1991**, *1075*, 93-101.
(40) Ochs, D.; Kaletta, C.; Entian, K.-D.; Beck-Sickinger, A.; Poralla, K. *J. Bacteriol.* **1992**, *174*, 298-302.
(41) Kelly, R.; Miller, S.M.; Lai, M.H.; Kirsch, D.R. *Gene*, **1990**, *87*, 177-183.
(42) Buntel, C.J.; Griffin, J.H. *J. Am. Chem. Soc.* **1992**, *114*, 9711-9713.
(43) Karlin, S.; Blaisdell, B.E.; Bucher, P. *Protein Eng.* **1992**, *5*, 729-738.
(44) McCurdy, A.; Jimenez, L.; Stauffer, D.A.; Dougherty, D.A. *J. Am. Chem. Soc.* **1992**, *114*, 10314-10321.
(45) Dougherty, D.A.; Stauffer, D.A. *Science* **1990**, *250*, 1558-1560.
(46) Kumpf, R.A.; Dougherty, D.A. *Science* **1993**, *261*, 1708-1710.
(47) Burley, S.K.; Petsko, G.A. *FEBS Lett.* **1986**, *203*, 139-143.
(48) Davies, D.R.; Metzger, H. *Ann. Rev. Immunol.* **1983**, *1*, 87-117.
(49) Sussman, J.L.; Harel, M.; Frolow, F.; Oefner, C.; Goldman, A.; Toker, L.; Silman, I. *Science* **1991**, *253*, 872-879.

RECEIVED May 2, 1994

Chapter 4

Molecular Asymmetry and Sterol Evolution

W. David Nes and M. Venkatramesh

Department of Chemistry and Biochemistry, Texas Tech University,
Lubbock, TX 79409–1061

Sterols are chiral natural products that form biomolecular complexes in cells. In this review, we examine the biomimetic and catalytic forces that may have acted to create -*ex nihilo suiet subjecti*- optical purity in sterol structure and stereochemistry. The causal factors that may have contributed to the maintenance and change in three-dimensional shape of sterols during evolution is also discussed.

The recognition that life and biochemical change on planet Earth is based on the biological properties of simple asymmetric natural products (such as, amino acids and sugars (monocyclic)) has been the subject of thoughtful speculation for over a century (*1,2*). The origin and phylogeny of optical purity in these molecules and proteins has generated much discussion (*3-6*). We now consider a more complex class of compounds - isopentenoid polycycles. Specific isopentenoids, like amino acids, are vital to the survival and reproductive fitness of the cell (*7*). Two groups of amphiphilic polycycles are examined here: sterols (tetracycles) and hopanoids (pentacycles). Their study is of interest for several reasons: first, hopanoids have been found to be the most abundant biomolecules on Earth (*8*); second, hopanoids share structural and functional (as membrane inserts) equivalence with sterols in bacteria (*9, 10*); third, there exist 512 stereoisomers (resulting from nine stereogenic centers) of tetrahymanol (a hopanoid) and 216 stereoisomers (resulting from eight stereogenic centers) of cholesterol, but only one stereoisomer is known to be synthesized of each compound in nature (*7*); fourth, sterols and triterpenes act as geomarkers and chemical fossils (*11*).

In this review we are also concerned with the causal forces, chemical and biochemical, that evolved specifically to make and break stereocenters in sterols. Several opposing views are examined as possible mechanisms for the origin of optically pure sterols and for introducing change in the sterol structure during evolution: Vitality (also referred to as the teleonomic or organization principle (*7,12*))- is the principle where the degree of variation in point mutations leading to a change in sterol enzyme architecture and catalytic efficiency is controlled by the type and amount of reactant and product (and product utilization) in the cell, and by the reaction mechanism, not chance; Vitalism- is the principle in which variation in structure results from non-physiochemical force (*13*); Lamarckism- is the principle that structural

0097–6156/94/0562–0055$10.70/0
© 1994 American Chemical Society

change results from heritable transmission of acquired characteristics and functions; Darwinism- is the principle that structural change in biomolecules is driven by random mutations in heritable material that give rise to new enzymes that catalyze different reactions. Natural selection of the biomolecules, according to Darwinism, operates at the population level by survival of the fittest organisms (*13-15*).

The Molecular Basis of Optical Purity

The most apparent and universal chemical characteristic of natural products is their intramolecular asymmetry which affects the size, shape and function of the compound. Cells apparently have selected particular stereoisomers from pairs of antipodes, because, one of the antipodes has special functions (resulting from its 3-dimensional shape) that maintains life, uniquely. The general phenomenon of asymmetry is sufficiently important to evolution that it is frequently considered with models (Figure 1) that examine the origin and early development of life (*3,5,16*). Phylogenetic trees that show the origin of progenotes (first living entities) and branching lines of descent are continuing to undergo revision (*17,18*). The molecular paradigm for evolution shows the formation of racemic mixtures of compounds precede the formation of optically pure compounds: prebiotic systems being primitive, produced racemic mixtures abiotically and use the antipodes indiscriminately; biological systems being sophisticated, use biochemical recognition systems, e.g., binding of the molecule to protein or lipid, that discriminate (thereby, serving as the mechanism for selection) the order and arrangement of the molecule's constitution. Thus we begin our examination of molecular asymmetry and sterol evolution with a brief review of chemical terms.

Asymmetry and molecular chirality (a term coined by Kelvin from the Greek word Khair meaning hand) are equivalent terms that refer to a geometrical property whereby an object cannot be superimposed on its mirror image by simple rotations and translations (cf. the example given in Human anatomy, Figure 2). That is, the object lacks specific symmetry properties- a plane, axis and center of symmetry (*19*). An

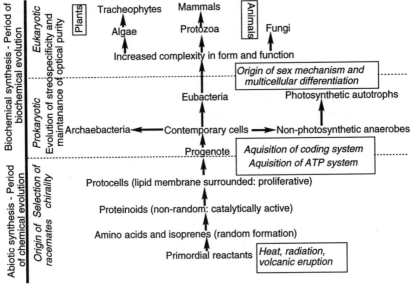

Figure 1. Evolution of stereochemistry

asymmetric object should not be confused with a dissymmetric (homochiral (*19,20*)) object. The two objects possess subtle differences in their symmetry properties; specifically, the latter lacks an alternating axis of symmetry. An achiral molecule is superimposable on its mirror image (Figure 2). Pairs of asymmetric molecules are structurally similar in that nothing is changed intramolecularly, except for the order of the groups that surround the asymmetric centers. As we will discuss, this change in order at the stereocenters may have a profound affect on the molecule's biological properties, even though the compound's physical properties may be rather unaffected.

In chemistry, compounds that possess one or more asymmetric centers are characterized into one of two divisions of isomers (Figure 3). The stereochemistry (i.e., chemistry in 3-dimensional space) of a carbon atom with four different atomic groupings attached to it is referred to as an asymmetric carbon atom. The classic experimental approach to detect and measure asymmetry is to use a polarimeter. This instrument detects the extent to which a substance will rotate a monochromatic plane of polarized light. The direction of rotation depends on: the nature of asymmetry (as influenced by the conformation and configuration of the molecule), the number of asymmetric groups and the order of the groups in the path of the light. Although generally not recognized in chemistry textbooks, the discovery of the polarimeter and its use by Pasteur and others to distinguish optical activity of natural products must be ranked as one of the scientific milestones of the 19th Century.

The relationship between optical activity and chirality is absolute. The term "optical activity" is an experimental one, and a sample that is optically active will rotate the plane of polarized light; by contrast, an inactive sample will not, although it may contain asymmetric centers. There is a special case of optical inactivity in which the equal and opposite rotations from a mixture of equal numbers of molecules of each enantiomer cancel. Such a mixture is said to be racemic and designated by the symbol (+/-).

Compounds that exist in mirror image form and exhibit optical activity are referred to as enantiomers. Every asymmetric molecule can take either of two opposite enantiomorphic forms, i.e., they may possess left-handedness or right-handedness. A compound made up of molecules of one-handedness will, because of its asymmetric electromagnetic field, rotate polarized light in one direction. A compound made of the same molecules, but of opposite handedness, will rotate polarized light in the reverse direction, though in equal amounts. Any substance that rotates polarized light clockwise (as you face the substance, with the substance between you and the light source), in a right-handed direction, is said to be dextrorotary (indicated by- d). If it rotates the light counter-clockwise, or in a left-handed direction, it is levorotary (l). Plus and minus signs may be used instead of d and l. The configurational prefix (capital) D and L is used to assign the absolute configuration (absolute streochemistry) of a molecule. The configuration refers to the arrangement in space of atoms of a molecule of a given constitution, without regard to differences in spatial arrangement that can be brought about without breaking bonds and reforming them in a different way. The relative configuration of a molecule of a given constitution defines the spatial arrangement of its atoms or groups relative to each other or to the corresponding atoms or groups in a molecule of a different compound. The absolute configuration of a molecule defines the absolute spatial arrangement of its atoms or groups when the orientation of the molecule is specified, i.e., the term may be used only with reference to the known actual positions of the atom(s) in space. A system used in modern bioorganic chemistry (but not used necessarily by biochemists) for the specification of the absolute configuration of molecules is the Sequence Rules (*19,20*).

When a compound is placed into the D- or L-series, its absolute stereochemistry is said to be known. Whether the molecule is placed into the D- or L-series is based on the configuration of a specific reference carbon atom somewhere in the molecule and whether this reference stereocenter projects similarly or opposite to the projection of D-

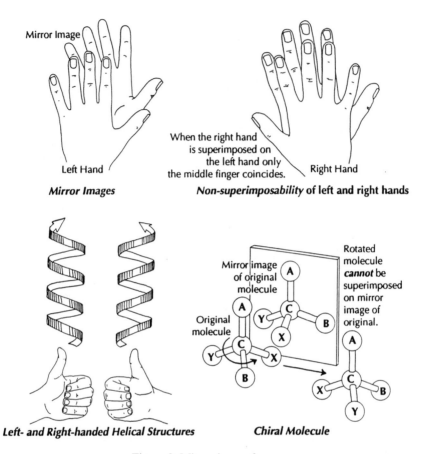

Figure 2. Mirror-image forms.

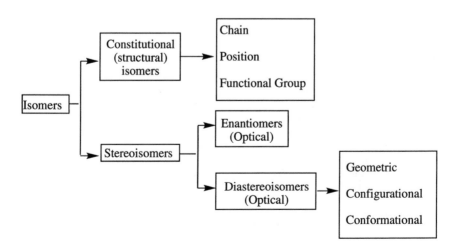

Figure 3. Classification of isomers

or L- glyceraldehyde. The pair of enantiomers D- and L- glyceraldehyde is shown in Figure 4. The configuration of groups around the chiral carbon (position 2) distinguishes D from L. The sequence rules may be used to determine the absolute configuration (either R (rectus) or S (sinistrus)) of glyceraldehyde. The R/S convention describes the order in which the substituents are arranged around an asymmetric carbon atom. Priority is based on the atomic numbers of the atoms to which the asymmetric carbon is attached, high atomic number having priority over low. Next is given priority to atomic weight, then to the number of substituents. Prochiral centers (pro-R and pro-S) often exist in isopentenoid polycycles. They are centers that are not asymmetric, but are potentially capable of reacting with an asymmetric active site to become chiral: that atom or group X, for instance, that would lead to an R stereochemistry, if preferred to the other is called pro-R. Sterols have several prochiral centers in their constitution, e.g., C-26/C-27 (*12*).

It follows from the preceding discussion that the sign of rotation may be determined experimentally, e.g., as d or l, or defined operationally by reference to the known absolute stereochemistry of glyceraldehyde, e.g., as D or L. R and S are defined by the Sequence Rules. The older literature is confusing in this respect because there was a time of transition when the lower case letters were used for configurational family as well as for actual rotation. It is also important to note that one may have complete knowledge of the relative configuration at each asymmetric center of a molecule (e.g., polycycle) without knowing its absolute stereochemistry. For example, the configuration at each asymmetric center of cholesterol, relative to that of the hydroxyl group at C-3, was established before the absolute configuration at any center had been deduced. Until then, the designation, for instance, of the C-3 hydroxyl group as β was arbitrary (though correct by chance) and hence it was not known whether the steric formula in Figure 5 was the correct representation of the cholesterol molecule, or that of its mirror image- ent-cholesterol (*21*). The generic structure and numbering system of sterols that sterol biochemists prefer to use is shown in Figure 5: this system deviates from the 1989 revised IUPAC-IUB system (natural product chemists' nomenclature) that describes a different numbering of carbon atoms and stereochemical nomenclature of select groups in the sterol structure (for a more detailed discussion cf. *12*).

In D-(+)-glucose there are four different asymmetric centers. The maximum number of optical isomers possible is related to the number (n) of asymmetric carbon atoms in the molecule (see also Figure 4). Thus for glucose there are $2^4 = 16$ stereoisomers. By convention for sugars and related hexoses, the absolute stereochemistry of the molecule, D or L, hinges on the configuration of the asymmetric carbon most distant from the carbonyl function, which is C-5 for glucose. The C-5 stereochemistry is then related to the absolute configuration of glyceraldehyde. The common enantiomer of the sugar fructose is termed D because of the stereochemical orientation about the appropriate carbon. However, this enantiomer actually rotates the plane of polarization to the left, hence D-(-)-fructose.

Similarly for amino acids, a single chiral group in the molecule is used to determine the absolute stereochemistry of the molecule as a whole. Here, the stereochemistry of the α-carbon and its surrounding groups, relative to the stereochemistry of glyceraldehyde, is used to assign the configuration of the molecule. Unfortunately, we cannot relate the molecular handedness of other natural products, particularly the isopentenoid polycycles, to the sugars and amino acids, since the absolute configuration of, for instance, the sterol molecule has not been defined.

In order to overcome this deficiency in biochemical nomenclature, we suggest that naturalo-(nat) sterols formed biosynthetically from the cyclization of 3(S)-oxido-squalene may be distinguished from unnatural synthetically or abiotically produced sterols (e.g., enantio-(ent)-lanosterol or ent-cholesterol) by examining their 3-

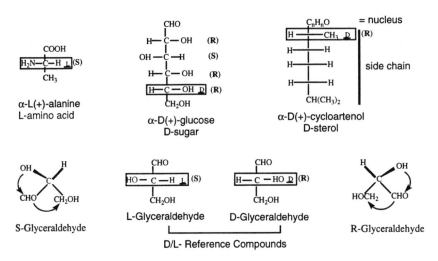

Figure 4. Absolute stereochemistry of molecules

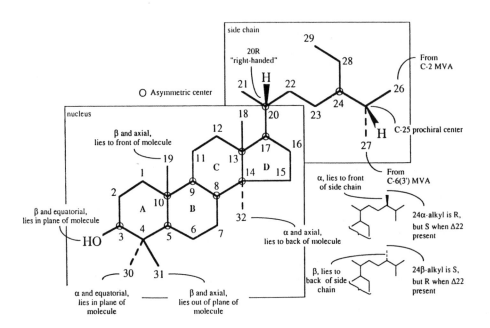

Figure 5. Numbering system of the sterol structure.

dimensional structure, then determining which configurational family, D or L, the molecule belongs to. For this purpose, we select two specific carbon atoms in the sterol constitution for stereochemical analysis and reference them to the stereochemistry of glyceraldehyde, C-20 and C-3; the arrangement about other asymmetric carbon atoms are then ignored. The carbon atom at C-20 has special significance in the sterol constitution in that it is central to the side chain and nucleus and C-3 possesses the equatorially oriented OH-group. Both carbon atoms are intimately involved in the formation of the protosteroid cation, a key intermediate in the cyclization of 3(S)-squalene oxide to sterols. The natural configuration for cholesterol at C-3 and C-20 is R (*21-23*). The configuration at C-20 and C-3 have been found chemically to conform to that of D-(+)-glyceraldehyde (*21*). All sterols that are produced from cyclization of 3(S)-oxidosqualene are known to possess the D-stereochemistry at C-3 and C-20 (*22*), therefore, they may be considered to be D-sterols (Figure 4). Sterols that possess the opposite stereochemistry at C-3 and C-20 to natural cholesterol may be considered L-sterols. The D-cholesterol exhibits a negative rotation and L-cholesterol (ent-cholesterol) exhibits a positive rotation of equal magnitude (*24*). The fully saturated sterol, cholestanol, also possesses a positive rotation (*22*), but its stereochemistry at C-3 and C-20 is similar to that of nat-cholesterol. Because of the structural differences in sterol intermediates and end products, the degree of optical activity may vary in "sterols". The variation in optical activity results from the extent of alkylation, double bond number and position in the side chain and nucleus. Nevertheless, as we noted for the D-sugars, D-sterols may have different experimentally determined optical activities, but they will always possess the same absolute configuration in their global stereochemistry when their constitution originates from the cyclization of 3(S)-oxidosqualene.

Basic to our discussion of sterol evolution is an operating definition for what is a sterol. The definition is confused somewhat by the insistence of natural product chemists to place sterol intermediates, such as lanosterol and cycloartenol, in with the class of natural products, the triterpenoids. Using the natural product chemists' nomenclature, all polycycles (tetracycles or pentacycles) that possess geminal methyl group at C-4 are grouped together as triterpenoids (*12*). We chose to define the term "sterol" by using a combination of chemical and biosynthetic reasoning as follows: a sterol is any chiral tetracyclic isopentenoid that is formed by the cyclization of 3(S)-oxidosqualene through the transition state possessing stereochemistry similar to the trans-syn-trans-anti-trans-anti-configuration, i.e., the protosteroid cation (Figure 6), and retains a polar group at C-3 (hydroxyl), an all trans-anti stereochemistry in the ring system and a side chain configuration of C-20 R. Also to be a sterol (distinct from sterol-like, biosynthetically, e.g., 10α-cucurbitacin, or functionally, e.g., tetrahymanol), the cyclization product must have the ability to undergo transformations to a 4,4-desmethyl product with a Δ^5-bond. The stereochemistry at C-3 and C-20 will remain a structural constant in each of the products of transformation, hence they may be considered D-sterols.

The Origin of Optical Purity in Isopentenoid Polycycles

Life processes operating at the molecular level involve chiral synthetic reactions performed within a chiral environment; presumably this has been so since the origin of the first living entity (*3-5*). However, in prebiotic times, before homochiral (the existence of one enantiomer) biochemistry, the probable product of a chemical reaction would have been the extracellular formation of an equal mixture of two enantiomers, one of which was sequestered selectively by a protocell. The question that has intrigued stereochemists since the time of Pasteur is what force designed optical purity in natural products originally and whether the same force continues to operate in living systems in one form or other.

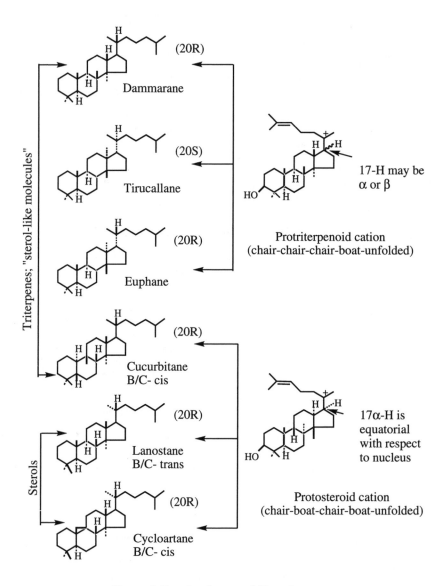

Figure 6. Sterol - triterpene bifurcation

As discussed elsewhere, all the "ingredients" to prepare racemic mixtures of sterol-like molecules anaerobically were available to the protocells (7). Squalene and, squalene surrogates, may have been synthesized in the primal soup from isoprenes that were formed (and then polymerized into different lengths of C-5 units) from hot volcanic eruptions. Squalene, a symmetrical olefin with six double bonds, is gyrosymmetric with respect to a rotational axis perpendicular to the C-12, C-13 bond. Squalene may, as a result of the hydrophobic effect, coil into a ball with the pi-lobes overlapping one another (Figure 7). This coiled structure is known to undergo anaerobic acid-catalyzed cyclization, spontaneously, in an aqueous-polar environment. The cyclization product(s) is expected to possess an hydroxyl-group (derived from water) connected to a structure with two or more rings. When the product is pentacyclic, the thermodynamically most stable structure is one that maintains an all trans-anti-stereochemistry arrangement of the rings. According to the Ruzicka hypothesis, the cyclization of all-trans squalene is determined by the conformation of the squalene backbone and on the geometry of the ring junctions; ring annulation is a concerted process, so that no stable transition state intermediates should be formed or neutralized by positive charges (25). Hence, no hydrogens from the medium are expected to enter the molecule during the cyclization or rearrangement reactions. The cyclization of squalene under abiotic conditions should yield a racemic mixture of products (Figure 7).

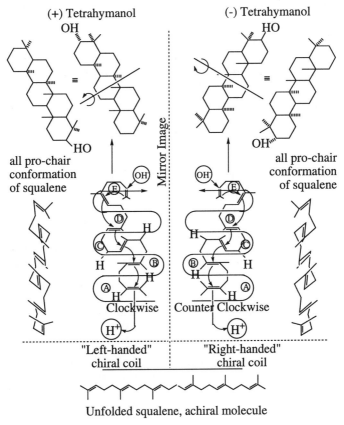

Figure 7. Enantiomeric forms of squalene and tetrahymanol

In the absence of mitigating factors, the probality of forming the D-molecule from a precursor possessing one or all the three pertinent symmetry elements is identical with the probability that the reaction will yield the enantiomer. In order for the reaction to yield unequal amounts of optical antipodes from an achiral substrate, an asymmetric synthesis must occur stereoselectively; some chiral agent must therefore be present with the reactant(s) to affect the product distribution. For polycyclic isopentenoids to form in an enantioselective manner abiotically, some chiral agent(s) must have been present during the abiotic period (so-called chemical evolution). Three chiral agents are often considered as likely candidates for asymmetric induction of amino acids during that period (they could act in a similar manner on polyene cyclizations). They are: (i) chiral surfaces that absorb the isomers stereoselectively, such as clay, quartz, etc., (ii) circularly polarized light derived from sunlight reflected from the sea, and (iii) electromagnetic radiation that affects β-decay of radionuclides (*3- 5*).

Some investigators believe that the adoption of the L-α-amino acid series in homochiral biochemistry of living organisms is based on their abiotic asymmetric synthesis resulting from the parity-violating energy difference between enantiomeric molecules (*3*). If this force was equally important in polyene cyclizations during prebiotic times, it certainly could be an evolutionary mechanism in the evolution of the first sterol. Another force that may have acted to design the optical purity in natural products is Vitalism. Japp in the late 19th Century (*2*) and, more recently Gonzalez (*26*), have considered this mechanism for the origin and introduction of monochirality in living systems. Both speculations point to an extant homochiral biochemistry that evolved to produce natural products from mechanisms that are not biochemical in origin, rather they are assumed to be of chemical or non-physiochemical origin.

An alternative scenario for the origin of optically active isopentenoid polycycles is as follows. They are formed by random abiotic synthesis, as racemates, then accumulated nonenantioselectively into the protocell (protocells are self-organized structures that possess some characteristics of living cells (*27*)) lipid membrane. Once in the plane of the monolayer, the isopentenoid polycycle could diffuse laterally, or undergo "flip-flop" to the other side of the membrane (producing membrane asymmetry) (*28*). The asymmetry of the isopentenoid antipodes was of less importance functionally (viz., with respect to forming biomolecular complexes in the membrane) than was their ability to possess amphiphilic character- which both enantiomers possessed equally. The lack of precise structural features of the first polycycle to act as a membrane insert may have allowed them to control the fluidity of the membrane and the size and shape of the protocell with greater degrees of freedom than in membranes of true living entities that are compositionally more complex. As Haldane speculated 40 years ago, any lipoid or sterol produced abiotically will at once migrate to the place of the strongest gradient and there form monomolecular layers, turning the gradient into a discontinuity (*29*). It follows then that the ability of lipids to self-assemble and use passive diffusion as an uptake mechanism, coupled with the thermodynamics of irreversible processes (*30*) and the hydrophobic effect (*31*) may have been the primal mechanisms that operated to promote the gradual change in molecular organization of the protocell membrane; ultimately producing an unequal distribution of enantiomers with time. The preferential accumulation of one enantiomer could seed for the further selection of optical purity in polycycle isopentenoid structure. A technique of a similar nature- equilibrium recrystallization of racemates, is used to produce unequal distribution of enantiomers (*20*). If there has been in the course of chemical evolution a free chiral product that could be the basis for selection in biochemical evolution (e.g., used in templating of the ancestral cyclase), then architectural suitability alone might account for the observed stereochemistry in tetrahymanol and related hopanoids. Another possibility that we assume more plausible is the stereochemistry of isopentenoid polycycles has been introduced into the molecule, then transformed by biochemical forces that mate the function of polycycles to the evolution of enzyme efficiency. The remainder of the review is concerned with this hypothesis.

The Hopanoid-Steroid Phylogenetic Bifurcation

The replacement of racemic compounds prepared abiotically with optically pure compounds produced biochemically is clearly the hallmark event that distinguishes life from non-life. Bacteria are considered to be the most primitive forms of life that are capable of biochemical dynamism and cell proliferation. They are thought to have originated in an anaerobic (absence of oxygen) environment (*32*); therefore, the "first sterols" must have been formed by anaerobic pathways. In cholesterol synthesis there are several steps involving oxygen, the first committed step is the epoxidation of squalene. Thus sterols, such as cholesterol, may not have been synthesized *de novo* by the first living entities. However, bacteria may have been capable of synthesizing pro-sterols anaerobically by the squalene pathway. Pro-sterols may be defined operationally as pentacycles or tetracycles that are sterol-like, 3-dimensionally. If eukaryotes are derived from prokaryotes and the hopanoid cyclase gave rise to the sterol cyclase, it follows the hopanoid cyclase is the likely candidate that gave rise to "pro-sterols". However, if as Woese speculates (*18*), the basic eukaryotic cell evolved in parallel with the two prokaryote lines, not from either of them (as the name prokaryote implies), then the basic sterol (tetracycle) structure may not have evolved directly from the line that gave rise to hopanoid cyclase descent; rather evolving directly from the progenote, implying different forces were at play acting on the evolution of cyclase efficiency.

The first consideration that the steroid and triterpene cyclase enzymes were related evolutionarily was reported in 1962 by W.R. Nes and coworkers (*33*). Nes and coworkers designed experiments using plants to test the idea that "ontogeny recapitulates phylogeny". They discovered that squalene was converted first to pentacycles (amyrins) then to sterols during pea seedling growth. Presumably, this ontogenetic switch in cyclase activity is related to an atmosphere that was transformed from one that was reducing to one that was oxidizing. Nes, in a series of subsequent papers further considered the role of pentacycles relative to tetracycles in evolution (*23, 33-35*); including hypothesizing first on the functional equivalence between hopanoids (tetrahymanol) and sterols and for the selection of the steroid cyclase relative to the hopanoid cyclase based on the ability of the sterol cyclase to select for a membrane insert that was as thin and flat a molecule as possible.

Ourisson and his coworkers discovered that pentacycles are widely occurring in bacteria. They confirmed (*36*) the earlier observation of Caspi and coworkers (*37*) that hopanoids are formed by anaerobic, proton-initiated attack of squalene. Ourisson and coworkers hypothesized that the hopane-producing squalene cyclase is an ancestral version of the epoxy squalene cyclase(s). They hypothesized (*9,10*) a more detailed phylogenetic sequence of cyclase evolution than Nes, who considered evolution of the isopentenoid pathway generally (cf. *34, 35*). As envisaged by Ourisson and coworkers, primitive biochemistry involves pentacycle cyclization of squalene to hopanoids (anaerobic bacteria) and of squalene-oxide (SO) cyclization to isoarborinol (aerobic bacteria). Advanced biochemistry involves tetracycle cyclization of SO to cycloartenol (in photosynthetic eukaryotes) and lanosterol (in non-photosynthetic eukaryotes). The mechanism by which the hopanoid cyclase evolved into the sterol cyclase is thought to be by point mutations (one amino acid substitution) of a specific functional group in the terminal end of active site, i.e., the nucleophile (basic group). Only a small change in the position of the nucleophile in the active site is required in this model to control the product distribution. The occurrence of Squalene oxide cyclase in bacteria is not well explained in the Ourisson model.

Insights into polyene cyclization mechanisms and structure of the products seemingly support the view that transformation of squalene in the all pro-chair conformation to hopanoids (tetrahymanol) is a "simple cyclization without rearrangements" (37), i.e., it is concerted and primitive. If the stereochemistry of the cyclization is found to be consistent with stereoelectronic requirements for a concerted reaction, it may nonetheless be stepwise. If, however, the stepwise mechanism is found to operate and operates as an evolutionary determinant for change (leading to the variation in cyclized product distribution), then the entropy-gradient involved in tetracyclic and pentacyclic cyclizations becomes phylogenetically significant. The entropy factor in enzyme kinetics reflects the importance of a particular orientation of reactant(s) with the catalyst. Therefore, in considerations of the transformation of a primitive (less efficient) enzyme into a more advanced (more efficient) enzyme, it is necessary to identify the conformational (binding forces) and mechanistic (kinetic and thermodynamic forces involved in bond-making and bond-breaking transformations) aspects of catalysis that have survived or evolved because they are better than conceivable alternatives.

Based on our conformational analysis of sterols, tetracyclic and pentacyclic triterpenes (38-42), we hypothesized a two-pathway model for step-wise polycyclization: one that proceeds through the energetically most favored pathway involving the C-20 prototriterpenoid (dammarenyl) cation, the other proceeding through the C-20 protosteroid cation (Figures 8 and 9). The trans-syn system in the ring systems of the protosteroid cation is energetically least favored compared with the trans-anti system of the prototriterpenoid cation because of diaxial and other steric interactions brought about by the boat form required of the ring B. It will also be seen from Figures 8 and 9 that stabilization of the C-20 cation by 1,2-hydride shift (17H to C-20) or ring enlargement (either from the 13,17 bond or 16,17 bond) will determine the product distribution. The initial folding orientation of the 5,6-, 6,7-, and 7,8- bonds (sterol numbering) determines the configuration and hence conformation of the B/C-ring junction (boat/chair or chair/chair) of the cyclized product, and similarly the orientation of the 16,17-, 17,20-, and 20,22-bonds (sterol numbering) determines the configuration of the 17(20)-bond. Since squalene may fold in any one of several chiral coils, due to rotation about the carbon-carbon bonds, the chirality observed in Nature must then be dependent on the asymmetric conformation induced by the cyclase (conformational control). The structure of the substrate and kinetic and thermodynamic control of the regulatory transition states (e.g., at C-20) will contribute secondarily to the product distribution (mechanistic control). Consistent with our hypothesis is the observations by Rohmer and coworkers that the bacterial pentacycle cyclase may cyclize racemic 2,3- squalene oxide to 3,21-hopane diols and cyclize 2,3-dihydrosqualene to the tetracyclic (20R)-dammar-13(17)-ene and (20R)-dammar-12-ene (43,44).

It is often assumed the (RS)2,3-epoxy squalene is a racemic mixture of enantiomers and orients in the squalene oxide cyclase in a different conformation than squalene (37,45,46). As shown in Figure 10, chiral substrates must compliment enzymes in a specific spatial orientation. Attachments between substrate and enzyme generally involves two complimentary sites. The reactivities of the substrate (distinct from their binding features) and enzymic sites must be different and asymmetric to one another. The substrate may have two but not more identical groups (Y and B in Figure 10) interacting with the enzyme and two dissimilar groups (A and X in Figure 10) associated with the central carbon atom. Functional group B and Y of the substrate may also be dissimilar. Regardless of their nature, there is only one fit on the active site of the enzyme surface. Generally, in enzymology we are concerned with a one-to-one relation between intermolecular asymmetry involving local functional groups on the substrate and enzyme; but, other interactions may exist that involve the 3-dimensional

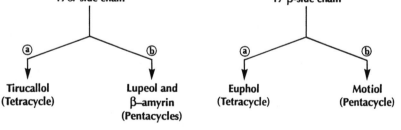

Figure 8. Cyclization of squalene oxide through the C-20 protosteroid cation to tetracycles and pentacycles.

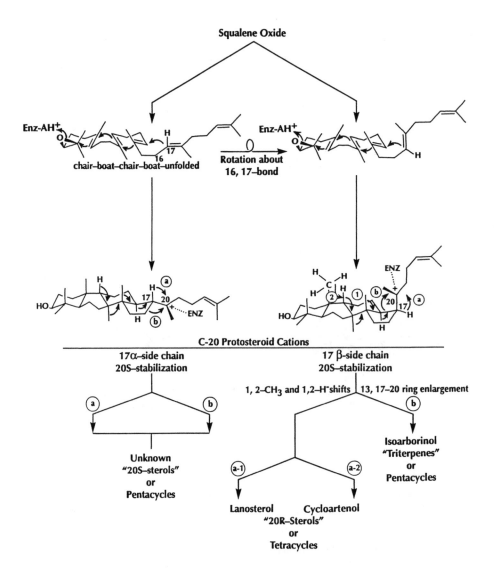

Figure 9. Cyclization of squalene oxide through the C-20
prototriterpenoid cation to tetracycles and pentacycles.

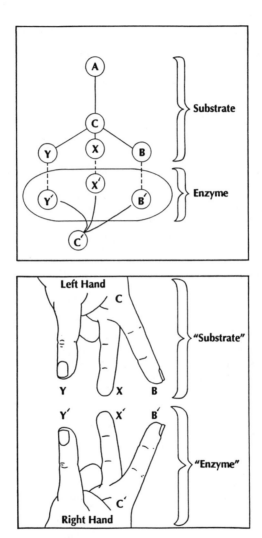

Figure 10. Asymmetric binding of enzyme to substrate.

shape and dipole momment of the structure as a whole (cf., our "steric-electric plug" model (*12*)).

Squalene is formally a symmetric molecule. However, upon binding, the substrate is forced into a conformation approximating the transition state that forms the initial activated complex (conformational control), after which proton-initiated attack on the nucleophilic center at the proximal end of the squalene coil promotes cyclization (mechanistic control). We hypothesize the nucleophilic centers at either end of coiled squalene control asymmetric binding of squalene to the cyclase. The double bonds at either end of squalene produce the necessary two points of attachment to the enzyme, they are similar to the identical functional groups B and Y shown in Figure 10. The optimal substrate for cyclase activity is one that fits the binding site in such a way that the catalytic reaction may proceed at a maximal rate for that enzyme. However, cyclases that have high affinities for their optimal substrates may also bind related compounds to some extent - this should not be viewed as showing the enzyme lacks a high degree of substrate specificity or is primitive. For instance, there may be productive binding between squalene oxide and the cyclase, but electronic effects that relate to the presence of the oxirane ring may enable squalene oxide to be catalyzed at a faster rate by the enzyme (a measure of increased enzyme efficiency). Moreover, squalene oxide variants that possess 25- or 35-carbon chains could be cyclized, the resulting sterols would possess a short or long side chain (*7*). The new sterol should be found to be a better membrane insert than hopanoids, which they are not (*7*).

3(S)- and 3(R)-Squalene oxide may orient in the cyclase in the same coiled conformation as squalene, i.e., relative to ring A, C-4 is up and the molecule spiraling in the left-handed helix (Figure 11). Proton-initiated attack on the three squalene substrates coiled in a similar enantiomeric conformation should produce products (tetracyclic and pentacyclic) that are mirror-image related. If only a single enantiomeric conformation is bound to the cyclase, then the stereochemical outcome of the subsequent cyclizations is fixed and thereafter determined relative to the absolute configuration of the A/B-trans ring junction, which all naturally occuring sterols and triterpenoids possess. The resulting absolute stereochemical control leading from the left-handed coiled structure is a natural consequence and would require no further conformational control (as shown by Johnson and coworkers (*47*)). In the natural coiled conformation, the 2,3(RS)-oxido squalenes become diastereoisomers, not enantiomers. The expected products of catalysis by a cyclase enzyme (primitive or otherwise) using the diastereoisomers of squalene oxide should be epimers (not antipodes) with the 3-α– and 3-β– oriented hydroxyl group. Interestingly, molecular antipodes are produced in some isopentenoid cyclizations that involve fewer numbers of isopentene units, viz., those producing the 10-carbon monoterpenes (*48*).

Since conformational and mechanistic control of cyclization is intimately associated with the quartenary structure of the cyclase, and the primary structure is intimately associated with the gene encoding for it, it follows there may be one gene for each natural product formed by squalene (and its oxide) cyclization: the "one gene-one triterpene/sterol" hypothesis. The mechanistic similarity between the prokaryote and eukaryote cyclases presents a strong circumstantial case for divergent evolution. In order to evolve, cyclase efficiency must have improved, but how? Many investigators (*49,50*) focus their attention on the evolution of enzyme function rather than the function of enzyme products. Koshland hypothesized that enzyme function must be subject to incremental improvements by small changes in enzyme structure, and the structure must be capable of incremental modification in a random manner (*49*). Classic Darwinism supports the role for randomness (that which effects the gene mutations). When this view is placed in terms of cyclase evolution, chance becomes the source of variation, and hence of product distribution (hopanoid versus sterol). However, variation in the structure of the cyclized product may not of necessity be a random process, nor one that is influenced by changes in the amino acid position that affects the

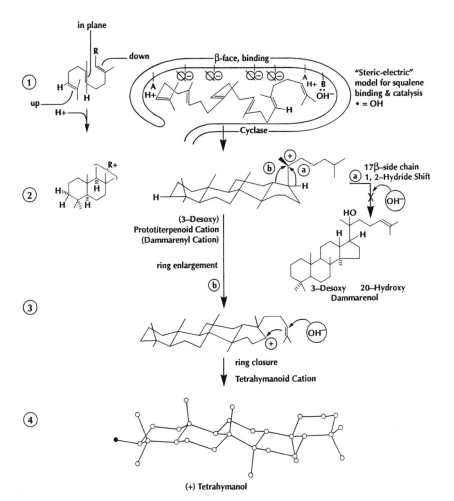

Figure 11. Hypothetical pathway for the cyclization of squalene to tetrahymanol. • in the X-ray structure represents the OH group. Ⓠ Represents an aromatic amino acid delivering a negative ⊖ point charge. The "OH⁻" may stabilize C-20 cation as a result of conformational changes in the cyclase resulting from allosteric modulation. The Base (B) delivering OH⁻ at the end of the cleft is "activated during catalysis to stabilize the carbenium ion(s) formed at C-20 or C-24.

juxtaposition of the substrate on the surface of the enzyme so that only the rate of catalysis is increased. Mutations may also affect transition state catalysis that control product distribution.

Based on the "biogenetic isoprene rule" (*25*), it is evident there are several, albeit fixed number of structural variants of squalene (and its oxide) pentacyclic- and tetracyclic-cyclizations. Clearly, there has been a change in the structure of polycycles during the evolution of bacteria to eukaryotes. However, the optical activity of the hopanoids may not have changed during this transition, suggesting that the ancestral cyclase may have already been fairly sophisticated in its ability to produce an unequal mixture of enantiomers. If a high degree of catalytic efficiency is slow in evolving, the essential steric features of the reaction are likely to be conserved (Doolittle has distinguished functional and mechanistic convergence from structural and genuine sequence convergence (*51*)- we infer the former meaning) in evolution once they have appeared, eventhough the substrate may be modified (e.g., squalene conversion to oxidosqualene) or a coenzyme changed. We hypothesize the ancestral cyclase activity has been slow in evolving, but nevertheless, modified with time. The cyclase may originally have possessed broad substrate specificity, allowing for a variety of cyclized structures to be formed (similar to the biomimetic cyclizations of squalene oxide and squalene where one to several rings are formed (*52*). This characteristic coupled with an undeveloped regulatory system (where the transition state and the ground state of the substrate bound to the enzyme are of equal magnitude (*53*)), may have allowed the first cells to produce a mixture of polycycles. Adaptive forces must have been at play to cause bacterial cyclases to produce pentacyclic products at the expense of tetracyclic products. We assume selection for increased catalytic efficiency resulted from a mating of cyclase activity to the function of the product(s). The function of the cyclized product to act as a membrane insert in bacteria may have been the primary force that selected for pentacycle products from energetically more favored tetracyclic products, e.g., 20-OH, 3-desoxy dammarene (Figure 11). In eukaryotes, the ability for sterols to play multiple roles (structural, hormonal, and as intermediates to other regulatory compounds) may have provided the selective advantage for sterols to be the primary products and triterpenes to be the secondary metabolites of squalene oxide cyclization.

   Ourisson *et al.*, have similarly considered the importance of function in their account of the molecular evolution of cyclases (*9,10*). They assume that sterols and hopanoids share functional equivalence in nature. We suggested another possibility. Pentacyclic triterpenoids are biogenetically related to the tetracyclic sterols, but they are structurally different (*39*). The variations in structure, while arising from mutations in the cyclase, are actively, rather than passively, retained because they serve a purpose, e.g., in one or another temporally expressed function(s) during development. If so, then we might expect that, in contrast to what is generally thought, sterols (which came after hopanoids in evolution) are alternatives to, rather than replacements of, hopanoids. As such, we should find distinguishing structural characteristics in sterols that hopanoids and related triterpenoids (isoarborinol) do not possess. These novel features should be found in sterols to increase their fitness compared with that of hopanoids. The different cyclized products of phylogenetic interest considered by Ourisson and coworkers (*9,10*) and ourselves (*38-42*) is shown in Figure 12. The interesting aspect of the 3-dimensional geometry of the pentacycles relative to the tetracycles (with the side chain to the right and in the staggered/pseudocyclic conformation) is that the former group is somewhat shorter and volumetrically smaller than sterols. In studies to test the suitability of triterpenoids and sterols as membrane components in fungi, we discovered that only sterols were effective membrane components (*54,55*). In vascular plants, we isolated many pentacyclic and tetracyclic triterpenoids, but they are not present in the cell membranes of actively proliferating cells, they are produced as secondary metabolites to form the surface wax (*7*). In fungi, pentacyclic triterpenoids may also be synthesized, but their occurrence is developmentally delayed relative to sterols (*7*). The switching mechanism from sterol to triterpene production may have evolved as a result of increased enzyme efficiency that caused structural refinement in the cyclization products, rather than a mechanism of replacement of one functional detail by another in the structure of the cyclase. The former mechanism is related to the organizational principle as the driving force of change whereas the latter mechanism is consistent with chance as the driving force.

   Griffin has recently hypothesized that aromatic amino acids in the active site may act as negative point charges (in accordance with the Johnson hypothesis) to affect cyclization (this volume). In keeping with this speculation (cf. Figure 11), we hypothesize further the replacement of key aromatic amino acids in the enzyme cleft with charged, neutral, or other aromatic acids may affect the tertiary and quarternary structure of the cyclase, thereby affecting the enzyme's conformational control on the ground state substrate-enzyme complex. This view differs from that of others that speculate changes in conformational control result from point mutations directed to the nucleophilic center that mediates stabilization of the final carbenium ion produced during cyclization (*9,10*).

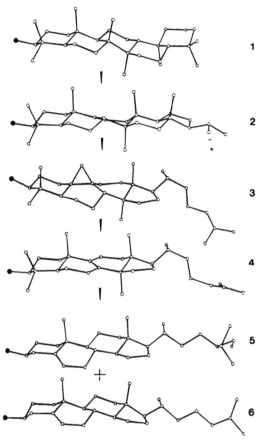

Figure 12. Hypothetical phylogenetic sequence of hopanoids giving rise to sterols. X-ray crystallographs obtained by Dr. Jane Griffin and coworkers of the Medical Foundation of Buffalo. (1: Tetrahymanol, 2: Isoarborinol, 3: Cycloartenol, 4: Lanosterol, 5: Sitosterol, 6: Cholesterol)

The Evolution of Sterol Stereochemistry

The sterol pathway is thought to originate with bacterial heterotrophs that developed enzymes that epoxidize squalene to squalene oxide (*56-58*). In keeping with the idea that ancestral enzymes lack a high degree of mechanistic specificity, the ancestral epoxidase probably produced a mixture of 3(S)- and 3(R)- squalene oxide. The squalene oxide diastereoisomers were then presented to the ancestral cyclase, which cyclized both substrates anaerobically to tetracyclic and pentacyclic products. They are not aerobic cyclizations, nor was the environment at this time likely aerobic, since photosynthesis came later. Thus the first sterols were not further transformed and were used structurally intact. The first cells likely produced pairs of C-3 OH epimers. The difference in configuration of the OH-group could influence sterol-lipid interaction in the membrane and this might be the source for selection. That squalene and squalene oxide coil in the same conformation, so that hopanoids occur in the same enantiomeric form as sterols explains why D-sterols predominate in nature. If however, the anti-universe were to exist such that L-sterols (Figure 13) were formed, would that matter? We have incubated cholesterol and ent-cholesterol in a sterol auxotroph, the yeast mutant GL-7, and found that both compounds were inserted into the membrane and supported growth equally well as the native fungal sterol ergosterol (*59*). Moreover, lanosterol, cholesterol and hopanoids are used equally well as growth support agents in bacteria (*9,10*). Thus, membrane function of these compounds fails to explain the selection of one enantiomer over the other, or the selection of 3-OH pentacycles (e.g.,

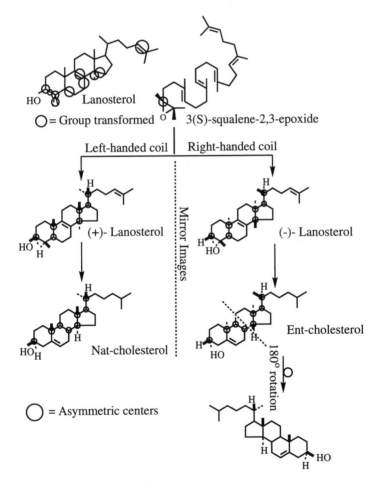

Figure 13. Cyclization of 3(S)-Squalene-2,3-epoxide to enantiomeric tetracycles

hopanoids) over tetracycles (e.g., lanosterol) by primitive bacteria (or even advanced bacteria that may synthesize both groups of polycycles). But membrane function may be important, nevertheless, to adaptation and speciation (the developement of new and novel organisms), particularly as changes in the composition of the membrane asymmetry may affect the size and shape of the lipid leaflet, which in turn, may promote changes in cell size, number, differentiation and growth rates (*7*), factors that impact on reproductive fitness.

Phylogenetic studies have indicated the existence of a two-pathway model of sterol biosynthesis, the so-called lanosterol-cycloartenol bifurcation (Figure 14) (*58*). Both the lanosterol and cycloartenol pathways evolved from bacteria- either non-photosynthetic heterotrophs or blue -green bacteria, respectively. The former pathway evolved in organisms that are completely non-photosynthetic, whereas the latter pathway evolved in organisms that used oxygenic photosynthesis at some point in their evolution. Thus organisms, such as green plants, slime molds (*56*), amoeba (*60*), algae

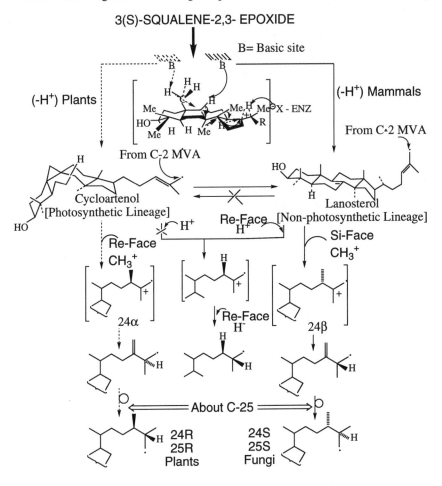

Figure 14. Cycloartenol - lanosterol bifurcation

grown in dark, and parasitic flowering plants that are "white" (*58*) all cyclize squalene oxide to cycloartenol. Alternatively, fungi and animals cyclize squalene oxide to lanosterol (*61*). The inability for the cyclase that gives rise to the lanosterol-based pathway to operate in organisms with a photosynthetic lineage may be interpreted to imply that the cycloartenol pathway evolved in parallel to the lanosterol-based pathway. Parallel evolution implies the lineages have changed in similar ways, so that the evolved descendents are as similar to each other as their ancestors were.

Which came first- the lanosterol- or the cycloartenol-based pathway? Ourisson *et al.*, suggest that the cycloartenol-based pathway evolved first (*9,10*). They reasoned that cycloartenol and lanosterol possess different shapes- bent versus flat. The different sterol shapes were discriminated by primitive cells. Both sterols possess a 14-methyl group protruding on its α-face. In the flat conformation (induced by the 8,9- bond in lanosterol), the 14-methyl group destroys planarity of the backside of the molecule. This structural feature alone is regarded as the harmful group that prevents beneficial sterol-lipid interactions (*62*). In the bent conformation (induced by the syn relationship of the 9,19-cyclopropyl group and 8-hydrogen in cycloartenol) the 14-methyl group is buried and shielded from its harmful affects (*62*).

The accepted mechanism of cyclizaion of squalene oxide to lanosterol and cycloartenol involves the intermediacy of the C-20 protosteroid cation, which maintains the chair-boat-chair-boat conformation. Based on the structure of the cyclization products and mechanistic reasoning regarding the necessary intervention of a nucleophile intermediate (X$^-$, enzyme) to allow for neutralization of the charge generated at C-9 during the subsequent backward rearrangement to stabilize the C-20 cation in cycloartenol synthesis, some investigators believe that the position of the base in the enzyme that abstracts the proton at C-9 (lanosterol synthesis) and the proton from the C-19 methyl group (cycloartenol synthesis) is spatially different (*63*). Hence, a single point mutation in the amino acid composition of the cyclase may explain the evolutionary progression of a cycloartenol-based pathway into a lanosterol-based pathway. It does not explain, however, the ordering of intermediates between lanosterol and cycloartenol and the functional end products, nor of the structural diversity of sterols.

In addition to the squalene oxide to cycloartenol cyclase, the set of phytosterol enzymes- the S-adenosyl-L-methionine C-24 methyl transferases (SMT) and the 9,19-cyclopropyl to 8- isomerase (COI) are considered as phylogenetically significant to plant evolution. The plant SMT is assumed to catalyze Re-face methylation of bent sterols, producing 24-α alkyl sterols (*64,65*), whereas the fungal SMT acts on flat sterols producing the C-24 alkyl epimer by a Si-face mechanism of attack (Figure 14). Plants (and marine organisms (*67*)) produce a variety of 24-alkyl sterols that are epimeric at C-24 (Figure 15), suggesting that several SMT enzymes may have evolved; each SMT may have a unique amino acid composition in the active site. Plants are also thought to possess a COI that is not similarly synthesized by animals or fungi. Hence, cycloartenol should not undergo transformation by organisms operating the lanosterol-based pathway (*68*). Another phylogenetic marker, which is recognized medically, is the C-24 alkylation (plant/fungi)- reduction (animals) bifurcation (*62*). It is assumed that neither plants nor fungi should transform cycloartenol or lanosterol to "animal" cholesterol, since these organisms are thought to lack the 24,25-reductase.

Recent mechanism studies involving incubations with stable isotopes (*66,69,70*) and studies by us on conformational analysis (*38-42*), distribution (*7*), structure-function/metabolism relationships (*54, 55*) and mechanistic enzymology of sterol transformation (*12,56*), negate the otherwise attractive explanation hypothesized by Ourisson *et al.*, (*9,10*) for evolution of the sterol pathway beginning with the cycloartenol-based pathway. The new data indicate that : the lanosterol-based pathway may have evolved before cycloartenol-based pathway, the squalene-oxide cyclase and

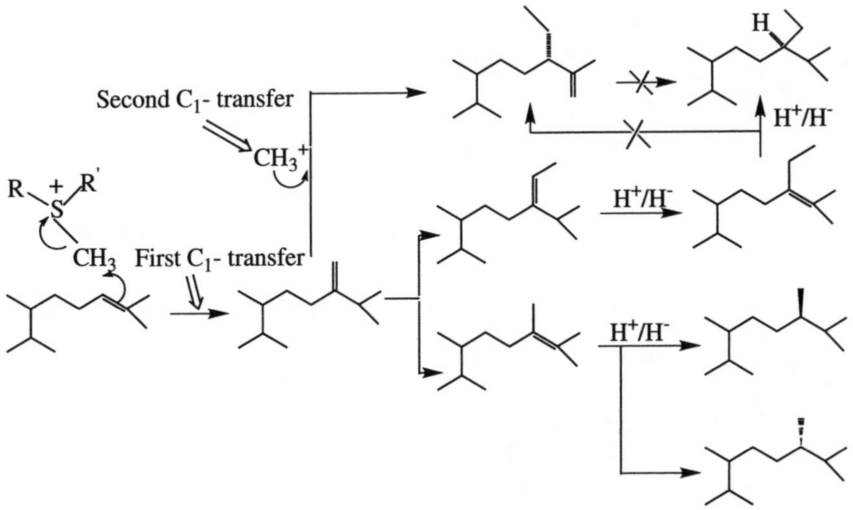

Figure 15. Transmethylation of sterol side chain

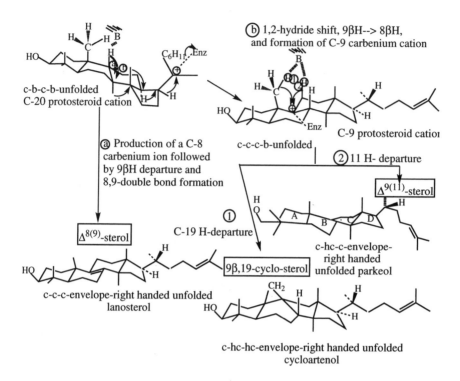

Figure 16. Hypothetical mechanism for the production of lanosterol, cycloartenol, and parkeol

COI may have a common ancestry, key enzymes that metabolize the sterol side chain (the SMT and the $\Delta^{24(28)}$ to 24(25) isomerase) may be the same enzyme and have evolved from a common genetic pool, and sterol allosterism of select sterol enzymes may account for some of the structural diversity of sterols in nature.

As shown in Figure 16, the production of cycloartenol requires a reaction step that is not required in the cyclization of squalene oxide to lanosterol. The additional transition state, the production of C-9 protosteroid cation, may involve the same base used in lanosterol synthesis to abstract the H-9. This follows from the observation of Arigoni that enzyme-catalyzed ring closure and acid- catalyzed ring opening of cycloartenol occurred with the same steric course, i.e., retention of configuration (cited in *19*). The steric course of the cyclopropane ring-opening during cycloartenol transformation is of interest because the retention mode indicates that the base in the active site which abstracts the proton is approximately situated over ring-C of the substrate (Figure 17). This is also the expected position for the basic group responsible for the loss of the 9-β hydrogen during lanosterol synthesis (*19*). Additional biomimetic studies involving acid-catalyzed ring opening of cycloartenol (71) have shown that several products may be formed, including the conformationally flat parkeol and the bent 10-α cucurbita-5,24-dienol (*40*). In plants that synthesize cycloartenol, squalene oxide has been found to cyclize to parkeol (*72*) and 10-α cucurbita-5,24-

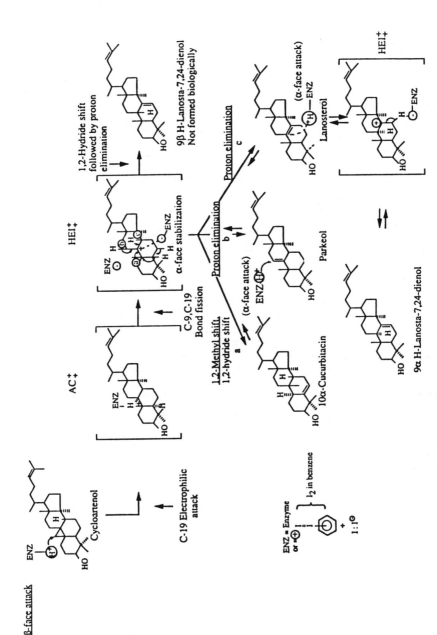

Figure 17. Hypothetical mechanism for the acid-catalyzed cyclopropyl ring opening of cycloartenol.

dienol (73), without the intermediacy of cycloartenol formation. The possibility exists that parkeol and 10α-cucurbitacin are produced by the same cyclase that produces cycloartenol. Allosteric effectors and/or competitive modulators might influence the cyclase conformation to shift from one producing cycloartenol to one producing parkeol or 10α-cucurbitacin. It follows, that subtle changes in the reaction progress (kinetic control) of cyclization may cause alternative tetracycles to be produced. The initial advantage of producing one structural isomer over the other by the ancestral cyclase may be related to the reactivity of the newly evolving enzyme to producing select sterols rather than to the effectiveness of one product to serve better as membrane insert. The ability of select products to be further transformed rather than to accumulate in the cell and act as competitive modulators of cyclization may also be significant.

If parkeol and 10α-cucurbitacin are alternative products, why not lanosterol from the cycloartenol-producing cyclase and vice versa? When a microsomal preparation from corn embryos was incubated with a squalene oxide variant that lacks one of the C-4 geminal methyl groups (2,3-trans-1'-nor-2,3-oxidosqualene), the cyclase produced 31-nor lanosterol and 31-nor cycloartenol (74). Thus, the cycloartenol cyclase of corn may produce either 8,9-sterols or 9,19-cyclopropyl sterol, depending on the nature of the substrate delivered to the enzyme. In plants, we speculate the kinetically favored outcome of cyclization is cycloartenol whereas in fungi and animals it is lanosterol. On thermodynamic grounds, the ancestral cyclase probably operated the lanosterol route first, since it involves one less transition state in its formation. Although lanosterol is likely the kinetically favored cyclized product in fungi, we expect that the native fungal cyclase may be found (when "pushed") to produce other products, such as parkeol and cycloartenol. Barton and coworkers have isolated parkeol from yeast (75) and we have demonstrated that parkeol and cycloartenol (apparent ring-opening to the 9(11)-position) are converted to ergosterol by yeast (55). Thus COI operates in fungi- or is it that the 8(9)- to 7(8)-isomerase catalyzes the opening of the cyclopropane ring? If our assumptions are correct, that parkeol and cycloartenol are produced by the same cyclase then yeast may be capable of producing cycloartenol. Interestingly our ability to transform cycloartenol and parkeol to ergosterol can only be demonstrated when using the sterol auxotroph GL-7 that is adapted to be heme-competent (55). In the wild-type cells endogenously formed lanosterol competes with and prevents uptake and transformation of structural isomers of lanosterol (59). Competition between pairs of structurally similar compounds may be an evolutionary strategy that operates to regulate expression of enzyme reaction coordinates and hence of product distribution.

The importance attributed to whether the structural isomers are flat or bent may have little to do with evolution of the sterol pathway. As shown in Figure 12, cycloartenol (and biosynthetically related 9,19-cyclopropyl sterols that lack C-4 methyl groups (Griffin, Allinger and Nes, unpublished)) and lanosterol are flat in the solid state. Moreover, we have shown they are similarly planar in solution (38), in artificial (76) and biological membranes (55), and when bound to enzymes (SMT) that catalyze sterols (42,77). The structural feature that is harmful to sterol-lipid/protein interactions is likely the C-4 methyl group(s). As shown by chromatography, the degree of substitution at C-4 of the sterol affects the hydrogen bonding strength of the polar group at C-3. For instance, in thin-layer chromatography (TLC) the mobility (R_f value) of C-4 methyl and C-4 desmethyl sterols is different (77). The 14-methyl group does not prevent the sterol serving as a competent membrane insert (55), nor is it important to the TLC behavior of sterols (77). Nevertheless, evolutionary pressures may have caused its removal to allow for the Δ^8-bond to be transformed into a Δ^5-bond (7).

The metabolism of the 24,25-double bond appears to be another phylogenetically significant area of sterol metabolism. Less advanced organisms produce either 24β-ethyl- (algae) or 24β–methyl- (fungi) sterols whereas vascular plants produce 24α–ethyl- sterols (*2,3*). However, some algae synthesize 24α-alkyl sterols (Patterson, this volume) and tracheophytes may, depending on their stage of development, produce a mixture of C-24 alkyl epimers (*78*). Recently, much new information has been obtained regarding the molecular details of the biomethylation-deprotonation step of 24-alkyl sterol biosynthesis. The mechanism may involve an ionic process of transmethylation, compared with the "X-group" mechanism postulated previously by Cornforth (*79*). Methylation of the 24,25-bond may proceed to one of several alternative products (Figure 15). Fungi and marine invertebrates may produce mixtures of C-24 alkylated products; 24α-sterols have not been detected in fungi and 24-methyl sterols have not been detected as the major 24-alkyl sterols in vascular plants, only in algae do 24-methyl sterols predominate the sterol mixture (Patterson, this volume and *80*).

In ergosterol biosynthesis, Arigoni hypothesized that a negatively charged counter-ion located next to the sulphonium center so as to facilitate its fixation at the active site of the SMT is identical with the basic group ultimately responsible for abstraction of the proton from the methyl group attacking the 24,25-bond (*66*). As a consequence, addition of the methyl group and removal of a proton must necessarily occur from the same face of the plane of the original trigonal bond of the substrate. It follows, the resulting carbenium ion formed during ergosterol transmethylation undergoes a 1,2-hydride shift on the opposite face followed by loss of a proton, producing an exocyclic methylene at C-24. These observations differ from biomimetic methylation studies of olefins where the favored product was a $\Delta^{24,25}$-methyl structure (*81*), which Arigoni indicated could not be produced in ergosterol synthesis for mechanistic reasons.

New data indicate that in phytosterol synthesis, the mechanism of biomethylation of the 24(25)-bond and the 24(28)-bond (second methylation) is mechanistically similar to fungal methylation, i.e., by Si-face attack (*69,70*). The α-configuration at C-24 introduced into phytosterols is the result of reduction of the 24(25)-bond, and the double bond is formed late in the transformation process by isomerization of a 24(28)-sterol (*12*). The isomerization of the 24(28)-bond may occur on the same active site where transmethylation proceeds. The observation that bioformation of $\Delta^{23(24)}$- and $\Delta^{25(27)}$- sterols also occurs during C-24 alkylation (*82*), suggests that in both single and double transmethylations some stabilization is received by the formation of carbenium ions at C-23, implying that the C-24 carbenium ion precursor is significantly long-lived and that the usual proton transfer from C-24 to C-25 is not necessarily concerted. The counter ion to capture the proton from the methyl group, or H-23 or H-27 may also be used to promote isomerizations of appropriately located double bonds in the side chain. Clearly, both reactions cannot occur simultaneously, unless the enzyme is polymeric with multiple subunits. We discussed elsewhere the possibility that the SMT is polymeric and the importance of kinetic and thermodynamic control of the reaction progress regulating product distribution (Figure 18) (*12,82*).

As shown in Figure 19, the cofactor S-adenosyl-L-methionine (AdoMet) may serve as the methyl cation donor and as the counter ion (as envisaged by Arigoni), particularly if the adenine portion of the cofactor is aligned next to an aromatic amino acid. Since the cofactor is on top of the sterol side chain and covers the terminal section, the enzyme itself is shielded from delivering the counter ion. It is well known in biochemistry that when aromatic rings are stacked side by side, the pi-electrons of the rings interact to form weak bonds. Some amino acids (e.g., tyrosine) have aromatic rings that bond in this way with each other or with flat resonant structures, e.g., the adenine portion of AdoMet, thereby serving as negative point charges to abstract

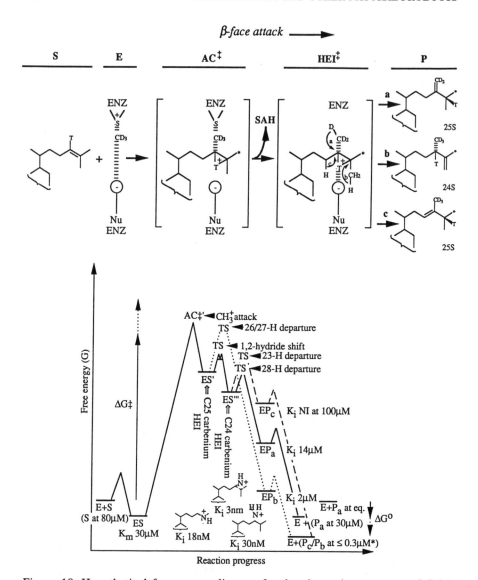

Figure 18. Hypothetical free-energy diagram for the alternative outcome of C-24 methylation. A first proximation of the relative energies of the reaction course intermediates are shown; ES- enzyme-substrate complex; AC - activated complex; TS - transition state; and EP - enzyme-product complex. Relative energies of the intermediate species were deduced from studies of product inhibition and the use of compounds designed as high energy intermediates (HEI) analogs. NI - not inhibitory. Endogenous microsomal concentrations are as indicated on the figures: S - substrate (cycloartenol), P_a - product (24(28)-methylenecycloartanol), P_b - (cyclolaudenol), and P_c - (cyclosadol), where * - 0.3 μM represents our limit of detection by GC/HPLC. Microsomal preparations contained circa 30 μM endogenous cycloartenol. A typical assay was performed with 50 μM cycloartenol, 50 μM SAM, and Tween 80 for 45 minutes at 30°C. Enzyme assays from references 42, 83.

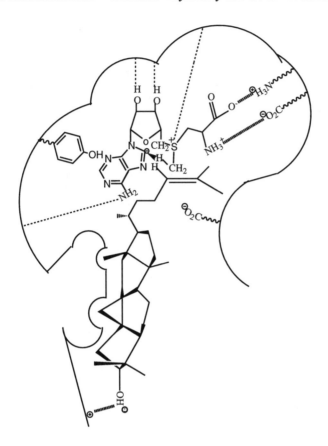

Figure 19. Proposed model for the binding of AdoMet and sterol to the SMT

protons from the methyl group during C-24 methylation. Our inability to demonstrate active transmethylation of 22(23), 24(25)- sterols (conjugated double bonds in side chain) by plant (*42,83*) and yeast (unpublished) microsomal preparations under conditions that cycloartenol and zymosterol, respectively, were actively methylated indicate that the conformation of the sterol side chain is important to catalysis (*42,83*). Another possibility we now consider is that, when the conjugated 1,4-butadiene system is placed in close contact with a negative point charge (from AdoMet) the nucleophilic character of the 24,25-bond is affected, so that, the necessary electronics for methylation to proceed are inhibited. This may explain the delayed introduction of the 22(23)-bond in the sterol side chain and of the requirement for the side chain to be flexible to allow 24(28)-products of methylation to be transmethylated by the same SMT that catalyzes the transmethylation of the 24,25-bond (*82*). Thus the SMT may play multiple roles in sterol transformation. For this reason, it is unclear what differences are to be expected in the amino acid composition of the SMT active site to explain the different 24-alkyl sterol compositions of plants, fungi and marine invertebrates.

Whereas several chiral centers of lanosterol and cycloartenol are transformed, suggesting the shape of the sterol may be affected, it is not. The global 3-dimensional shape of the molecule remains unchanged. Structural changes in the molecule may result from transformations that affect the tilt of the 3-OH group, and the tilt of the 17(20)-bond. They are features that influence the hydrogen-bonding interactions of the OH-group and the flexibility of the side chain, respectively (Figure 12). Sterol transformation likely occurred at different rates in the photosynthetic and non-photosynthetic lineages because cofactor availability (e.g., oxygen) was limiting early in evolution and the electronics and substrate specificity involved in each reaction mechanism were different. Obviously, the lanosterol pathway evolved at a faster rate than the cycloartenol pathway. This follows from the observation that the cyclopropyl group interferes with C-14 demethylation, so that, plants redesigned the ordering of the intermediates in the pathway to open the cyclopropyl ring to produce 8(9)-sterols that could serve as more efficient substrates for C-14 demethylation.

Several steps in the sterol pathway are anaerobic, e.g., C-24 methylation, C-24 and C-14(15) reductions and isomerization of double bonds in the nucleus and side chain. The existence of anaerobic steps in a pathway have been used to indicate the primitive nature of enzyme systems (e.g., hopanoid cyclase). Advanced steps are those that involve mixed-function oxygenase reactions, e.g., desaturations at C-5 and C-22 and methyl removal at C-4 and C-14. Interestingly, the first step in phytosterol biosynthesis is the transmethylation of cycloartenol (83), but this may be fortuitous in terms of sterol evolution, since less-advanced and more-advanced fungi use zymosterol as the preferred substrate for C-24 methylation (reviewed in 82 and 84) and the second step in phytosterol transformation appears to be C-4 demethylation, rather than by the anaerobic acid-catalyzed opening of the cyclopropane ring. Advanced pathways should also possess enzymes that exhibit a high degree of substrate specificity. But the current dogma in the sterol field is that the sterol pathway of plant, fungi and animals use enzymes that possess a broad substrate specificity, so that major and minor pathways are available to transform whatever intermediates are produced to Δ^5-end products (85-88). Because of this broad substrate specificity, the rate-limiting enzyme of sterol biosynthesis must be associated with an enzyme that does not catalyze the formation of the tetracycle structure, i.e., HMGCoA reductase (85-88).

We challenged the central dogma recently (12,82,83). Our hypothesis is that the enzymes that transform the tetracyclic products of squalene oxide cyclization possess a high degree of substrate specificity for sterol, and kinetically favored taxa-specific pathways operate in nature, so that only certain intermediates may ultimately be transformed to Δ^5-sterols under normal physiological conditions (12,82,83). A possible kinetically favored sterol pathway that we hypothesize may exist in yeast is shown in Figure 20 (12). The genes for each of the enzymes that have been mapped are located on several different chromosomes, rather than clustering on a single chromosome. It is unclear how many genes must change to produce a new sterol enzyme. Do these genes occur on consistent chromosomal positions in nature, and if not what does that mean in terms of the expression and ordering of the pathway enzymes? Unfortunately, neither the gene mapping data nor mechanistic enzymology data provide a clue to the ordering of the pathway intermediates, nor do they explain why yeast add a methyl group to the sterol side chain (it is energy expensive, costing 12 ATP equivalents). Since the 24-desalkyl sterol, cholesterol, is found to serve equally well as the native end product in the bulk membrane role (55), membrane function cannot explain the evolution of the transmethylation reaction in yeast. We have shown that fungi (90), including yeast (91) synthesize hormonal levels of cholesterol, which means the 24,25-reduction step is highly regulated relative to methylation of that double bond system. We confirmed that yeast may reduce the 24,25-bond by a kinetic isotope effect resulting from 24-deuteriated lanosterol incubated with GL-7. It is transformed to 24-D cholesta-5,7,22-trienol (unpublished). In another ascomycetous

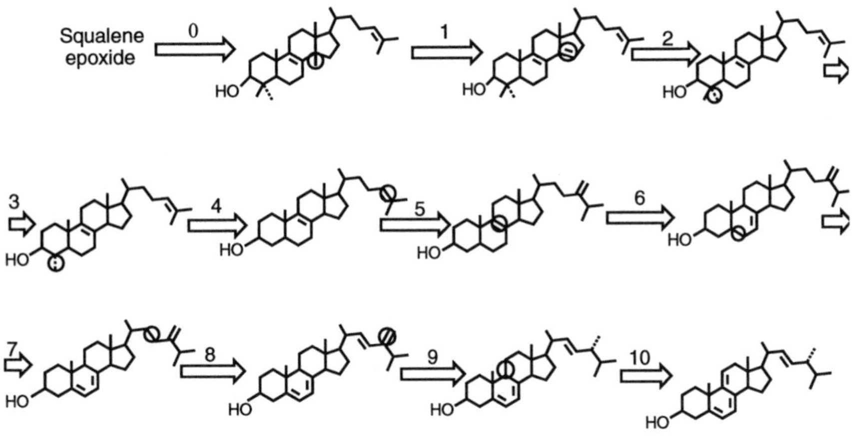

YEAST STEROL TRANSFORMATION PATHWAY

STEP	ENZYME	FAVORED SUBSTRATE	ACTIVITY	GENE	CHROMOSOME
0	SO cyclase	Squalene oxide	6.0	ERG7	8
1	C-14 demethylase	lanosterol	5.0	ERG11	8
2	Δ14(15)- reductase	14-desmethyllanosterol	N.D.	ERG24	N.D.
3,4	C-4 demethylase	4α-methylzymosterol	0.5	N.D.	N.D.
5	C-24 methyl transferase	zymosterol	3.3	ERG6	13
6	Δ8(9) to Δ7(8)- isomerase	24(28) methylenezymosterol	0.2	ERG2	13
7	C-5 desaturase	7,24(28) cholestadienol	10.2	ERG3	12
8	C-22 desaturase	5,7,24(28) cholestatrienol	0.18	ERG5	N.D.
9	Δ24(28)- reductase	5,7,22,24(28) cholestatetraenol	33.0	ERG4	7
10	Δ9(11)- desaturase	ergosterol	N.D.	N.D.	N.D.

N.D. indicates not determined. We thank Dr. Martin Bard for the key to the gene mapping data (cf. 97 for a key to the literature). Activity refers to the reaction rate in microsomal preparations and is expressed as nmoles/hr (V_{max}) and were obtained from the literature (reviewed in 44,55,82,84,97).

Figure 20. Hypothetical kinetically favored sterol transformation pathway in yeast

fungus we demonstrated isomerization of the 24(28)-bond to the 24(25)-position, after which the double bond was reduced, producing 24β-methyl cholesterol (*92*). Additionally, we discovered using radiotracer methods that vascular plants synthesize cholesterol (*93*). Thus many, if not all, of the sterols produced by plants may be synthesized by fungi, and cholesterol which is thought to be a unique zoosterol may be produced by plants and fungi (*90*). The newly observed regulatory role of sterol (*62,90,94-96*) in triggering growth and reproduction may be an agent for survival and promoting change. The sterol functioning in the hormonal role may involve sterol-protein binding that discriminates stereochemistry of the molecule that cannot be distinguished when the sterol is used in the bulk membrane role. Thus 24-alkylation may serve a purpose in fungi when the sterol in functioning hormonally.

Maintenance and Change in Optical Activity of Sterols

The key to evolutionary change is variation. Life as we know it operates within a set of physical laws in which there are opposing forces that cause variation: one that promotes the system to disorder, the other promotes the system to a higher and more complex order of organization. The shift from a balanced cellular economy (away from physiological homeostasis) is the start point for change in one direction or the other. As often expressed in Biology textbooks, genetic change is a prerequisite for evolutionary change, and the mutation process is the ultimate source of this variability (*13*). It follows that transmutations in the sterol genotype lead to changes in the sterol phenotype.

Darwinism provides one explanation for change in optical activity of sterols. If use and disuse of sterols regulated enzyme activity and enzyme activity affected gene mutations, then Lamarckism might be the mechanism of change. The "Organizational Principle" is another mechanism for change. Here, we assume that biosynthesis and function of sterols are mated to environmental pressures, so that mutations are possible. In our model (*7*) there are limits to the amount of change that can be introduced into the enzyme and sterol structure. Thus when a more catalytically efficient enzyme is produced, the new gene(s) remains encoded only if the new sterol is more suitable than the original product.

Chemotaxonomic studies indicate that the sterol phenotype differs qualitatively and quantitatively in Nature (*7,23*). From these studies it is evident that enantiomeric purity has been maintained in the sterol constitution, but structural variants, constitutional isomers and diastereoisomers, are produced. The changes in sterol stereochemistry apparently evolved to produce superior compounds that could play multiple roles.

Acknowledgments. We thank the support of the Asgrow Seed Company. The X-ray crystallographic pictures shown in Figure 12 were supplied to us by our collaborator, Dr. Jane Griffin (Medical Foundation of Buffalo). The art-work for some figures was done by Ms. Lesley Keneipp (Texas Tech University, Engineering Services).

Literature Cited

1.   Pasteur, L. C.R. *Acad. Sci. Paris* **1848**, *26*,535.
2.   Japp, F.R. *Nature* **1898**, *58*, 452-460.
3.   Mason, S.F. *Nature* **1984**, *311*, 19-23.
4.   Brewster, J.H. *J. Chem. Educ.* **1986**, *63*, 667-670.
5.   Hegstrom, R.A.; Konderpudi, D.K. *Sci. Amer.*, 108-115.
6.   Ulbricht, T.L.V. In *Comparative Biochemistry;* Florkin, M. Ed.; Academic Press: New York, NY, 1962, pp 1-25.
7.   Nes, W.D. *Rec. Adv. Phytochem.* **1990**, *24*, 283-327.
8.   Ourisson, G. *Nature* **1987**, *326*, 126-127.
9.   Ourisson, G.; Albrecht, P.; Rohmer, M. *Trend. Biochem. Soc.* **1982**, *7*, 236-239.
10.   Ourisson, G.; Rohmer, M.; Poralla, K. *Ann. Rev. Microbiol.* **1987**, *41*, 301-333.
11.   Mackenzie, J.R.; Brassell, S.C.; Eglinton, G.; Maxwell, J.R. *Science* **1982**, *217*, 491-504.
12.   Parker, S.R.; Nes, W.D. In *Regulation of Isopentenoid Metabolism* ; Nes, W.D.; Parish, E.J.; Trzaskos, J.M. Eds.; ACS Symp. Series 497; American Chemical Society: Washington, D.C., 1992, 110-145.

13. Mayr, E. *The Growth of Biological Thought.* The Belknap Press; Cambridge, 1982, 1-571.
14. Landman, D.E. *Sci. Amer.* **1993**, 150.
15. Gould, S.J. *Science* **1982**, *216*, 380-387.
16. *The Origin of Life and Evolutionary Biochemistry*; Dose,K.; Fox, S.E.; Deborin, G.A.; Pavlovskaya, T.E. Eds.; Plenum Press: New York, NY, 1974, 476pp.
17. Wheelis, M.L.; Kandler, O; Woese, C.R. *Proc. Natl. Acad. Sci.* **1992**, *89*, 2930-2934.
18. Woese, C.R. In *Evolution From Molecules to Men*; Bendal, D.S. Ed.; Cambridge University Press, 1983, 209-233.
19. Retey, J.; Robinson, J.A.; *Stereospecificity in Organic Chemistry and Enzymology*; Verlag Chemie: Weinheim, Germany, 1982, 1-333.
20. March, J. *Advanced Organic Chemistry,* John Wiley & Sons: New York, NY, 1992, 94-130.
21. Myant, N.B. *The Biology of Cholesterol and Related Steroids*; William Heinemann Medical Books Ltd.: London, England, 1981, 1-227.
22. Fieser, L.F.; Fieser, M. *Steroids;* Reinhold: New York, NY, 1959, 1-100.
23. Nes, W.R.; McKean, M.L. *The Biochemistry of Steroids and Other Biologically Significant Isopentenoids*; University Park Press: Baltimore, MD, 1977, 35-70.
24. Rychnovsky, S.C.; Mickus, D.E. *J. Org. Chem.* **1992**, *57*, 2732-2736.
25. Ruzicka, L. *Proc. Chem. Soc.* **1959**, Nov., 341-359.
26. Gonzales, O.J. *J. Chem. Educ.* **1985**, 503-509.
27. Bernal, J.D. *The Origin of Life,*;Weindenfield and Nicilson: London, 1967, 3-75.
28. Singer, S.J. *Ann. Rev. Biochem.* **1974**, *43*, 805-833.
29. Haldane, J.B.S. *New Biol.* **1953**, *16*, 12-40.
30. Prigogine, I.; Gehehiau, J. *Proc. Natl. Acad. Sci.* **1986**, *83*, 6245-6249.
31. Tanford, C. *Science* **1978**, *200*, 1012-1018.
32. Whittaker, R.H.Q. *Rev. Biol.* **1959**, *34*, 210-226.
33. Baisted, D.J.; Capstack, E.; Nes, W.R. *Biochemistry* **1962**, *1*, 537-543.
34. Nes, W.R.; Baisted, D.J.; Capstack, E.; Newschwander, W.W.; Russell, P.T. In *Biochemistry of Chloroplasts;* Goodwin, T.W.; Ed.; Academic Press: New York, NY, 1967, 273-282.
35. Nes, W.R. *Lipids* **1974**, *9*, 596-612.
36. Rohmer, M.; Bouvier, P.; Ourisson, G. *Proc. Natl. Acad. Sci.* **1979**, *76*, 847-851.
37. Caspi, E. *Acc. Chem. Res.* **1980**, *13*, 97-104.
38. Nes, W.D.; Benson, M.; Lundin, R.E,; Le, P.H. *Proc. Natl. Acad. Sci.* **1988**, *85*, 5759-5763.
39. Nes, W.D.; Wong, R.Y.; Griffin, J.F.; Duax, W.L. *Lipids* **1991**, *26*, 649-655.
40. Nes, W.D.; Wong, R.Y.; Benson, M.; Akihisa, T. *Chem Commun.* **1991**, 18, 1272-1274.
41. Nes, W.D.; Wong, R.Y.; Benson, M.; Landrey, J.R.; Nes, W.R. *Proc. Natl. Acad. Sci.* **1984**, 81, 5896-5900.
42. Nes, W.D.; Janssen, G.G.; Bergenstrahle, A. *J. Biol. Chem.* **1991**, 266, 15202-15212.
43. Abe, I.; Rohmer, M. *Chem. Commun.* **1991**, 13, 902-903.
44. Abe, I.; Rohmer, M.; Prestwich, G.D. *Chem. Rev.* **1993**, 2189-2206.
45. Barton, D.H.R.; Jarman; Watson, K.C.; Widdowson, D.A. *J.Chem. Soc. Perkin Trans. 1* **1975**, 1134-1138.

46. Rohmer, M.; Anding, C.; Ourisson, G. *Eur. J. Biochem.* **1980**, *112*, 541-547.
47. Johnson, W.S.; Telfer, S.J.; Cheng, S.; Schubert, U. *J. Amer. Chem. Soc.* **1987**, *109*, 2517-2518.
48. Alonso, W.R.; Rajaonarivony, J.I.M.; Gershenzon, J.; Croteau, R. *J. Biol. Chem.* **1992**, *267*, 7582-7587.
49. Koshland, D.E. Jr. *Fed. Proc.* **1976**, *35*, 2104-2111.
50. Jenson, R.A. *Ann. Rev. Microbiol.* **1976**, *30*, 409-425.
51. Doolittle, R.F. *TIBS*, **1974**, *19*, 15-18.
52. Johnson, W.S. *Bioorgan. Chem.* **1976**, *5*, 51-98.
53. Kraut, J. *Science* **1988**, *242*, 533-540.
54. Nes, W.D.; Saunders, G.A.; Heftmann, E. *Lipids*, **1982**, *17*, 178-183.
55. Nes, W.D.; Janssen, G.G.; Crumley, F.G.; Kalinowska, M.; Akihisa, T. *Arch. Biochem. Biophys.* **1993**, *300*, 724-733.
56. Nes, W.D.; Norton, R.A.; Crumly, F.G.; Madigan, S.J.; Katz, E.R. *Proc. Natl. Acad. Sci.* **1990**, *87*, 7565-7569.
57. Cavalier-Smith, T. *Ann. New York Acad. Sci.* **1987**, *503*, 17-54.
58. Nes, W.R.; Nes, W.D.; *Lipids in Evolution*, Plenum Press: New York, NY, 1980, 71-81.
59. Nes, W.D.; Venkatramesh, M.; Rychnovsky, S.C. (manuscript submitted).
60. Raederstorff, D.; Rohmer, M. *Eur. J. Biochem.* **1985**, *164*, 422-434.
61. Nes, W.D.; Le, P.H.; Berg, L.R.; Patterson, G.W.; Kerwin, J.L. *Experientia* **1986**, *42*, 556-558.
62. Bloch, K.E. *CRC Crit. Rev. Biochem.* **1983**, *14*, 47-92.
63. Hanson, K.R. *Ann. Rev. Plant Physiol.* **1972**, *23*, 335-366.
64. Rahier, A.;Genot, J.C.; Schuber, F.; Benveniste, P,; Narula, A.S. *J. Biol. Chem.* **1984**, *259*, 15215-15223.
65. Nicotra, F.; Ronchetti, F.; Russo, G.; Lugara, G.; Castellato, M. *J.Chem. Soc. Perkin Trans. 1* 1990, 105-109.
66. Arigoni, D. *Ciba Found. Symp.* **1978**, *60*, 243-260.
67. Zimmerman, M.P.; Djerassi, C. *J. Amer. Chem. Soc.* **1991**, *113*, 3530-3533.
68. Goodwin, T.W. In *The Biochemistry of Plants*; Stumpf, P.K., Ed.; Volume 4; Academic Press: New York, NY, 1980, 485-507.
69. Seo, S.; Uomori, A.; Yoshimura, Y.; Seto, H.; Ebizuka, Y.; Noguchi, H.; Sankawa, U.; Takeda, K. *J.Chem Soc. Perkin Trans. 1*, 1990, 105-109.
70. Nes, W.D.; Norton, R.A.; Benson, M. *Phytochemistry* **1992**, *31*, 805-811.
71. Shimizu, N.; Itoh, T.; Saito, M.; Matsumoto, T. *J.Org. Chem.* **1984**, *49*, 712-714.
72. Akhila, A.; Gupta, M.M.; Thakur, R.S. *Tetrahedron Letts.* **1987**, *28*, 4085-4088.
73. Balliano, G.; Caputo, O.; Viola, F.; Delprino, L.; Cattel, L. *Phytochemistry* **1983**, *22*, 915-921.
74. Cattel, L.; Anding, C.; Benveniste, P. *Phytochemistry* **1976**, *15*, 931-935.
75. Barton, D.H.R.; Kempe, U.M.; Widdowson, D.A. *J. Chem. Perkin Trans. 1* **1972**, 513-522.
76. Collins, J.M.; Nes, W.D.; Quinn, P.J.; Wolfe, D.H.; Cunningham, B.A.; Kucuk, O.; Westerman, M.P.; Lis, L.J. *FASEB J.* **1992**, *6(A)*, 241.
77. Xu, S.; Norton, R.A.; Crumley, F.G.; Nes, W.D. *J. Chromatogr.* **1988**, *452*, 377-398.
78. Kalinowska, M.; Nes, W.R.; Crumley, F.G.; Nes, W.D. *Phytochemistry* **1990**, *29*, 3427-3434.
79. Cornforth, J.W. *Angew. Chem. Int. Ed. Engl.* **1968**, *7*, 903-911.

80. Akihisa, T.; Kokke, W.C.M.C.; Tamura, T. In *Physiology and Biochemistry of Sterols*; Patterson, G.W.; Nes, W.D., Eds.; American Oil Chemists Society Press: Champaign, IL, 1991, 172-228.
81. Julia, M.; Marazano, C. *Tetrahedron Lett.* **1985**, *41*, 3717-3724.
82. Nes, W.D.; Parker, S.R.; Crumley, F.G.; Ross, S.R. In *Lipid Metabolism in Plants*; Moore, T.S., Ed.; CRC Press: Boca Raton, 1993, 389-426.
83. Janssen, G.G.; Nes, W.D. *J. Biol. Chem.* **1992**, *267*, 25856-25863.
84. Janssen, G.G.; Kalinowska, M.; Nes, W.D. In *Biochemistry and Physiology of Sterols*; Patterson, G.W., Nes, W.D., Eds.; American Oil Chemists Society Press: Champaign, IL, 1992, 88-122.
85. Benveniste, P. *Ann. Rev. Plant Physiol.* **1986**, *37*, 275-308.
86. Schropfer, Jr., G.J. *Ann. Rev. Biochem.* **1981**, *50*, 585-603.
87. Parks, L.W. *CRC Crit. Rev. Microbiol.* **1978**, *6*, 300-341.
88. Bach, T.J. *Lipids* **1986**, 21, 82-88.
89. Lees, N.D.; Arthington, B.A.; Bard, M. In *Regulation of Isopentenoid Metabolism;* Nes, W.D.; Parish, E.J.;Trzaskos, J.M. Eds.; ACS Symp. Series 497, American Chemical Society: Washington, D.C., 1992, 246-259.
90. Nes, W.D. *ACS Symp. Series* **1987,** *325*, 304-328.
91. Xu, S.; Nes, W.D. *Biochem. Biophys. Res. Commun.* **1988**, *155*, 509-517.
92. Nes, W.D.; Le. P.H. *Biochim. Biophys. Acta* **1990**, *1042*, 111-125.
93. Heupel, R.C.; Sauvaire, Y.; Le, P.H.; Parish, E.J.; Nes, W.D. *Lipids* **1986**, *21*, 69-75.
94. Nes, W.R. *ACS Symp. Ser.* **1987**, *325*, 252-267.
95. Nes, W.D.; Heupel, R.C. *Arch. Biochem. Biophys.* **1986**, *244*, 211-217.
96. Rodriguez, R.J.; Taylor, F.R.; Parks, L.W. *Biochem. Biophys. Res. Commun.* **1983**, *112*, 47-53.
97. Corey, E.J.; Matsuda, P.T.M.; Bartsel, B. *Proc. Natl. Acad. Sci.* **1994**, *91*, 2211-2215

RECEIVED April 21, 1994

Chapter 5

Phylogenetic Distribution of Sterols

Glenn W. Patterson

Department of Botany, University of Maryland, College Park, MD 20742

Sterols have extremely widespread distribution in living organisms
... in bacteria, protozoa, algae, fungi, higher plants and animals.
Organisms that are unable to synthesize sterols are not uncommon,
but many require external sources of sterol. Only a few species
contain no sterols and they have other isopentenoids which
substitute for sterols (1). Although sterols have multiple roles in
living organisms, in most organisms, the role requiring the largest
amount of sterol is as a structural constituent of membranes. It is
no surprise then, that of the two great divisions of living
organisms, sterols are relatively abundant in eukaryotes and are
present in relatively small amounts in prokaryotes (2).

Sterols of Prokaryotes

For instance, sterols were identified in *E. coli* (3,4) and *Azotobacter chroococcum*
(4) at 0.0004% and 0.01% of dry weight, respectively, while *Streptomyces
olivaceus* contained sterols at 0.0035% of dry weight (5). The identification of
sterols in such small amounts without the use of isotopes does not clearly indicate
that sterol synthesis is taking place in these organisms. In fact, in numerous
other bacteria, no sterols were detected when the limit of detection was 0.0001%
(4). However, in *Cellulomonas dehydrans*, the incorporation of ^{14}C-glucose into
squalene and sterols was demonstrated (6). There have been several reports of
sterols in the cyanobacteria (blue green algae) (7), where the results are similar
to those of other prokaryotes in that the amount of sterol present is usually small.
Some of the sterols of cyanobacteria resemble those of green algae (7). It is
difficult to grow many cyanobacteria axenically, therefore there is still a question
whether cyanobacteria actually synthesize sterols. Some cyanobacteria are
reported to synthesize triterpenoids of the hopane series, and these triterpenoids
may be present in many bacteria as well (8,9). The sources and roles of both the
sterols and hopanoids in prokaryotes need further study. A member of the

0097–6156/94/0562–0090$08.00/0

primitive Archeobacteria, *Methylococcus capsulatus* is unique in its sterol composition in that its quantity of sterol is higher than that of other bacteria (0.22% of dry weight) (10). The sterols of *Methylococcus* also are unusual in that they contain the rare $\Delta^{8(14)}$-sterols (11) and an 8-carbon side chain with one or two methyls at C-4 (Figure 1). The sterols of *Methylococcus* reflect a sterol biosynthetic pathway that is very primitive and does not metabolize sterols beyond the 4-methyl $\Delta^{8(14)}$-stage.

Sterols of Protoctista

Eukaryotic organisms are usually divided into four kingdoms Protoctista, Fungi, Animals, and Plants. The Protoctista is defined by exclusion (12). Its members are eukaryotes that are neither fungi, plants or animals and are simple organisms with many being unicellular. A large number of Protoctista belong to the algae. Several hundred species of algae have been examined for sterol composition, and the kinds of sterols present are very closely related to taxonomic position (7). For example, of the 152 species of red algae examined, all except two had Δ^5 -sterols as the principal sterol component. Cholesterol is by far the most common sterol in the red algae. By the same token, fucosterol, another Δ^5 sterol, is the principal sterol of every species of brown algae examined (Figure 2). In other algal classes fewer species have been examined, but the predominant sterols in the following classes are: charophyta, 24-ethylcholesterol and isofucosterol; chrysophyta, 24-ethylcholesta-5,22-dienol; xanthophyta, 24-ethylcholesterol; cryptophyta, 24-methylcholesta-5,22-dienol (Table I) (7). The major sterols in all of the above algal classes are Δ^5-sterols, but other types of sterol are important in some algae. The dinoflagellates have a sterol composition unlike that of any algal group and are characterized by dinosterol (Figure 3) and other sterols with a 4-methyl group and frequently additional carbons attached to C-23 and C-24 (7). Although dinosterol has no nuclear double bonds, many dinoflagellates contain cholesterol as a principal or a secondary sterol. Most diatoms contain Δ^5-sterols but they may be C_{27}, C_{28} or C_{29} sterols and no individual sterol appears to be dominant. The green algae contain primitive, single-celled members and more advanced multicellular algae which are believed to be the origin of higher plants. The orders containing multicellular algae, contain primarily 24-ethylcholesterol and isofucosterol (Figure 2), while the more primitive, primarily unicellular orders contain a wide range of sterols including Δ^5, Δ^7 and $\Delta^{5,7}$-sterols. Over a dozen chlorophyte species contain Δ^7 sterols, and ergosterol($\Delta^{5,7}$) is a major sterol in nearly a dozen species. Within the algae, ergosterol has been reported only from unicellular chlorophyte species.

Algal sterols have been thought to differ from those in higher plants by having 24β-oriented alkyl groups while higher plants are 24α (13). However, the recent isolations of sitosterol (24α-ethyl) from *Hydrodictyon reticulatum* (14); 24α sterols from *Nitzschia* (15), *Phaeodactylum* (16) and, *Amphora* (17); Campesterol (24α-methyl) from *Tetraselmis* (18); several 24α-sterols from the Prymnesiophyta (7); epibrassicasterol (24α-methyl) from the Cryptophyta (19,20); and 24α-ethyl sterols from several Chrysophytes (21,22), demonstrate that 24α-sterols are present in at least seven algal classes. The relationship

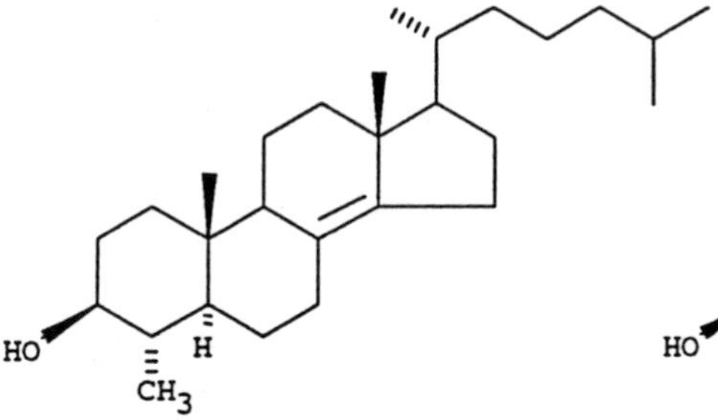

4-METHYLCHOLEST-8(14)-ENOL 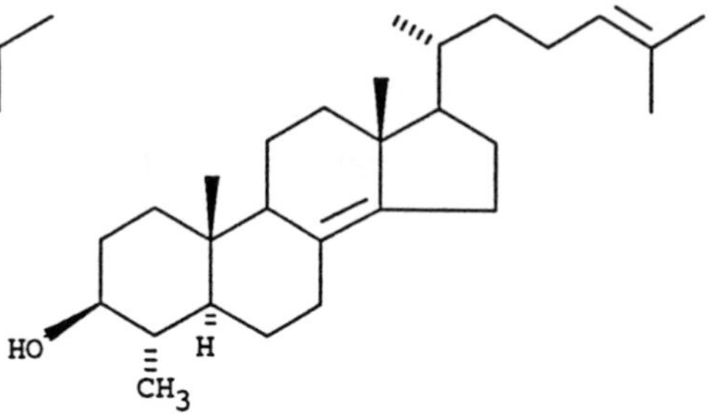   4-METHYLCHOLESTA-8(14),24-DIENOL

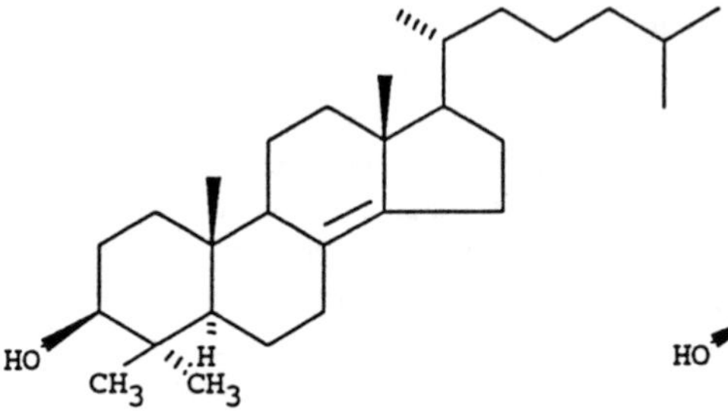

4,4-DIMETHYLCHOLEST-8(14)-ENOL 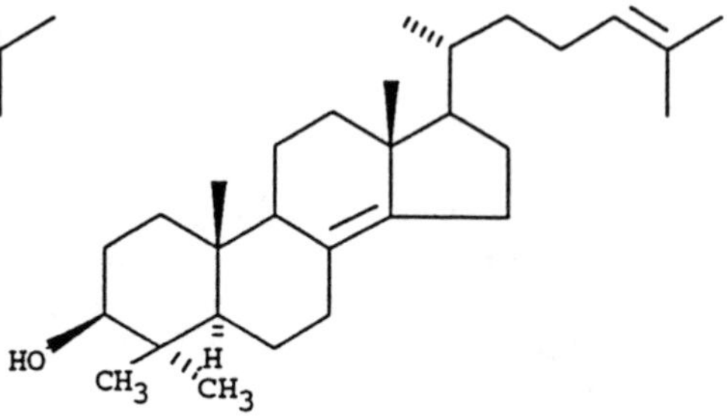   4,4-DIMETHYLCHOLESTA-8(14),24-DIENOL

Figure 1. Sterols of methylococcus

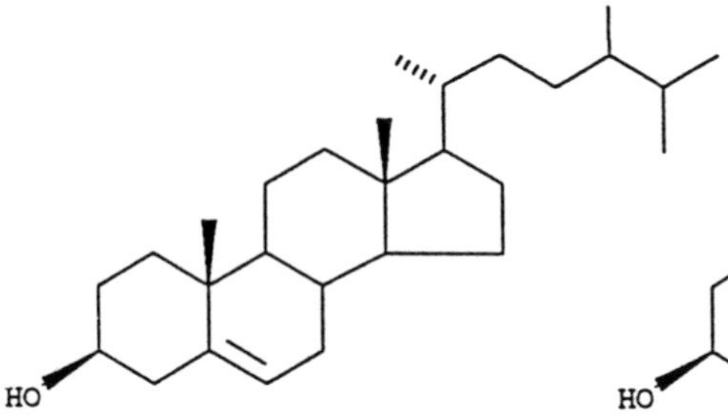

24-METHYLCHOLESTEROL

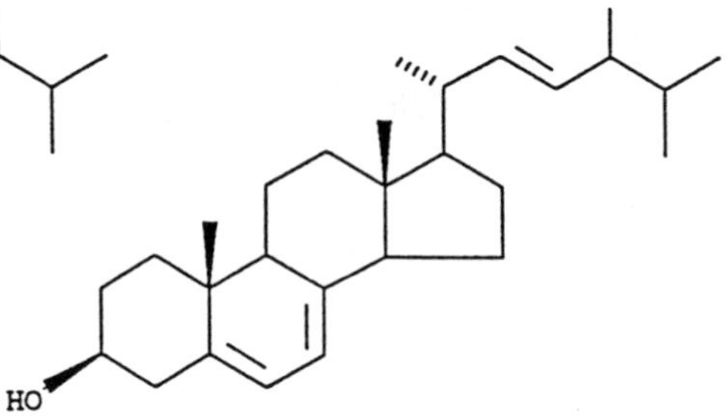

ERGOSTEROL

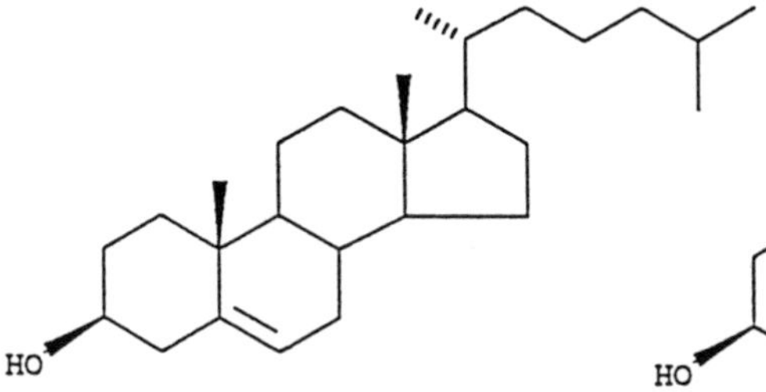

CHOLESTEROL

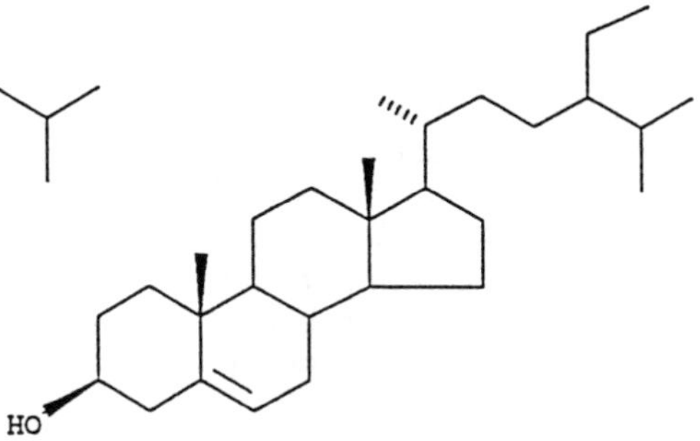

24-ETHYLCHOLESTEROL

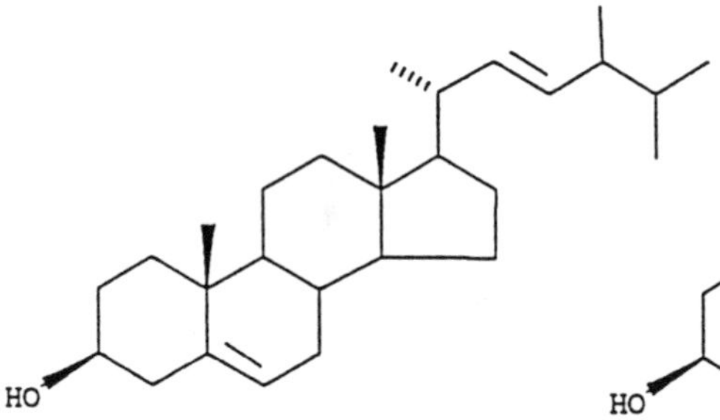

24-METHYLCHOLESTA-5,22-DIENOL

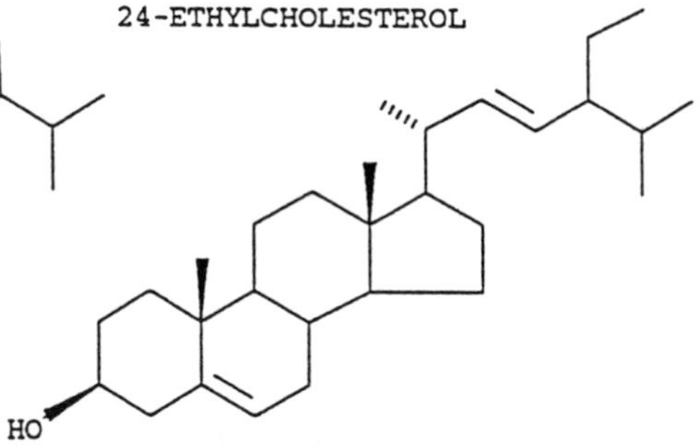

24-ETHYLCHOLESTA-5,22-DIENOL

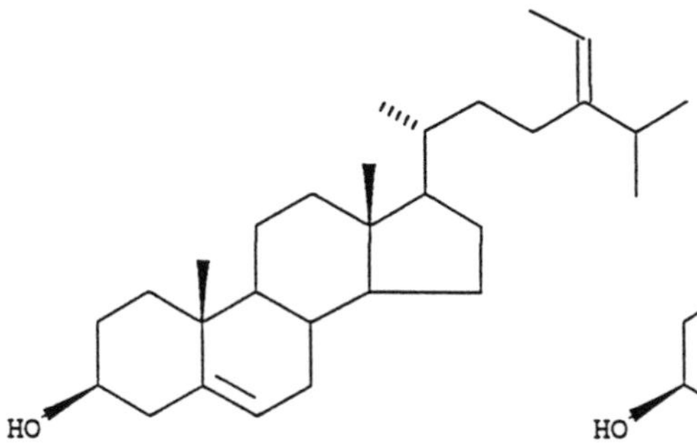

FUCOSTEROL

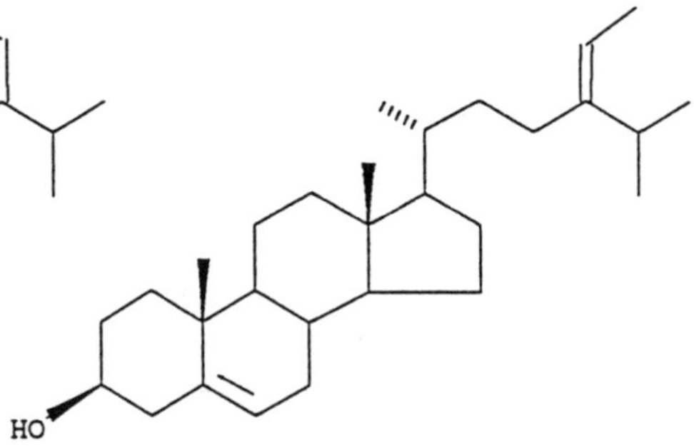

ISOFUCOSTEROL

Figure 2. Major sterols of algae

Table I. Dominant Sterols* of Several Algal Phyla

	CHOL	FUCO	EC5,22	MC5,22	DINO	24EC	ISOFUCO
Rhodophya	X						
Phaeopha		X					
Crytophyta			X				
Charophyta						X	X
Xanthophyta						X	
Chrysophyta		X					
Dinophyta					X		

* CHOL = cholesterol
  FUCO = fucosterol
  EC5,22 = ethylcholesta-5,22-dienol
  MC5,22 = 24-methylcholesta-5,22-dienol
  DINO = 4,23,24-trimethylcholest-22-enol (dinosterol)
  24EC = 24-ethylcholesterol
  ISOFUCO = isofucosterol

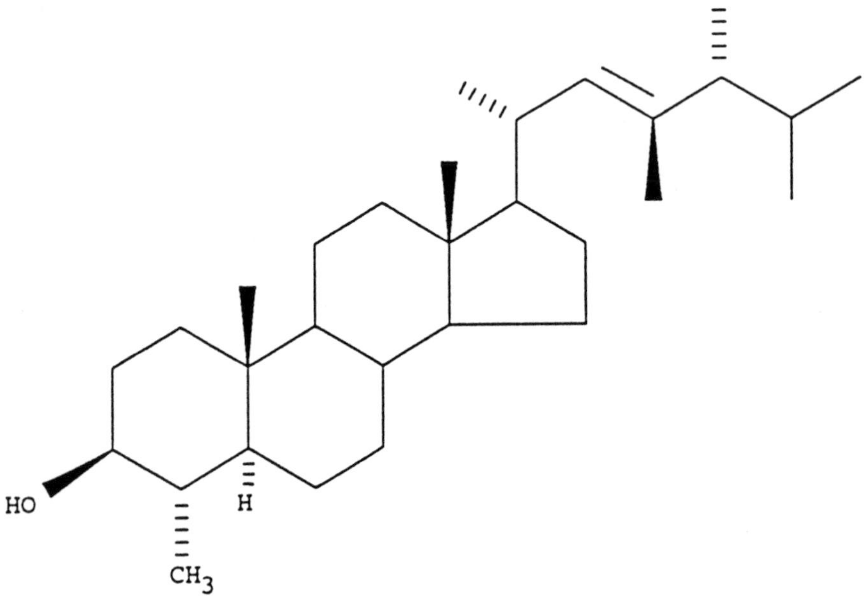

DINOSTEROL

Figure 3. Structure of dinosterol

between sterol composition and algal phylogeny has been strengthened by recent analyses and by improvements in our understanding of algal taxonomy.

Ergosterol is a major component in a few primitive green algal orders, Volvocales and Chlorococcales and in a few members of the Euglenophyta (7). Outside the algae, however, ergosterol is the most common sterol of the protozoa. It is the major sterol of all of the flagellates and amoebas examined (Table II). The third major group of protozoans examined (ciliates) do not synthesize sterols but may absorb them from the environment (Table II). *Tetrahymena pyriformis* is unusual in that it does not synthesize sterols, but in the absence of sterols it does synthesize the pentacyclic triterpenoids tetrahymanol and diploptenol (Figure 4). A recent publication on marine ciliates showed that all eight species examined contained tetrahymanol (38). Five of the above species also contained hopanol (Figure 4) but none were reported to contain sterols. Most sterol analyses in the protozoa were conducted before the advent of capillary gas chromatography. It is probable that protozoan sterol composition is more complex than it appears to be at present. Only one species has been examined in the myxomycota (plasmodial slime molds) and in the Acrasiomycota (cellular slime molds). *Physarum* (Myxomycota) contains C_{28} and C_{29} Δ^5-sterols (39). *Dictyostelium* (Acrasiomycota) contains nuclear saturated sterols (24R-24-ethylcholest-22-enol) as major components, some with 4-methyl groups (40). The rare sterol mixture in *Dictyostelium* is structurally related to that in the Prymnesiophyte alga *Pavlova* (41) and to that of the dinoflagellates (7).

Three groups of organisms that once were considered fungi and now are classified in the Protoctista (12) are the Hypochytrids, Chytrids and Oomycetes. The most common sterols in the chrytrids and hypochytrids are cholesterol, 24-methylcholesterol and 24-ethylcholesterol (Table III). Cholesterol is the principal sterol of at least four chrytrid genera. The Oomycetes also contain primarily Δ^5-sterols (Table IV) and cholesterol is a major component in many species. The C_{28} and C_{29} sterols of the Oomycetes are fucosterol and 24-methylenecholesterol, indicating that the Oomycetes lack the ability to reduce the $\Delta^{24(28)}$-double bond. All species of the order Peronosporates examined (Table IV) lack the ability to synthesize sterols (52,54). They do not require sterol for growth but do require sterol for sexual reproduction (54,55,56).

Sterols of Fungi

The assessment of sterol composition in fungi has been much simplified by the classification of the Hypochytrids, Chrytrids, and Oomycetes in the Protoctista (12). The fungi are composed of three major groups, the Zygomycota, Ascomycota, and Basidiomycota (12). Well over a hundred species have been examined for sterol composition and ergosterol and related $\Delta^{5,7}$-sterols appear to be the dominant sterols in most species. However, in the Zygomycota, some species have ergosterol and related C_{28} sterols as principal sterols (50,57-59) and others contain 24-methylcholesterol and cholesterol (60), or 24-ethylcholesterol and cholesterol (61). The principal sterol in the overwhelming number of Ascomycota examined is ergosterol, but in the orders Tuberales and Tafrinales brassicasterol is sometimes the principal sterol (62,63). The sterols of most

Table II. Sterols of Protozoa

Species	Major Sterol*	Reference
Phylum Zoomastigina (Flagellates)		
Trypanosoma cruzi	ERG 5,7;ERG,STIG 5,7,22	26
T. ranarum	ERG	30
T. rhodesiense	ERG	29,31
T. mega	ERG	31
T. lewisi	ERG	31
Leptomonas culicidarum	ERG	28
Crithidia fasiculata	ERG	26
C. oncopelti	ERG	29
Blastocrithidia	ERG	1
Herpetomonas	ERG	1
Leishmania tarentolae	ERG	27
Phylum Rhizopoda (Amoebas)		
Acanthamoeba polyphaga	ERG, STIG 5,7,22	23
A.sp.	ERG	25,37
Hartmannella rhyodes	ERG	25,37
Mayorella palestinensis	ERG	25,37
Naegleria gruberi	ERG	24
N. lovaniensis	ERG	24
Phylum Ciliophora (Cilates)		
Paramecium aurelia	REQ	32,34
Tetrahymena pyriformis	TETRA	1
Tetrahymena setifera	REQ	35
T. paravorax	REQ	36
Cyclidium sp.	TETRA	38
Pleuronema sp.	TETRA	38
Parauronema acutum	TETRA	38
Anophryoides soldoi	TETRA	38
Uronema nigricans	TETRA	38
Metanophrys sp.	TETRA	38
Diplogmus sp.**	TETRA	38
Protocruzia sp. **	TETRA	38

* ERG = ergosterol
    ERG5,7 = ergosta-5,7-dienol
    STIG 5,7,22 = Stigmesta-5,7,22-trienol
    REQ = sterol required, none synthesized
    TETRA = tetrahymenol
** Tentative identification

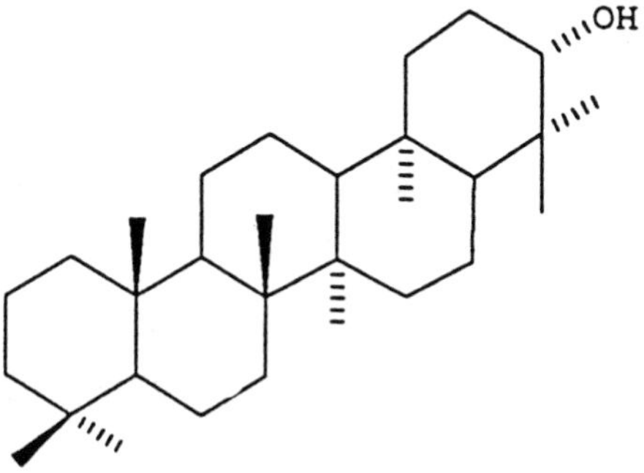

TETRAHYMANOL

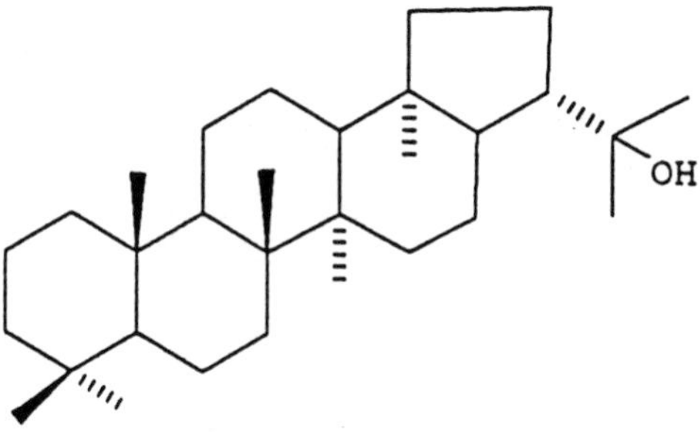

DIPLOPTEROL

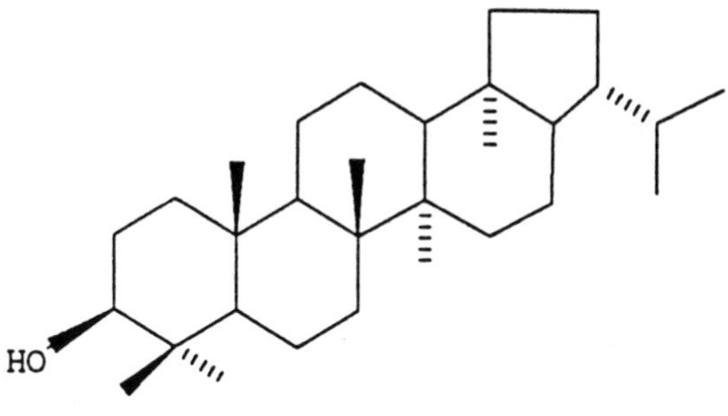

HOPANOL

Figure 4. Tetracyclic triterpenoids of protozoa

Table III. Sterols of Chytrids and Hypochytrids

Species	Major Sterol*	Reference
Hypochytridiomycota		
Hypochochytridium catenoides	24EC5,22;24MC;24EC	42,44,51
Rhizidiomyces apophysatus	24EC;24MC	42
Chytridiomycota		
Catenaria anguillulae	LAN,24EC,CHOL	42
Blastocladiella emersonii	CHOL;CHOL7	42
B. ramosa	CHOL7;DILAN;14-8;14-7	43
Monoblepharella sp.	24EC;24MC	42, 43
Allomyces macrogynus	CHOL;24EC	42,43,44
A. arbuscula	CHOL	43
A. javanicus	CHOL	43
Spizellomyces punctatum	24MC;24EC	42
Rhizophlycis rosea	CHOL	42,44
Chytridium confervae	CHOL	42
Rhizophydium spaeotheca	24MEC;24EC	42

* CHOL = cholesterol
  24MC = 24-methylcholesterol
  24EC = 24-ethylcholesterol
  LAN = lanosterol
  CHOL7 = cholest-7-enol
  DILAN = 24-dihydrolanosterol
  14-8 = 14α-methylcholest-8-enol
  14-7 = 14α-methylcholest-7-enol
  24MEC = 24-mentylenecholesterol

Table IV. Sterols of Oomycetes

Species	Major Sterol*	Reference
Saprolegniales		
Achyla bisexualis	FUCO, 24MEC	45
A. hypogyna	FUCO, 24MEC, CHOL	51
A. caroliniana	CHOL, DES	50
A. americana	FUCO, 24MEC, CHOL	47,49
Dictyuchus monosporus	FUCO, FUCO7, 24MEC, CHOL7	46
Saprolegnia ferax	24MEC, DES, FUCO	48,50
S. megasperma	24MEC, FUCO, CHOL	50
Leptolegnia caudata	24MEC, FUCO, DES	50
Aplanopsis tennestris	FUCO, CHOL	50,51
Pythiopsis cymosa	24MEC, CHOL	50
Leptomitales		
Apodachlyella completa	FUCO, 24MEC	47,50,51
Apodachyla minima	CHOL	50
A. brachnema	CHOL, FUCO	50
Lagenidiales		
Lagenidium callinectes	DES, FUCO, CHOL	47
L. giganteum	None	51
Peronosporales		
Phytophthora infestans	None	50
P. cactorum	None	54
P. cinnamoni	None	52
Pythium ultimum	None	50
P. debaryanum	None	50
P. graminicola	None	53

* FUCO = fucosterol
CHOL = cholesterol
24MEC = 24-methylenecholesterol
DES = demosterol
CHOL7 = cholest-7-enol
FUCO7 = 24-ethylcholesta-7,24(28)(Z)-dienol

Basidiomycota are either ergosterol or related C_{28} sterols, although some species contain closely-related sterols with a Δ^7-monoene nucleus (64). It is clear that ergosterol is the principal sterol in most fungi, although occasionally a species will contain Δ^5-sterols similar to those of the Oomycetes (65) or Δ^7-sterols similar to those of some species of green algae. Most sterols of fungi are C_{28}, although C_{29} sterols occur in some species (65). Ergosterol is the principal sterol in some species of green algae and euglenophytes and is the major sterol of most protozoa and fungi. It is not the major sterol of any higher plant or animal.

Sterols of Plants

Sterols have been analyzed from hundreds of plants and early thoughts were that sitosterol and related sterols stigmasterol and campesterol were typical of plants (66). The composition of sterols in the plant kingdom does not appear quite so simple at the present time. The most simple plants, bryophytes, contain primarily Δ^5 sterols, although several species contain small quantities of $\Delta^{5,7}$-sterols (67). A major difference between plant sterols and sterols algae and fungi is the configuration of the alkyl group at C-24 in the side chain. In plant sterols "sitosterol" is normally 100% 24α-ethyl and "campesterol" is usually a 2:1 mixture of 24α-methyl and 24β-methylcholesterol, respectively (1). In algae and fungi most sterols examined have the 24-β-oriented alkyl group (7,65). In terms of their C-24 orientation, bryophytes are intermediate between algae and fungi on one hand and higher plants on the other. In six Bryophytes analyzed for C-24 orientation, 3 species contained 24-ethylcholesterol as a mixture with 10-40% of the 24β-isomer and 6 species contained 24-methylcholesterol as a mixture with 28-77% of the 24β-isomer. In the pteridophytes, all of the 24-ethylcholesterol and 24-ethyl-5,22-cholestadienol is 24α. However, in the lyocopods, all species examined had more 24β-methylcholesterol than 24α-methylcholesterol. In the more advanced horsetails, whisk ferns, and true ferns 24α-methyl was more plentiful than 24β-methyl in all species examined (68). At the apex of plant evolution, angiosperms contain pure sitosterol (24α) and stigmasterol (24α) and most species contain campesterol (24α-methyl) and 24-dihydrobrassicasterol (24β-methyl) at a 2:1 ratio, respectively. From these data it appears that during evolution of higher plants from green algae, sterols have changed from being 100% 24β in most green algae to being 80-100% 24α in the angiosperms. All higher plant species except angiosperms appear to have Δ^5-sterols as principal components. Early reports indicated that Δ^5-sterols were overwhelmingly dominant in the angiosperms as well, although individual members of the Chenopodiaceae, Leguminosae, Theaceae, Sapotaceae and apparently all members of the Curcurbitaceae contain Δ^7-sterols (66). Recent research indicates that in the order Caryophyllales, Δ^7-sterols are much more common than was believed previously. Of the 12 generally recognized families in the Carophyllales, seven families appear to have major and sometimes dominant quantities of Δ_7 sterols with lesser quantities of Δ^5 and Δ^0 sterols (69). The other five families have predominantly Δ^5-sterols. Sterols with the $\Delta^{5,7}$ nucleus have not been found in major amounts in any angiosperm. The majority of angiosperm species examined to date are those containing sitosterol, stigmasterol, and campesterol (70), but a substantial number of angiosperms have a composition dominated by spinasterol and 7-stigmastenol (Figure 5) (69).

Sterols of Animals

Studies of sterol composition have been complicated by the fact that (a) many primitive animals have complex compositions and (b) in many animal phyla it has not been conclusively determined whether or not sterols can be synthesized from smaller molecules. Much of the early work on primitive animals was conducted by Bergmann with sponges (71). Sponges were found to have great diversity of sterol composition which were usually such complex mixtures that they were not totally resolvable. Sponges were shown to contain C_{27}, C_{28} and C_{29} sterols with nuclear double bonds primarily in the Δ^5 or Δ^7 positions (71). Although Bergmann examined over 50 species of sponges, due to inadequate analytical technology, limited detail could be obtained on sponge sterol composition. More recent work has confirmed the complexity and diversity of sterols in sponges (72,73). Some sponges have been reported to synthesize sterols (74), while in others sterol biosynthesis was not observed (74,75).

Complex mixtures of mostly Δ^5-sterols, with cholesterol usually dominant, are found in the Coelenterates (1). Some species are of interest because they possess the unusual C_{30} sterol gorgosterol. At present there is very little evidence to support biosynthesis of sterols in Coelenterates.

The Phyla Platyhelminthes and Nemathelminthes also apparently are unable to synthesize sterols from smaller molecules (1,76), but many nematodes are plant parasites and are able to dealkylate phytosterols to produce cholesterol (76). Both phyla contain cholesterol as their principal sterol (1).

Annelids contain cholesterol and a mixture of other sterols. Marine annelids have been shown to synthesize cholesterol but biosynthesis has not been demonstrated in terrestrial forms (1).

The phylum Arthropoda is the largest of the animal kingdom. Sterol composition and sterol metabolism have been studied extensively, especially in economically-important species. Sterols do not appear to be synthesized by Arthropods, although in a few cases the data are not conclusive (77-79). Cholesterol is the principal sterol in most but not all species (1), and occasionally it is accompanied by various phytosterols. Most Arthropods require cholesterol, which is obtained directly in the diet or, in many cases, by dealkylation of dietary phytosterols to cholesterol (80).

Led by the work of Bergmann, the sterols of molluscs were recognized very early to be complex mixtures (71) which could not be resolved adequately at that time. More recent studies have confirmed the complexity of sterol composition of mollusks (81-83) and using liquid chromatography and gas chromatography, 17 sterols were identified in one species of scallop (82) and nearly forty were identified in the eastern oyster (83). These studies indicate that, except for the chitons which contain Δ^7-sterols (84), the sterols of molluscs are primarily Δ^5 and may contain from 26 to 30 carbons. Although examinations of many mollusc species indicates that sterol biosynthesis from acetate occurs, it is sometimes slow (81). In the eastern oyster, the most frequently studied

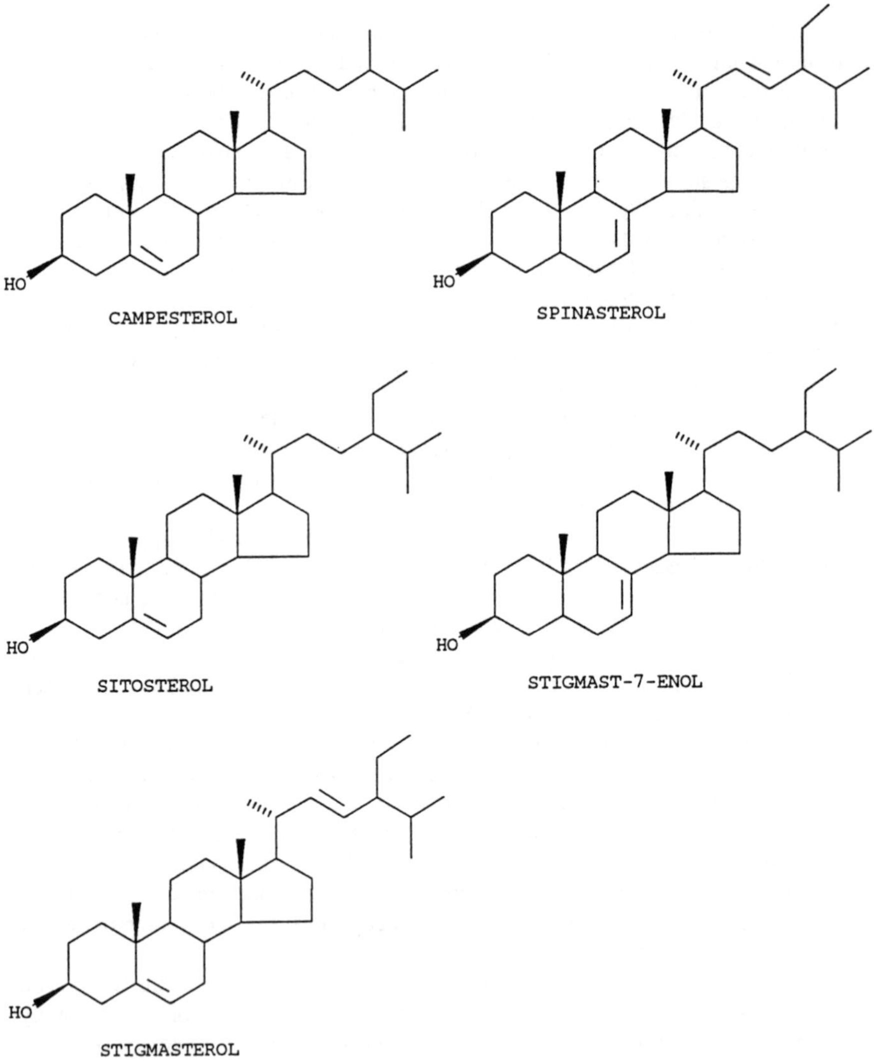

CAMPESTEROL

SPINASTEROL

SITOSTEROL

STIGMAST-7-ENOL

STIGMASTEROL

Figure 5. Sterols of plants

mollusk, the net result from three laboratories is that no synthesis occurs (85-87). Occasional positive results appear to be attributable to the activity of associated microorganisms (88).

Echinoderms can be divided into two groups based on sterol composition. Starfish and sea cucumbers contain predominantly Δ^7-sterols while the sea lilies, brittle stars, and sea urchins contain primarily Δ^5-sterols (89). Two early studies indicated that sterols are not synthesized in echinoderms (85,90), but later studies demonstrated biosynthesis in most echinoderm species studied (89,91). One of the latter studies (89) also provided data suggesting that echinoderms synthesize C_{27} sterols, but obtain C_{28} and C_{29} sterols from the diet.

The sterols of vertebrate animals are almost pure cholesterol, usually with only traces of biosynthetic precursors present. All vertebrates apparently synthesize cholesterol but unlike most of the more primitive animals, vertebrates generally do not absorb C-24 alkylated sterols from their diets (1).

Primitive organisms appear to produce sterol end-products that are only a few steps from cyclization, while more advanced organisms produce sterols biosynthetically more distant from cyclization (Figure 6).

Conclusions

Sterols are present in all major groups of living organisms. Many prokaryotes, however, contain sterols in small amounts or have sterols replaced by pentacyclic triterpenoids such as the hopanoids. A very primitive prokaryote, *Methylococcus,* synthesizes a "primitive" sterol mixture, one in which the sterol nucleus has been little altered after cyclization. Most members of the Protoctista synthesize sterols in an extremely wide range of C_{27} to C_{30} sterols with Δ^5, Δ^7 and $\Delta^{5,7}$ nuclear double bonds. There is good correlation between sterol structure and taxonomic position especially in the algae and protozoa. Ergosterol is easily the most common sterol in the fungi, but some fungi do contain Δ^7-sterols or mixtures of Δ^5-sterols. Sitosterol is the most common plant sterol. However, a large number of plants in the order Caryophyllales contain primarily Δ^7-sterols or have complex mixtures of Δ^5 and Δ^7-sterols. All animals contain sterols but a surprising number of animals do not synthesize sterols. Lower animals absorb sterols from the diet or the environment and contain cholesterol and complex mixtures of other sterols. In most lower animals, conclusive proof of the existence of sterol biosynthesis rarely has been attained. Vertebrate animals all synthesize cholesterol and rarely contain significant amounts of any other sterol. It appears that in the sterol composition of living organisms, we see small amounts of sterols or sterol-like molecules in prokaryotes, an enormous variety of sterols in the Protoctista, and established major sterols of ergosterol in fungi, sitosterol in plants, and cholesterol in animals. The wide variety of sterols in early eukaryotes supports Marguiles' concept of this group as an evolutionary experiment in symbiosis. Very distinctly different sterol composition in fungi, plants, and animals seem to suggest separate "lines" leading to these groups from primitive eukaryotes. Sterols also provide clues to possible precursor organisms to plants, animals, and fungi.

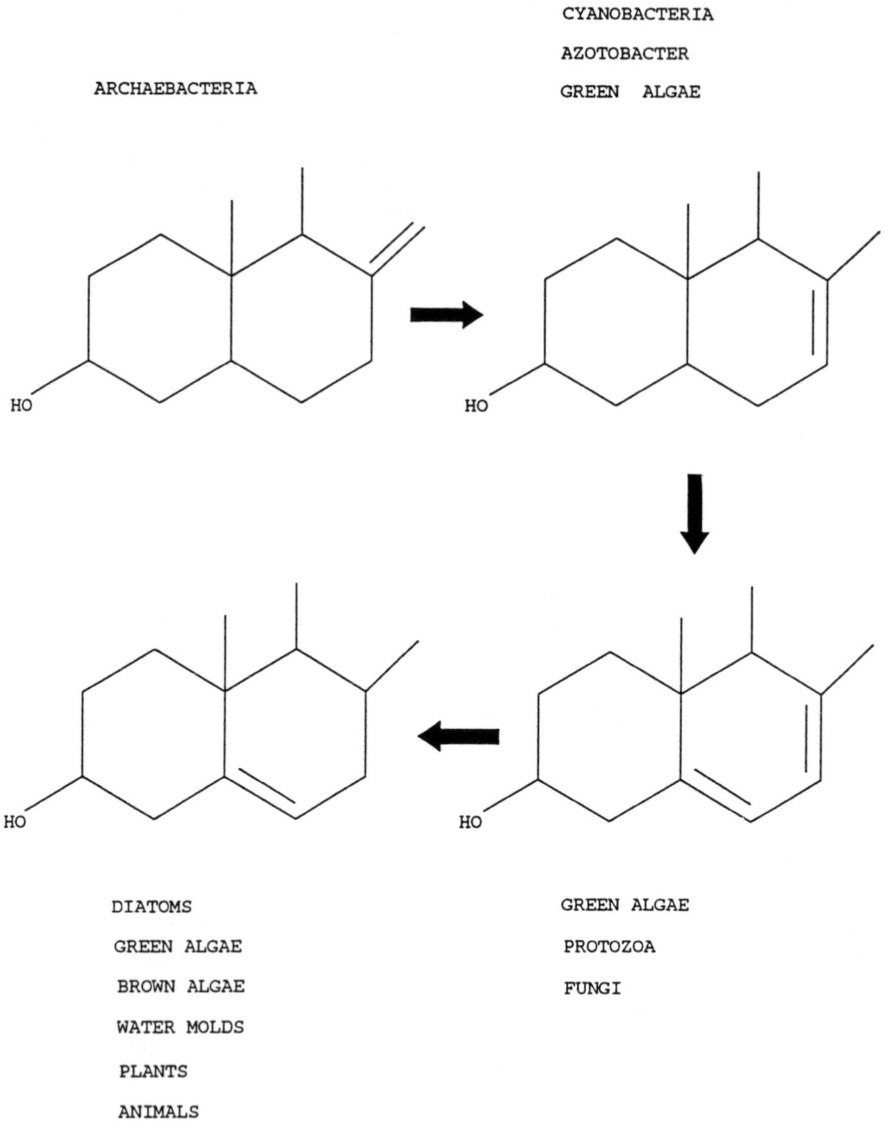

Figure 6. Distribution of sterols in living organisms

Literature Cited

1. Nes, W.R. and McKean, M.L. The Biochemistry of Steroids and Other Isopentenoids; University Park Press, Baltimore, MD, 1977; pp 37-57.
2. Stanier, R.Y., Organisation and Control in Prokaryotic and Eukaryotic Cells, 1, Twentieth Symp. Soc. Gen. Microbiol., Cambridge Univ. Press, 1970.
3. Schubert, K., Rose, G., Tummler, R., and Ikekawa, N., Z. Physiol. Chem. 1964, 339, 293-295.
4. Schubert, K., Rose, G., Wachtel, H., Harhold, C., and Ikekawa, N., Eur. J. Biochem. 1968, 5, 246-251.
5. Schubert, K., Rose, G., and Harhold, C., Biochem. Biophys. Acta. 1967, 137, 168-171.
6. Weeks, O.B. and Francesconi, M.D., J. Bact. 1978, 136, 614-624.
7. Patterson, G.W., The Physiology and Biochemistry of Sterols; American Oil Chemists' Society, Champaign, Illinois. 1992. Patterson, G.W. and Nes, W.D. eds., 1992, pp.118-157.
8. Rohmer, M. (This book)
9. Ourisson, G. and Rohmer, M. 1982. In: Current Topics in Membranes and Transport, Academic Press, New York, 1982, Vol. 17, pp. 153-182.
10. Bird, C.W., Lynch, J.M., Pirt, F.J, Reid, W.W., Brooks, C.J.W., and Middleditch, B.S., Nature 1971, 230, 473-474.
11. Bouvier, P., Rohmer, M., Benveniste, P., and Ourisson, G., Biochem. J. 1976, 159, 267-271.
12. Margulis, L. and Schwartz, K.V., Five Kingdoms, W.H. Freeman and Company, New York 1988, 376 pp.
13. Patterson, G.W., Lipids 1971, 6, 120-127.
14. Yokata, T., Kim, S.K., Fufui, Y., Takahashi, N., Takeuchi, Y. and Takematsu, T., Phytochemistry 1987, 26, 503-507.
15. Gladu, P.K., Patterson, G.W., Wikfors, G.H., Chitwood, D.J. and Lusby, W.R., Phytochemistry 1991, 30, 2301-2303.
16. Rubinstein, I. and Goad, L.J. 1974. Phytochemistry 13, 485-487.
17. Orcutt, D.M. and Patterson, G.W., Comp. Biochem. Physiol. 1975, 50B, 579-583.
18. Patterson, G.W., Tsitsa-Tzardis, E., Wikfors, G.H., Gladu, P.K., Chitwood, D.J., and Harrison, D., Comp. Biochem. Physiol. 1993, 105B, 253-256.
19. Goad, L.J., Holz, G.G., and Beach, D.H., Phytochemistry 1983, 22, 475-476.
20. Gladu, P.K., Patterson, G.H., Wikfors, G.H., Chitwood, D.J. and Lusby, W.R., Comp. Biochem. Physiol. 1990, 97B, 491-494.
21. Rohmer, M., Kokle, W.C.M.C., Fenical, W., and Djerassi, C., Steroids 1980, 35, 219-231.
22. Kokke, W.C.M.C., Shoolery, J.N., Fenical, W., and Djerassi, C., J. Org. Chem. 1984, 49, 3742-3752.
23. Raederstorff, C. and Rohmer, M., C.R. Acad. Sci. Paris 1984, 299, 57-60.

24.   Raederstorff, C. and Rohmer, M., Eur. J. Biochem. 1987, 164, 421-426.
25.   Halevy, S. and Finklestein, S., J. Protozool. 1965, 12, 250-252.
26.   Kern, E.D., Von Brand, T., and Tobie, E.J., Comp. Biochem. Physiol. 1969, 30, 601-610.
27.   Halevy, S., and Aviva, L., Ann. Trop. Med. Parasitol. 1966, 60, 439-444.
28.   Halevy, S., and Sarel, S., J. Protozool. 1965, 12, 293-296.
29.   Williams, B.L., Goodwin, T.W., and Ryley, J.F., J. Protozool. 1966, 13, 227-230.
30.   Halevy, S. and Gisry, O., Proc. Soc. Exp. Biol. 1964, 117, 552-555.
31.   Dixon, H., Gingu, C.D., and Williamson, J., Comp. Biochem. Physiol. 1972, 41B, 1-18.
32.   Conner, R.L., Landrey, J.R., Burns, C.H., and Mallory, F.B., J. Protozool. 1968, 15, 600.
33.   Mallory, F.B., Gordon, J.T., and Conner, R.L., J. Am. Chem. Soc. 1963, 85, 1362.
34.   Conner, R.L., VanWagtendonk, W.J., and Miller, C.A., J. Gen. Microbiol. 1953, 9, 434-439.
35.   Holz, G.G., Jr., Erwin, J.A., Wagner, B., and Rosenbaum, N., J. Protozool. 1962, 9, 359-363.
36.   Holz, G.G., Jr., Erwin, J.A. and Wagner, B., J. Protozool. 1961, 8, 297-300.
37.   Halevy, S., Avivi, L. and Katan, H., J. Protozool. 1966, 13, 480-483.
38.   Harvey, H.R. and McManus, G.B., Geochim. Cosmochim. Acta 1991, 55, 3387-3390.
39.   Lenfant, M., LeCompte, M.F., and Farrugia, G., Phytochemistry 1970, 9, 2529-2935.
40.   Nes, W.D., Norton, R.A., Crumley, F.G., Madigan, S.J., and Katz, E.R., Proc. Natl. Acad. Sci. U.S.A. 1990, 87, 7565-7569.
41.   Gladu, P.K., Patterson, G.W., Wikfors, G.H., and Lusby, W.R., Lipids 1991, 26, 656-659.
42.   Weete, J.D., Fuller, M.S., Huang, M.Q., and Gandi, S., Experimental Mycol. 1989, 13, 183-195.
43.   Southall, M.A., Motta, J.J., and Patterson, G.W., Amer. J. Bot. 1977, 64, 246-252.
44.   Bean, G.A., Patterson, G.W., and Motta, J.J., Comp. Biochem. Physiol. 1972, 43B, 935-939.
45.   Popplestone, C.R. and Umrau, A.M., Phytochemistry 1973, 12, 1131-1133.
46.   Berg, L.R., Patterson, G.W. and Lusby, W.R. Phytochemistry, 24, 616-618.
47.   Berg, L.R. and Patterson, G.W. Exptl. Mycol. 1986, 10, 175-183.
48.   Berg, L.R., Patterson, G.W., and Lusby, W.R., Lipids 1983, 18, 448-452.
49.   Fox, N.C., Congiglio, J.G., and Wolf, F.T. Exptl. Mycol. 1983, 7, 216-226.

50. McCorkindale, N.J., Hutchinson, S.A., Pursey, B.A., Scott, W.T., and Wheeler, R., Phytochemistry 1969, 8, 861-867.
51. Warner, S.A., Sovocool, G.W. and Domnas, A.J., Mycologia 1983, 75, 285-291.
52. Wood, S.G. and Gottlieb, D., Biochem. J. 1978, 170, 343-354.
53. Schlosser, E., Shaw, P.D. and Gottlieb, D., Arch. Microbiol. 1969, 66, 147-153.
54. Elliott, C.G., Hendrie, M.R., and Knights, B.A., and Parker, W., Nature 1964, 203, 427.
55. Haskins, R.H., Tulloch, A.P., and Micetich, R.G., Can. J. Microbiol. 1964, 10, 187-195.
56. Hendrix, J.W., Science 1964, 144, 1028.
57. Weete, J.D., Laseter, J.L., and Lawler, G.C. Arch. Biochem. Biophys. 1973, 155, 411.
58. Goulston, G., L.J. Goad, and Goodwin, T.W., Biochem. J. 1967, 102, 15C.
59. Blank, F., Shortland, F.E., and Just, G., J. Invest. Dermatol. 1962, 39:91.
60. Beilby, J.P. and Kidby, D.K., Lipids 1980, 15, 375-378.
61. Beilby, J.P., Lipids 1980, 15, 949-952.
62. Weete, J.D., Kulifaj, M., Montant, C., Nes, W.R., and Sancholle, M. Can. J. Microbiol. 1985, 31, 1127-1130.
63. VanEijk, G.W. and Roeymans, H.J., Antonie van Leeuwenhoek 1982, 48, 257-264.
64. Weete, J.D. and Laseter, J.L., Lipids 1974, 9, 575-581.
65. Weete, J.D., Phytochemistry 1973, 12, 1843-
66. Bergmann, W., Ann. Rev. Plant Physiol. 1953, 4, 383-426.
67. Chiu, P.L., Patterson, G.W. and Fenner, G.P., Phytochemistry 1985, 24, 263-266.
68. Chiu, P.L., Patterson, G.W. and Salt, T.A. Phytochemistry 1988, 27, 819-822.
69. Salt, T.A., Xu, S., Patterson, G.W., and Adler, J.H., Lipids 1991, 26, 604-613.
70. Akihisa, T., Kokke, W.C.M.C., and Temura, T., The Physiology and Biochemistry of Sterols, American Oil Chemists' Society, Champaign, Illinois, Patterson, G.W. and Nes, W.D., eds., pp. 172-228.
71. Bergmann, W., In Comparative Biochemistry; Florkin, M. and Mason, H.S., eds., Academic Press, New York, 1962, Vol. 3, pp. 103-162.
72. Erdman, T.R. and Thompson, R.H. Tetrahedron 1972, 28, 5163-5173.
73. Shikh, Y.M. and Djerassi, C., Tetrahedron 1974, 30, 4095-4103.
74. Walton, M.J. and Pennock, J.G., Biochem. J. 1972, 127, 471-479.
75. DeRosa, M., Minale, L. and Sodano, G., Comp. Biochem. Physiol 1973, 45B, 883.
76. Chitwood, D.J., In Physiology and Biochemistry of Sterols; Patterson, G.W. and Nes, W.D., eds.; American Oil Chemists' Society, Champaign, Illinois, 1992, pp. 257-293.
77. Clayton, R.B., J. Biol. Chem. 1960, 234, 421.

78.   Clayton, R.B., Edwards, A.M., and Bloch, K., Nature 1962, 195, 1125-
79.   Kaplanis, J.N., Robbins, W.E., Vroman, H.E., and Bryce, B.M.,
       Steroids 1963, 2, 547-
80.   Svoboda, J.A., Weirich, G.F., and Feldlaufer, M.F., In Physiology and
       Biochemistry of Sterols; Patterson, G.W. and Nes, W.D., eds.; American
       Oil Chemists' Society, Champaign, Illinois, pp. 294-323.
81.   Voogt, P.A. Thesis, University of Utrecht, 1970.
82.   Idler, D.R., Khalil, M.W., Gilbert, J.D. and Brooks, C.J.W., Steroids
       1976, 27, 155-166.
83.   Teshima, S., Patterson, G.W., and Dutky, S.R., Lipids 1980, 15, 1004-
       1011.
84.   Idler, D.R. and Wiseman, P., Int. J. Biochem. 1971, 2, 516-528.
85.   Salaque, A., Barbier, M., and Lederer, E., Comp. Biochem. Physiol.
       1966, 19, 45-51.
86.   Trider, D.J. and Castell, J.D., J. Nutrition 1980, 110, 1303-1309.
87.   Teshima, S. and Patterson, G.W., Lipids 1981, 16, 234-239.
88.   Holden, M.J. and Patterson, G.W., Lipids 1991, 26, 81-82.
89.   Goad, L.J., Rubinstein, I., and Smith, A.G., Proc. R. Soc. London B
       180, 223-246.
90.   Nomura, T., Tsuchiya, Y., Andre, D. and Barbier, M., Bull. Jap. Soc.
       Sci. Fish. 1969, 35, 299-302.
91.   Kanazawa, A., Teshima, S., and Tomita, S., Bull. Jap. Soc. Sci. Fish
       1974, 40, 1257-1260.

RECEIVED May 2, 1994

Chapter 6

Evolution of the Oxysterol Pathway

Edward J. Parish

Department of Chemistry, Auburn University, Auburn, AL 36849

Oxysterols can be defined as sterols bearing a second oxygen function, in addition to that at carbon-3. In man, and possibly other organisms, certain oxysterols may have a key role in the regulation of sterol biosynthesis. Side chain oxgenated sterols, a class of oxysterol, are widely distributed in nature. They have been shown to be produced through biosynthesis in plants, animals, and microorganisms. The occurrence of this oxysterol biosynthetic pathway, producing sterols oxygenated at carbon-24 and/or-25, can be found in many classes of organisms including man. In this report, we have focused on the evolutionary origins of this biosynthetic pathway of which we are the final recipients.

The topic of evolution is a popular subject for discussion. Unfortunately, as one investigates the topic in some detail, he will discover that in certain areas of evolution much is based on extrapolation or assumption and that clear and dependable supporting data may not exist or is not available. One could surmise that the topic itself could be redefined as a function of time and, although firmly established, is in a state of evolution.

In the present case, if we are to examine the evolution of a biosynthetic pathway, we will have to make some assumptions. First, we must assume that the major forms of life that existed in the past still exist and that the steroid biochemistry we observe in the present is a reflection of what has existed previously. Although earlier forms of life evolved first, and are preceived as less advanced than those which came later, all major forms exist today and can be studied in the present. Second, the steroid

0097–6156/94/0562–0109$08.00/0

biochemistry found in each type of organism studied today has existed in the past and that by examining the differences or similarities in these pathways we can trace or observe the origins and evolution of sterol pathways.

An excellent example of this is the comparative study of the biosynthesis of sterols and biologically significant steroids in various organisms and the origins of this biosynthesis in a variety of organisms in the books by W. R. Nes and M. L. McKean (1) and W. R. Nes and W. D. Nes (2).

Studies on the fossil record with respect to molecular artifacts, which would include steroids, have been conducted (3-7). Although some steroids which have undergone diageneses in ancient sediments have observed steriochemistry which is similar to the original molecule, the vast majority are found as rearranged derivatives or in a reduced form (including many aromatic structures) which make it difficult to identify the original steroid precursor. In the present study, oxysterols are labile substances which commonly contain expoxy, hyxroxyl, and ketone functional groups. Chemically, epoxides may easily undergo ring opening or rearrangement to ketones, hydroxyl groups or their derivatives may be subject to dehydration or elimination to form alkenes which may be further reduced to aromatic structures by dehydrogenation, and ketones may be reduced to hydroxyl groups or be a direct route to more unsaturated or aromatic structures (8). Oxysterols would be easily transformed and degraded, initially by microbial activity and later by geochemical processes, into thermodynamically more stable saturated and aromatic hydrocarbons in mature sediments and petroleums. Also, to further complicate and undermine the usefulness of the fossil record, most common sterols of biological origin are known to undergo autoxidation (air oxidation) which is known to produce many oxysterols which might also be produced by biological processes (9,10). Under these circumstances, the fossile record is of very limited valve in the study of oxysterol biosynthesis in ancient species.

In the course of his studies on the composition of fats, the French Chemist Chevreul (1815) observed that a white crystalline compound could be obtained from the unsaponifiable material of certain animalfats. He found it to be identical with the compound isolated from gallstones some 45 years earlier by Poulletier de la Salle and in recognition of the initial source material named it cholesterine (Greek: *chole*, bile, and *steros*, solid). This name is still used in the continental literature although the more descriptive cholesterol is used in the English literature. Investigations into the occurrence of cholesterol and other sterols in the unsaponifiable portion of fats followed rapidly and research on the struture of these compounds soon followed (1,11-13).

Other unsaponifiable compounds similar to cholesterol have been found in the lipids of a variety of plant and animal sources. The collective name sterols has been adopted for all crystalline unsaponifiable alcohols with properties resembling those of cholesterol. In general, these compounds are 3-mono-hydroxysteroids having 27, 28 or 29 carbon atoms and nearly all have one or more double bonds. The double bond is most commonly found at position 5 with double bonds at C-22 and C-7 also being prevalent. All the naturally occurring sterols have 3β-hydroxyl groups.

As a class of compounds, oxysterols can eb defined as sterols bearing a second oxygen function, in addition to that of carbon-3, and having an iso-octyl or modified iso-octyl side chain (14-17). These compounds have demonstrated a variety of diverse biological properties, which include cytotoxicity, atherogenicity, carcinogenicity, mutagenicity, hypocholesterolemia, and effects on specific enzymes (14-22). Widely distributed in nature, they have ben found in animal tissues and food stuffs (9-10) and have been isolated from drugs used in folk medicine for the treatment of cancer (13-25). Other sutudies have shown that certain oxysterols have significant activity in the inhibition of DNA synthesis in cultured cells (20-27). A number of oxygenated derivatives of cholesterol and sterol intermediates in cholestrol biosynthesis have been found to be potent inhibitors of sterol biosynthesis in animal cells in culture (21,22,28). The specific inhibition of cholesterol biosynthesis in mammalian cells by oxygenated derivatives of cholesterol and lanosterol has been shown in many cases to devrease cellular levels of HMG-CoA reductase activity.

This response has been attributed to a decreased rate of HMG-CoA reductase synthesis (28-30) and in some instances to an increase in enzyme degradation (28-30). Other oxysterols are known to depress the rate of cholesterol synthesis from lanosterol in rat liver homogenates and may inhibit the 14-demethylation of lanosterol (31-33).

A large number of oxysterols have been evaluated for their abilities to repress HMG-CoA reductase activity in cultured mammalian cells (21, 22, 34). In general, potency has been found to vary over a wide range depending on the structural features of the oxysterol. As a general trend, inhibitory activity increases as the distance between carbon-3 and the second oxygen group becomes greater. Steroids with oxugen groups in ring D and the side chain have been shown to have the greatest activity. An intact side chain is a requirement for potent activity; a decrease in the length of the (iso- octyl) side chain results in decreased activity (35). Other noticeable trends indicate a relationship between inhibitory activity and the extent to which the second oxygen function is sterically hindered. In general, axial hydroxyl groups are more hindered and possess lower activities than the less hindered equatorial conformation (21, 24, 35). Stearic hindrance from othe rparts of the steroid moelcule can also result in diminished activity (i.e., effect of carbon-14 alkyl substituents on teh carbon-15 hydroxyl group (36). It has been suggested that oxygen functions in conformationally flexible positions such as those in ring D and in the side ore inhibitory steroids due to increased effective hydrogen bonding or hydrophilic interactions with receptor molecules (22).

These observations suggest a regulatory mechanism which, by analogy to steroid hormone receptors and bacterial induction-repression systems, reqiures a binding protein to recognize oxysterols and mediate subsequent cellular events. Evidence for the existence of a specific cytosolic receptor protein for oxysterosl has been presented (34, 37). After the activities of a number of sterols were evaluated, a good correlation was found between the actions of certain oxysterols on HMG-CoA reductase in L cells and their affinity for a oxysterol binding protein (34). The actions of oxysterols which depress the rate of cholesterol biosynthesis from lanosterol and possibly inhibit 14-demethylation of lanosterol are also postulated to exert their actions by an oxysterol binding protein (31-33).

It has been shown that mammalian systems produce oxysterols. Derivatives of cholesterol by hydroxylated in the 7 α-, or 25-, or 26-positions are produced in liver during bile acid biosynthesis and in side-chain hydroxylation at the 20 α–and 22R-positions in the initial step in the conversion of cholesterol in steroid hormones in endocrine organs (38-40). Also, all cells produce 32-hydroxylanosterol during the conversion of lanosterol to cholestrol (41-42). The oxygenated intermediates lanost-8-en-3β,32-diol and 3β-hydroxylanost-8-en-32-al have been isolated from rat liver microsomes incubated with 24,25-dihydrolanosterol (43). A number of C-32 hydroxylated derivatives of cholesterol and lanosterol were shown to be potent inhibitors of HMG-CoA reductase, sterol biosynthesis, and possess a high affinity for the oxysterol binding protein mammalian systems (34,44,45). Another mode of oxysterol biosynthesis has been described which utilizes the isopentenoid pathway to produce side-chain oxygenated derivatives of cholesterol and lanosterol (46-47). This mode of oxysterol biosynthesis is termed the "oxysterol pathway" and results in the formation of sterols oxygenaged at carbon-24 and/or 25. Such compounds are derived from squalene 2,3-epoxide by the introduction of a second oxygen function to form squalene 2,3,22,23-dioxidosqualene prior to cyclization. Thus, this intermediate has been shown to form 24(S),25-epoxylanosterol, 24(S), 25-epoxycholesterol, and 25-hydroxycholesterol in mammalian systems (46-49). 24(S),25-epoxycholesterol has been isolated from cultured mouse L cells, Chinese hamster lung fibroblasts, and human liver (50). These oxygenated side-chain derivatives have been shown to be potent inhibitors of HMG-CoA reductase, sterol biosynthesis, and possess a high affinity for the oxysterol binding protein (46-50). These results add support to the hypothesis that oxysterols may be natural regulators of cholesterol biosynthesis in mammalian cells (17). The striking regulatory effects of side chain oxysterols, and the lack of effect of pure cholesterol, has led to this hypothesis which suggests these compounds are the natrual feedback regulators of the cholesterol biosynthetic pathway. These points are illustrated in Figure 1 (17,51).

In addition to sterol biosynthesis in vertebrate animals, the biosynthesis of sterols seem to be a ubiquitous property of all photysynthetic plants from the prokaryotic blue-green algae to climax angiosperms (52-57). In addition, biosynthesis occurs in most fungi and some protozoa (1). Little attention has been paid to the biosynthesis of oxysterols in plants and microorganisms. However, a study and review of the biosynthesis of oxysterols in the total biosphere is necessary if one is to document and demonstrate the evolution of oxysterol biosynthesis (58,59).

Discussion

The occurrence of side chain oxysterols has been reported in plants, animals, and microorganisms (58,59). In this report, the evolution of oxysterols resulting from primary metabolism via the "oxysterol pathway" will be discussed. this includes sterols oxygenated in the side chain at carbon-24 and/or-25 (Figure 2).

It is assumed that the reported occurrence, in the literature, of these side chain oxysterols result from biosynthesis and that samples have been handled correctly to prevent significant autoxidation (air autoxidation). This could be problematic since sterol side chains (e.g. cholesterol) are known to slowly undergo autoxidation to produce C-24 and/or hydroxy products (9,10). Also, fucosterol is known to yield 24-ketocholesterol as an autoxidation product (9,10,60). However, site specific or exclusive autoxidation at only one site is usually the exception since this phenomenon is reported to produce complex mixtures of a variety of oxidized products which may be difficult to resolve (9,61).

Also included in this report, are two documented alternative biochemical pathways which are known to introduce oxygen functionaly at C-25 (and C-26) directly without using the "oxysterol pathway". In the first, sterols may be hydroxylated at C-25 in the liver during bile acid biosynthesis (1). In the second, vitamin D_3 is known to be hydroxylated directly at C-25 in mammalian systems (1). These documented instances occur in the animal kingdom. Plant metabolism, especially the biosynthesis of secondary plant metabolites, has not been examined in sufficient detail to indicate the extent to which direct introduction of a hydroxyl group at C-25 might occur. However, a great deal of evidence is available to indicate that many of these side chain oxysterols are produced via the "oxysterol pathway".

The "oxysterol pathway", from which side chain oxysterols are derived, is that which involves the cyclization of 2,3,22,23-dioxidosqualene (DOS) to form sterols oxygenated at C-24 and 25. Enzymatic conversions of some terminally modified squalene 2,3-oxide analogs into the corresponding lanosterol analogs have shown that the substrate specificity of the enzyme is low (62-65). A quantitative comparison of the relative efficiency of transformation of squalene 2,3-oxide and its terminally modified analogs (including DOS) was studied using the cyclase system prepared from hog liver (66). Subsequently, a number of investigators have

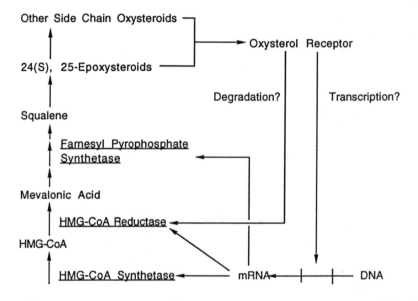

Figure 1.   Model for the oxysterol regulation of sterol biosynthesis.

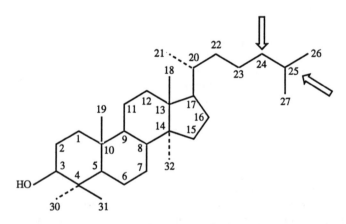

Figure 2.   Designations of carbon atoms in the sterol nucleus and side chain and indicated sites of primary metabolism under discussion.

shown that 24(S),25-epoxylanosterol, 24(S),25-epoxycholesterol and 24(S), 25- epoxycholesterol have been found in cultured fibroblasts (50) and the latter oxysterol was isolated from human liver (49) giving support to the suggestion that it may participate in the regulation of hepatic cholesterol metabolism in vivo.

A schematic comparison of the oxysterol or dioxidosqualene pathways found in plants and animals is presented in Figure 3. As early as 1970 DOS was shown to be cyclized by microsomes from bramble tissues grown in vitro, to form 24,25-epoxycycloartenol (66,67). In Ononis spinosa, DOS is the precursor of α-onocerin under anaerobic conditions. (68). In a related study, the investigation of the squalene-2,3;22,23-diepoxide-α-onocerin cyclase enzyme system from Ononis spinosa root has provided evidence for the formation of the intermediate pre-onocerin, which contains a C-24,25 epoxide function originating from DOS (69). Agaiol, the major steroidal triterpenoid of Aglaia ordorata has the 24(S) configuration at the C-24,25 epoxide and may have been formed from DOS with the (22S) configuration (71). 25(27)-Dehydrolanost-8-en-3β-ol, isolated from Cerus giganteus, was proposed to arise from the DOS pathway via opening the C-24,25 epoxide to give the C-25 hydroxy derivative which is then dehydrated to form the Δ 25(27) double bond (71). DOS has been found in rat liver homogenates (72) and isolated from yeast in 1977 (73) together with the cyclized product 24,25-epoxylanosterol (73). In a related study (Figure 4), ti was shown that 2(R,S),3-epiminosqualene was oxidized and cyclized to 24(R,S),25-epiminolanosterol in Gibberella Fujikuroi (74) and, furthermore it was subsequently found that the 24(RS),25-epiminolanosterol was metabolized via opening of the aziridine ring to 25-aminolanosterol in the same organism (75). this latter study demonstrates that the enzymes involved in the DOS pathway are non-specific.

The literature contains many examples of steroids and steroidal triterpenoids, isolated from plants and microorganisms, which contain an oxygen function on C-24 and/or C-25 (76-90). The presence of these compounds in such numbers, indicates the possible widespread occurrence and use of the "oxysterol pathway" in these organisms. Recently, it has been shown that several species of red algae contain a series of side chain oxysterols including C-24 and 25 hydroxy derivatives and a C-24,25 epoxide (Figure 5) (60). In addition, brown algae contains small quanties of 24-ketocholesterol (Figure 6) in addition to fucosterol (60). It was noted that care must be taken when handling and processing samples of brown algae since the fucosterol present is known to undergo outoxidation to 24-ketocholesterol. Certain soft corals (Figure 7) are also known to contain 25-hydroxy sterols which appear to result from metabolic processes (60). Two new polyhyrdoxysterols, (24S)-24-methylcholest-5-ene-3β,25 ξ,26-triol and (24S)-24-methylcholestane-3β,5β, 6α,25-tetrol were isolated from two Indian Ocean soft coral species of Sclerophytum

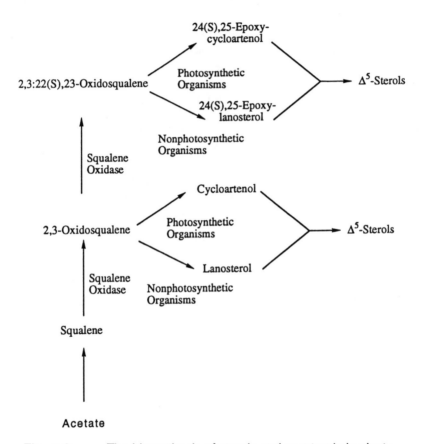

Figure 3.   The biosynthesis of sterols and oxysterols in plants, animals and microorganisms.

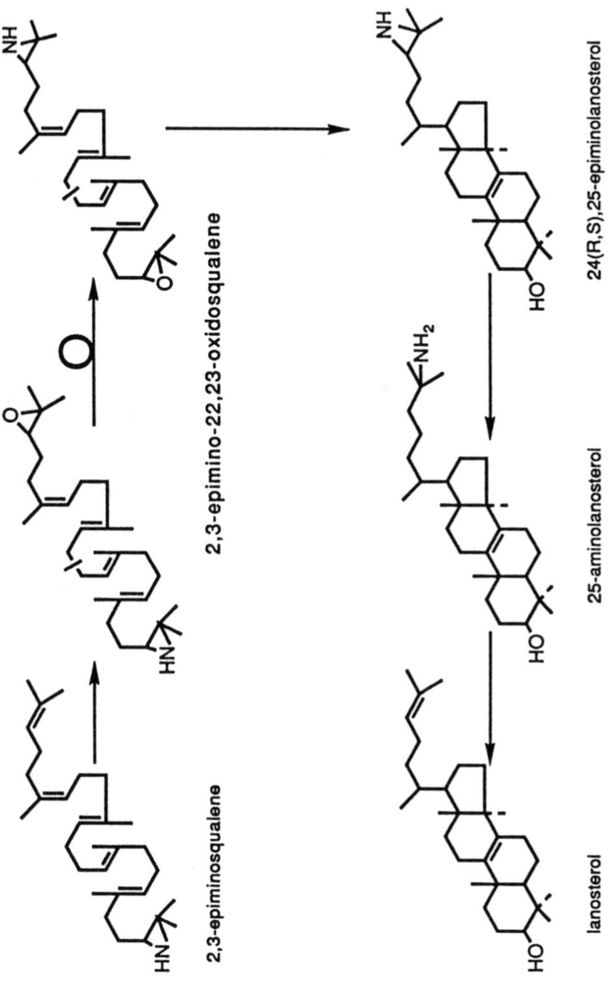

Figure 4. The metabolism of 2(R,S),3-epiminosqualene in *Gibberella fujikuroi*.

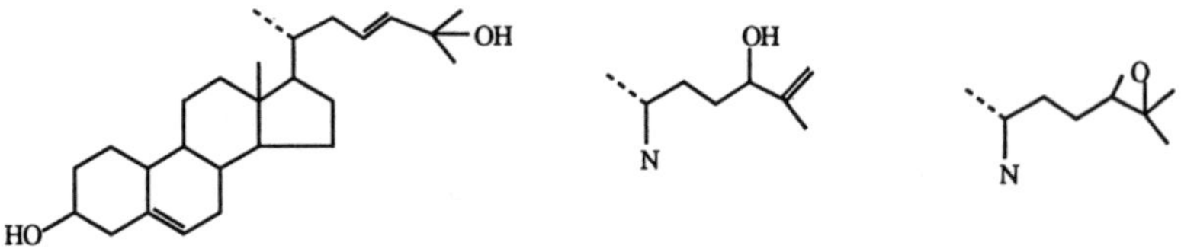

Asparagopsis armata	Same	Same
Rissoella verruculosa	Same	Same
Rhodymenia palmata	Same	
Liagora distenta		
Scinaia furcellata		

Figure 5.   The distribution of C-24 or 25 oxysterols in several red algae species.

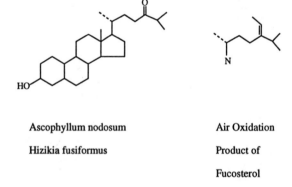

Ascophyllum nodosum	Air Oxidation
Hizikia fusiformus	Product of
	Fucosterol

Figure 6.   The known distribution of 24-ketocholesterol in brown algae and the structure of fucosterol, its autoxidation precursor.

Recently, a series of articles has described the occurrance of 24(R),25-dihydroxy vitamin D_3 in plasma (Figure 8) which was shown to originate, through metabolic labelling sutdies, from vitamin D_3 by the direct introduction of oxygenated function onto the side chain (92).

Hydroxylation of the side chain of cholesterol at C-24, C-25 and at C-26 is known to occur in rat liver mitochondra in vitro (93,94). 26-Hydroxylation is believed to constitute the initial step in a minor pathway in bile acid formation (95) whereas the importance of 24- and 25-hydroxylation of cholesterol is unknown (96). Steroids related to cholesterol such as sitosterol adn 5β-cholestan-3β-ol have been reported to be bile acid precursors (97,98). Purified cytochrome P-450, prepared from rat liver microsomes, was found to catalyze efficient 25-hydroxylation of 5β-cholestan-3α, 7α, 12α-triol as well as vitamin D_3 (99,100). A number of steroids and steroidal triterpenoids have been isolated from plants and microorganisms which contain a single oxygenated functional group at C-25 (90, 101-103).

The presence of these substances, in large numbers, in plants, animals and microorganisms might suggest a regulatory role in the life cycle of these organisms. A recent study found that 24,25(R,S)-exoxylanosterol has an inhibitory effect on sterol biosynthesis in Saccharomyces cerevisiae (strain G204, HEM1 HIS3) (104). These preliminary results indicate the potential for future studies in this area, especially in light of the regulatory role oxysterols have demonstrated in mammalian systems.

Conclusion

The extensive studies cited above clearly demonstrate the abundance of these compounds in nature. Apparently, plants, animals and microorganisms possess the biosynthetic ability to produce C-24 and 25 oxysterols in abundance. As this report has demonstrated, the origins of these biosynthetic pathways can be found in lower organisms and in some instances have been known before their discovery in animal systems. A great deal of effort has been spent on the study of biosynthesis of oxysterols and their regulatory role in mammalian biochemistry. The extensive study of the biosynthesis of oxysterols and their regulatory role in animal metabolism and impact it might have in the control of human heart disease has been underlying goals of these studies.

Ample evidence has been presented to show a clear lineage of C-24 and/or 25 oxysterol biosynthesis in the biosphere via the "oxysterol pathway". Those results would seem to indicate the evolution of this pathway from lower organisms and finally its occurrence and function in mammalian systems. Certainly, we are the fortunate final recipients of this evolutionary event.

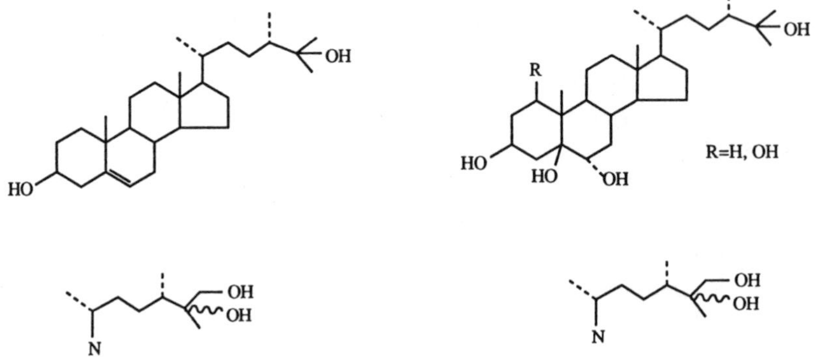

Sclerophytum sp.

Figure 7.   Examples of C-25 oxysterols found in soft corals.

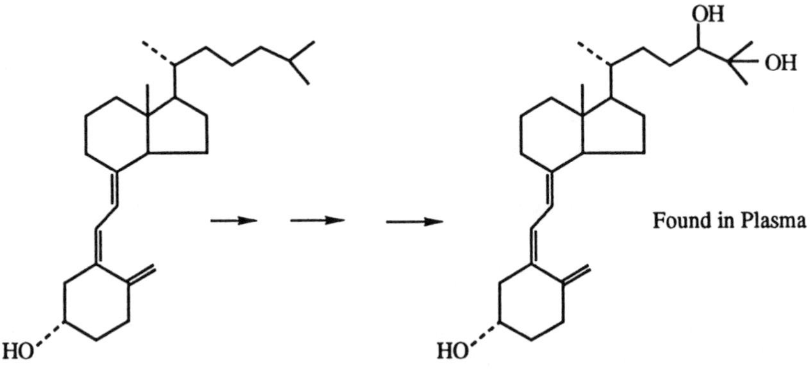

Vitamin D$_3$ 24R, 25-Dihydroxy vitamin D$_3$

Figure 8.   The structure and metabolism of vitamin D$_3$ to 24R,25-dihydroxy vitamin D$_3$ in plasma.

References

1. Nes, W. R.; McKean, M. L. *The Biochemistry of Steroids and Oher Isopentenoids*, University Park Press, Baltimore, 1977.
2. Nes, W. R.; New, W. D. *Lipids in Evolution*, Plenum Press, New York, 1980.
3. Eglinton, G.; Murphy, M. *Organic Geochemistry*, Springer-Verlag, New York, 1969.
4. Albrecht, P.; Ourisson, G., *Angew, Chem.* 1971 10, 209.
5. Bjoroy, M. *Advances in Organic Geochemistry 1981*, John Wiley, New York, 1983.
6. Mackenzie, A. S.; Brassell, S. L.; Eglington, G.; Maxwell, J. R. *Science* 1982 *217*, 491-502.
7. Brassell, S. L.; Eglington, G.; Maxwell, J. R. 1983 *Biochem. Soc. Trans. 11*, 575-586.
8. Larock, R. C. *Comprehensive Organic Transformations*, VCH Publications, New York, 1989.
9. Smith, L. L. *Cholesterol Autoxidation*, Plenum Press, New York, 1981.
10. Smith, L. L. *Chem. Phys. Lipids* 1987 44, 87-105.
11. Fieser, L. F.; Fieser, M. *Steroids*, Reinhold. New York, 1959.
12. Fieser, L. F.; Fieser, M. *Natural Products Related to Phenanthrene*, 3rd ed., Reinhold, New York, 1949.
13. Cook, R. P. *Cholesterol;* Academic Press, New York, 1958.
14. Kandutsch, A. A. and Chen, H. W. *J. Biol. Chem.* 1973 *248*, 8108-8114.
15. Kandutsch, A. A.; Chen, H. W. *J. Biol. Chem.* 1974, *249*, 6057-6063
16. Chen, H. W., Kandutsch, A. A.; Waymouth, C. *Nature* 1974, *251*, 419.
17. Kandutsch, A. A.; Chen, J. W.; Heiniger, H. J. *Science*, 1978, *201*. 498-504
18. Parish, E. J.; Nanduri, N. B. B.; Kohl, H. H.; Taylor, F. R. *Lipids* 1980, *21*, 27-34
19. Parish, E. J.; Nanduri, N. N. N.; Kohl, H. H.; Nusbaum, K. E. *Steroids*, 1986, *48*, 407-413
20. Parish, E. J.; Chitrakorn, S.; Luu, B.; Schmidt, G.; Ourisson, G. *Steroids*, 1989, *53*, 579-586.
21. Schroepfer, G. J.; Jr. *Ann. Rev. Biochem.* 1981, *51*, 585-112.
22. Gibbons, G. F. *Biochem. Soc. Trans. LONDON*, 1983, *11*, 649.
23. Cheng, K.-P; Nagano, N.; Bang, L.; Ourisson, G. *J. Chem. Res.*, 1977 (S)218; (M)2501-2506.
24. Nagano, H.; Poyser, J. P.; Cheng, K.-P.; Bang, L.; Ourisson, G. *J. Chem. Res.* 1977, (S)218; (M)2522-2571.
25. Zander, M; Patrick, K.; Bang, L.; Ourisson, G. *J. Chem. Res.* 1977, (S) 219; (M)2572-2577.
26. DeFay, R.; Astruc, M. E.; Soussillion, S.; Decomps, B.; Crastes de

Paulet, A. *Biochem. Biophys. Res. Commun.*, 1982, *106*, 362-373.
27. Astruc, M.; Laporte, M.; Tabacik, C.; Crasted de Paulet, A. *Biochem. Biophys. Res. Comm.* 1978, *83*, 691-704.
28. Sinensky, M.; Target, R.; Edwards, P. A. *J. Biol. Chem.*, 1981, 256, 11774-11783.
29. Faust, J. R.; Luskey, K. L.; Chin, D. J.; Goldstein, J. L.; Brown, M. S. *Proc. Natl. Acad. Sci. USA*, 1982, *79*, 5205-5209.
30. Tanaka, r. D.; Edwards, P. A.; Lan, S.-F; Fogelman, A. M., *J. Biol. Chem.* 1983, *258*, 13331.
31. Sonoda, Y.; Sato, Y. *Chem. Pharm. Bull. JAPAN 31*, 1983, 1698-1706.
32. Saton, Y; Sonoda, Y.; Morisaki, M.; Ikekawa, N. *Chem. Pharm. Bull. JAPAN*, 1984, *32*, 3305-3314.
33. Morisaki, M.; Sonoda, Y.; Makino, T.; Ogihara, H.; Ikekawa, H.; Sato, Y. *J. Biochem.* 1986, *99*, 597.
34. Taylor, F. R.; Saucier, S. E.; Shown, E. P.; Parish, E. J.; Kandutsch, A. A. *J. Biol. Chem.* 1984, *259*, 12382-12393.
35. Gibbons, G. F. in 3-Hydroxy-3-methylglutaryl Coenzyme A *Reductase* (Sabine, L, ed.) CRC Press, pp. 153-232 W. Palm Beach, FL, 1983.
36. Schroepfer, G. J., Jr.; Parish, E. J.; Tsuda, M.; Raulston, D. L.; Kandutsch, A. A.; Taylor, F. R.; Shown, E. P. *J. Biol. Chem. J.* Lipid Res. 1979, 20, 994-1005.
37. Kandutsch, A. A.; Taylor, F. R.; Shown, E. P. *J. Biol. Chem.* 1984, *259*, 12388-12399.
38. Vlahcevic, Z. R.; Schwartz, C. C.; Gustafsson, J.; Halloran, L. G.; Danielson, H.; Swell, L. *J. Biol. Chem.* , 1980, *255*, 2925-3006.
39. Pederson, J. I,; Bjorkhem, I.; Gustafson, J. *J. Biol. Chem.* 1979, *254*, 6464-6471
40. Smith, L. L.; Teng, J. I,; Lin, Y. Y.; Seitz, P. K.; McGehee, M. F. *J. Steroid Biochem.* 1981, *14*, 889-895.
41. Gibbons, G. F.; Pullinger, C. R.; Chen, H. W.; Carnee, W. K.; Kandutsch, A. A. *J. Biol. Chem.* 1980, *255*, 395-407.
42. Tabacik, C.; Aliau, S.; Serrou, B.; Crastes de Paulet, A. *Biochem. Biophys. Res. Commun.*, 1981, *101*, 1087-1098.
43. Shafiee, A.; Trzaskos, J. M.; Paik, Y.-K; gaylor, J. L. *J. Lipids, Res.*, 1988, *27*, 1-74
44. Schroepfer, G. J., Jr.; Pascal, R. A., Jr.; Shaw, R.; Kandutsch, A. A. *Biochem. Biophys. Res. Commun.*, 1978, *83*, 1024-1031.
45. Schroepfer, G. J., Jr.; Parish, E. J.; Pascal, R. A., Jr.; Kandutsch, A. A. *J. Lipid Res.*, 1980, *21*, 571-582
46. Nelson, J. A.; Stackbeck, S. R.; Spenser, T. A. *J. Biol. Chem.* 1981, 256, 1067-1078.
47. Panini, S. R.; Sexton, R. C.; Rudney, H. *J. Biol. Chem.* 1984, 259, 7767-7784.
48. Panini, S. R.; Sexton, R. C.; Gupta, A. K.; Parish, E. J.; Chitrakorn, S.; Rudney, H. *J. Lipid. Res.* 1986, *27*, 1190.

49. Spencer, T. A. Gayen, A. K. Phirwa, S. ; Nelson, J. A. Taylor, F. R.; Kandutsch, A. A.; Erickson, S. *J. Biol. Chem.* 1985*260*, 13391.
50. Saucier, S. F.; Kandutsch, A. A.; Taylor, F. R. Spenser, T. A.; Phirwa, S.; Gayen, A. K. *J. Biol. Chem.* 1985, *260*, 14571-14581.
51. Taylor, R. F. in *Regulation of Isopentenoid Metabolism* (Nes, W. D.; Parish, E. J.; Trzaskos, J. M., eds.) ACS Symposia Series, Washington D.C. 1992, Vol. 497, pp. 81-93.
52. Reitz, R. G.; Hamilton, J. G., Comp. *Biochem. Physiol.*, 1969, *25*, 401-409.
53. de Souza, N. J.; Nes, W. R., *Science*, 1968, *162*, 363.
54. Forin, M. C.; Maume, B.; Baron, C., *C. R. Acad. Sci., Paris*, 1972, 2*74D*, 113-120.

55. Paoletti, C.; Pusparaj, B.; Florenzano, G.; Capella, P.; Lercher, G. *Lipids*, 1976, *11*, 266-272.
56. Nes, W. R. *Adv. Lipid Res.*, 1977, *15*, 233-242.
57. Goad, L. J. and Goodwin, T. W. in *Progress in Phytochemistry* (Reinhold, L. and Liwschitz, Y., eds.) Interscience Publishers, New York, 1972, Vol. 3, pp. 113-129.
58. Parish, E. J. in *Physiology and Biochemistry of Sterols* (Patterson, G. W.; Nes, W. D., eds.) Am. Oil Chem. Soc., Champaign, Illinois, 1991, pp. 324-336.
59. Parish, E. J. in *Regulation of Isopentenoid Metabolism* (Nes, W. D.; Parish, E. J.; Trzaskos, J. M., eds.) ACS Symposia Series, Washington, D.C. 1992, Vol. 497, pp. 146-161.
60. Faulkner, D. J. *Nat. Prod. Reports* 1992, *9*, 323-354; references therein.
61. Parish, E. J. in *Analysis of Sterols and Other Biologically Significant Steroids* (Nes, W. D.; Parish, E. J., eds.) Academic Press, Inc., New York, 1989, pp. 133-149.
62. vanTamelen, E. E., Sharpless, K. B., Willett, J. D., Clayton, R. B. and Burlingame, A. L. B. 1967 *J. Am. Chem. Soc. 89*, 3920-3923.
63. Corey, E. J. and Gross, S. K. 1967 *J. Am. Chem. Soc. 89*, 4561-4563.
64. Anderson, R. J. Hanzlik, R. P., Sharpless, K. B., Van Tamelen, E. E. and Clayton, R. B. 1969 *J. Am. Chem. Soc. 91*, 53-56.
65. Freed, J. H. 1971 *Dissertation Abst. Intern.* 32, 1447-B.
66. Shishibori, T., Fukui, T. and Suga, T. 1973 *Chem. Letters*, 1289-1292.
67. Heintz, R., Schaefer, P. C. and Benueniste, P. 1970 *Chem. Commun.*, 946-949.
68. Rowan, M. G., Dean, P. P. G. and Goodwin, T. W. 1971 *F. E. B. S. Letters* 12, 229-233.
69. Rowan, M. G. and Dean, P. D. G. 1972 *Phytochemistry 11*, 3111-3116.
70. Boar, R. B. and Damps, K. 1973 *Chem. Comm.*, 115-119.
71. new, W. D. and Schmidt, J. O. 1988 *Phytochemistry 27*, 1705-1708.

72. Corey, E. J., Ortiz de Montellano, P. R., Lin, P. R. and Dean, P. D. G. 1967 *J. Am. Chem. Soc. 89*, 2797-2798.
73. Field R. B. and Holmlund, C. E. 1977 *Arch. Biochem. Biophys. 180*, 465-471.
74. Nes, W. D. and Parish, E. J. 1988 *Lipids 23*, 375-376.
75. Nes, W. D., Xu S. and Parish, E. J. 1989 *Arch. Biochem. Biophys. 273*, 323-331.
76. Elliger, C. A., Benson, M. Haddon, W. F., Lundin, R. E., Waiss, A. C., Jr. and Wong R. Y. 1988 *J. Chem. Soc. Perkin I*, 711-717.
77. Connolly, J. D. and Honda, K. L. 1989 *J. Chem. Soc. (C)*, 2435-2440.
78. Hui, W. H., Luk, K., Arthur, H. R. and Loo, S. N. 1971 *J. Chem. Soc. (C)*, 2826-2829.
79. Kutney, J. P., Eigendorf, G., Swingle, R. B., Knowles, G. D. Rowe, J. W. and Magasampagi, B. A. 1973 *Tetrahedron Lett.*, 3115-3119.
80. Butruille, D. and Dominguez, X. A. 1974 *Tetrahedron Lett.*, 639-642.
81. Shiengthong, D., Kokpol, U., Karntiang, P. and Massy-Westropp, R. A. 1974 *Tetrahedron 30*, 2211-2216.
82. Sakakibara, J., Hotta, Y., Yasue, M. Iitaka, y., and Yamazaki, K. 1974 *Chem. Commun.*, 839-844.
83. Sakakibara, J., Hotta, Y. and Yasue, M. Chem. *Pharm. Bull. JAPAN 23*, 400-459.
84. Boar, R. B. and Damps, K. 1977 *J. Chem. Soc. Perkin I*, 510-518.
85. Ikeda, M., Sato, Y., Izawa, M. Sassa, T. and Miura, Y. 1977 Agric. *and Biol. Chem. JAPAN 41*, 1539-1543.
86. Ikeda, M., Watahabe, H., Hayakawa, A., Sato, K. Sassa, T. and Miura, Y. 1977 *Agric. and Biol. Chem. JAPAN 41*, 1543-1548.
87. Ikeda, M., Niwa, G., Tohyama, K. Sassa, T. and Miura Y. 1977 Agric. *and Biol. Chem. JAPAN 41*, 1803-1807.
88. Okorie, D. A and Taylor, D. A. H. 1977 *Phytochemistry 16*, 2029-2036.
89. Sherman, M. M., Borris, R. P., Cordell, G. A. and Farnsworth, N. R. 1980 *Phytochemistry 19*, 1499-1506.
90. Kutney, J. P., Eigendurf, G., Worth, B. R., rowe, J. W., Conner, A. H. and Nagasampagi B. A. 1981 Hely. Chim. Acta 64, 1183-1189.
91. Kobayashi, M., Kanda, F., Rao, C. V. L., Kumar, G., Trimurtulu, G. Rao, C. B. 1990 *Chem. Pharm. Bull. JAPAN 38*, 1724-1732.
92. Yamada, S., Shimizu, M., Fukushima, K., Niimura, K., Maeda, Y. 1989 *Steroids 54*, 145-154.
93. Bjorkhem, I. and Danielsson, H. 1974 *Mol. Cell. Biochem. 4*, 79-95.
94. Aringer, L. Eneroth, P. and Nordstrom, L. 1976 *J. Lipid Res. 17*, 263-272.
95. Daniellson, H. in The Bile Acids: *Chemistry, Physiology and Metabolism* (Nair, P. P. and Kritchevsky, D., eds.) Vol. 2, plenum Press, New York, 1973; pp. 1-32.
96. Aringer, L. and Nordstrom, L. 1981 *Biochem. Biophys. Acta 665*, 13-21.

97. Salen, G., Ahrens, Jr., E. H. and Grundy, S. M. 1970 *J. Clin. Invest. 49* 952.
98. Bell, R. G., Hisa, S. L., Matschiner, J. T., Doisy, E. A., Jr., Elliott, W. H., Thayer, S. A. and Doisy, E. A. 1965 *J. Biol. Chem. 240*, 1054-1058.
99. Hansson, R., Holmberg, I., and Wikvall, K. 1981 *J. Biol. Chem. 256*, 4345-4349.
100. Anderson, S., Holmberg, I. and Wilkuall, K. 1983 *J. Biol. Chem. 258*, 6777-6781.
101. Tomita, Y. and Sakvrai, E. 1974 *Chem. Comm.* 434-436.
102. Hirata, T., Murai, K. Suga, T. adn Christensen 1977 *Chem. Letters*, 95-99.
103. Han, B. H., Kulshreshtha, D. K. and Rastogi, R. P. 1977 *Phytochemistry 16*, 141-147.
104. Casey, W. M., Burgess, J. P. and Parks, L. W. 1991 *Biochem. Biophys. Acta 1081*, 279-284.

RECEIVED May 2, 1994

Chapter 7

Evolutionary Aspects of Steroid Utilization in Insects

James A. Svoboda, Mark F. Feldlaufer, and Gunter F. Weirich

Insect Neurobiology and Hormone Laboratory, Agricultural Research Service, U.S. Department of Agriculture, Beltsville, MD 20705

Since insects are unable to biosynthesize the steroid nucleus, they require dietary sterols for structural and physiological (hormonal) purposes. Cholesterol will satisfy this dietary need in most cases, but since phytophagous insects ingest little or no cholesterol from dietary materials, they must convert dietary C_{28} and C_{29} phytosterols to cholesterol or other sterols. Through evolutionary development, certain insects have acquired the ability to metabolize dietary sterols in unique ways and to produce and utilize a variety of ecdysteroids (molting hormones) for hormonal control of development and reproduction. Thus, insects are able to flourish in virtually every conceivable ecological niche. Certain comparative studies that illustrate these evolutionary processes will be discussed in this chapter.

Steroid biochemistry research has rarely been used to demonstrate phylogenetic relationships between insects. Relatively few species belonging to only eight of the 29 or so orders of insects have been examined with respect to sterol metabolism (Figure 1). Phylogenetic relationships of families and orders have been taken from Ross, H. H. *et al.* (*1*). Listed from most primitive to most advanced, these include the orders: Thysanura, Orthoptera, Hemiptera, Homoptera, Hymenoptera, Coleoptera, Lepidoptera, and Diptera (Table 1). Over the years, a number of interesting differences in sterol utilization and metabolism between species have been discovered and some of these can be related to phylogenetic relationships; the rationale for some other differences is not so obvious. These studies have made it clear that it is difficult to generalize about insect sterol utilization and metabolism. We will discuss results of our work, and the work of a number of colleagues, relevant to the topic at hand.

Primitive insects date from the Carboniferous period of the Paleozoic era, or several hundred million years ago (*2*). Arthropods, which include the Class

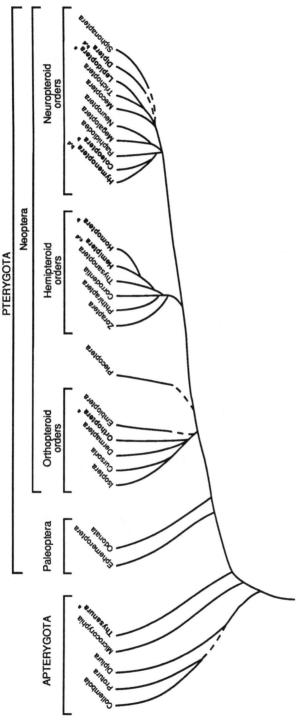

Figure 1. Suggested family tree of the orders of insects. Orders in which sterol utilization and metabolism have been investigated.

[a] Orders in which all species examined were capable of dealkylation of C_{28} and C_{29} phytosterols.

[b] Order includes species capable of dealkylation and species which are incapable.

[c] No species reported to be capable of dealkylation.

[d] Includes species capable of producing C_{28} or C_{29} ecdysteroids.

Insecta, could have arisen from annelids, which never acquired the ability to biosynthesize sterols. To date, no insect has been found to be capable of biosynthesizing the steroid nucleus (3). Thus, insects require a dietary source of sterol to support development and reproduction, and this dietary requirement for sterol is the only known difference in nutritional requirements between insects and

Table 1. Orders* of insects in which sterol metabolism has been examined.

Ametabolous - no metamorphosis

Thysanura - Primitively wingless - *e.g.* firebrat

Hemimetabolous - incomplete metamorphosis

Orthoptera - straight-winged - *e.g.* grasshoppers, cockroaches
Homoptera - winged and wingless - *e.g.* aphids, cicadas, leaf-hoppers
Hemiptera - half-winged - the true bugs, *e.g.* stink bugs

Holometabolous - complete metamorphosis

Hymenoptera - membrane-winged - *e.g.*, bees, wasps, ants
Coleoptera - Forewings thickened - includes only beetles
Lepidoptera - scale-covered wings - *e.g.*, moths, butterflies
Diptera - one pair of wings - *e.g.*, flies, mosquitoes

*Listed in phylogenetic order; most primitive first.

mammals (4). Because cholesterol generally supports normal growth, development, and reproduction in insects, insects that feed on animals or animal products readily utilize host or dietary cholesterol. However, the diet of plant feeding or phytophagous insects often contains little or no cholesterol. Most phytophagous insects have overcome this problem by evolving the ability to convert C_{28} and C_{29} phytosterols to cholesterol by dealkylating the C-24 alkyl group of the sterol side chain. In addition, many omnivorous species of insects have been shown to possess similar capabilities. Much of the discussion in this chapter will center on the question whether or not various species are capable of side chain dealkylation, on species-specific variations from the usual metabolism of phytosterols and on how these differences might be related from a phylogenetic or evolutionary standpoint.

Phytosterol Dealkylation

The usual pathways of dealkylation and conversion of the most common 24-alkyl phytosterols (sitosterol, campesterol, and stigmasterol) to cholesterol are illustrated in Figure 2. The major portions of these pathways were elucidated in detailed studies with two lepidopteran species, *Manduca sexta* and *Bombyx mori* (*3,5*). Whether a 28- or 29-carbon sterol is involved, the initial metabolic step is an oxidation to produce a $\Delta^{24,28}$-bond followed by a subsequent epoxidation prior to the dealkylation of the 1- or 2-carbon unit to form a Δ^{24}-bond. Desmosterol ($\Delta^{5,24}$-cholestadienol) is the common terminal intermediate in the conversion of a number of phytosterols to cholesterol. Certain azasteroid inhibitors have been used to demonstrate the presence or absence of a 24-reductase enzyme in sterol metabolism studies in a number of species of insects (*6*). Similar pathways have been found in most phytophagous as well as omnivorous species that have been examined in detail.

The conversion of these 24-alkyl sterols to cholesterol provides the necessary precursor to the 27-carbon compound ecdysone (Figure 2), which, along with 20-hydroxyecdysone, is one of the two most commonly occurring molting hormones (ecdysteroids) in insects. Although some of the most fascinating aspects of insect steroid biochemistry involve studies with insects that are unable to dealkylate, and their adaptation to produce C_{28} or C_{29} ecdysteroids, most insects that have been examined in detail use ecdysone and 20-hydroxyecdysone as molting hormones. Certainly, steroid metabolism in insects does provide a unique target for insect control. By appropriately interfering with the pathways of phytosterol conversion to cholesterol or by limiting the availability of utilizable sterols, the supply of precursor for molting hormones may be depleted to a level that would not support normal development.

Thysanura - Primitively Wingless Insects

The firebrat, *Thermobia domestica* of the order Thysanura, is the most primitive insect found to be capable of dealkylating phytosterols to cholesterol (*6*). In this study, dietary sitosterol was shown to be converted to cholesterol, and with the inclusion in the diet of an azasteroid that inhibits Δ^{24}-reductase and causes an accumulation of desmosterol, it was also determined that desmosterol was an intermediate in the conversion of sitosterol to cholesterol. To date, this is the most primitive insect that has been examined with respect to sterol metabolism.

Orthoptera - Cockroaches and Locusts or Grasshoppers

The first demonstration of dealkylation of the sterol side chain resulted from studies with the German cockroach, *Blattella germanica*, which converted [^{14}C]ergosterol to [^{14}C]22-dehydrocholesterol (*7*). This insect is incapable of reducing the Δ^{22}-bond. Subsequently, the conversion of [^{3}H]sitosterol to [^{3}H]cholesterol was shown to occur in this same species (*8*). Also, this conversion has been shown to occur in at least two other cockroach species, *Eurycotis floridana* (*9*) and *Periplaneta americana* (*6*). Studies with two locust species,

Locusta migratoria (*10*) and *Schistocerca gregaria* (*11*), provided evidence that these orthopterans were able to dealkylate phytosterols, such as sitosterol, via pathways similar to those found in other species (Figure 2). No unusual aspects of sterol metabolism have been reported for Orthoptera. Recently, the effect of dietary plant sterol composition on insect sterol metabolism was demonstrated in *L. migratoria.* When fed wheat seedlings treated with a fungicide (fenpropimorph) that greatly reduced the content of normal Δ^5-sterols while increasing the 9β,19-cyclopropylsterols in the plants (*12-14*), the locusts were unable to produce sufficient cholesterol to support normal development, and ecdysteroid titers were greatly reduced.

Hemiptera - True Bugs

The next more advanced order of insects that has been examined to any extent with respect to sterol metabolism capabilities is that of the Hemiptera, or true bugs. Interestingly, none of the phytophagous species investigated, to date, is capable of dealkylating the phytosterol side chain. In fact, the milkweed bug, *Oncopeltus fasciatus*, was the first phytophagous insect found incapable of dealkylating and converting C_{28} and C_{29} phytosterols to cholesterol (*15*). Also, makisterone A, a C_{28} ecdysteroid, was discovered in *O. fasciatus*, where it is the major molting hormone in the eggs of this insect (*16*). Since this early work with the milkweed bug, studies with a number of other phytophagous hemipterans likewise unable to dealkylate indicate that these insects have also adapted to producing makisterone A as their major molting hormone. In addition to *O. fasciatus*, these include *Nezara viridula* and *Dysdercus cingulatus* (*17,18*), *D. fasciatus* (*19*), and *Megalotomus quinquespinosus* (*20*). The recent discovery of makisterone C, a 29-carbon ecdysteroid, as the predominant ecdysteroid of the embryonic stage of *D. fasciatus*, provided the first evidence that a C_{29} steroid can serve as a molting hormone (*21*). It is interesting to note that a predacious hemipteran, *Podisus maculiventris*, also produces makisterone A as its major ecdysteroid even though there is adequate cholesterol in the diet of this species and cholesterol comprises nearly 90% of its total sterols (*17,18*). Thus, the mechanism for ecdysteroid biosynthesis is strongly conserved in this species, which evolved from phytophagous ancestors and is secondarily predacious. Several blood-feeding Hemiptera (*Rhodnius prolixus, Arilus cristatus*, and *Cimex lectularius*) were found to have only C_{27} ecdysteroids (ecdysone and 20-hydroxyecdysone) (*17*), which reflects the fact that cholesterol is the predominant sterol in the diet of these species (*18*).

Homoptera

The order Homoptera used to be considered as a suborder of Hemiptera, so it is appropriate to consider some sterol-related studies with a homopteran of the family Aphididae at this point. In a significant and thorough study, [^{14}C]sitosterol was converted to radiolabeled cholesterol in the aphid, *Schizaphis graminum* (*22*), and no sterol biosynthesis occurred in the aphid or its symbiotes. These results cast doubts on the validity of several previous reports suggesting sterol biosynthesis by aphid symbiotes provided necessary sterols to their host(s) (see Ref. 3 for discussion of these studies). It is interesting to note that phytosterol dealkylation occurs in a homopteran species, whereas it has not been demonstrated in the phylogenetically more advanced Hemiptera.

Hymenoptera - Bees, Ants, Wasps

One of the earliest biochemical studies on phytosterol dealkylation was done with the Virginia pine sawfly, *Neodiprion pratti* (23). This research demonstrated that dietary [³H]sitosterol was converted to [³H]cholesterol by this insect and that no cholesterol was produced from labeled acetate or mevalonate, the typical precursors in organisms that can biosynthesize sterols.   To date, this rather primitive hymenopteran is the only known phytophagous hymenopteran capable of dealkylating phytosterols.

More recently, the utilization and metabolism of dietary sterols in the honey bee, *Apis mellifera*, has been examined in detail utilizing a chemically defined diet (24).   Experiments with radiolabeled sterols demonstrated unequivocally that this important beneficial insect is unable to remove C-24 alkyl groups from the sterol side chain (25).   The nurse bees have the ability to selectively transfer certain sterols to the brood food they produce for the larvae. As a result, 24-methylenecholesterol is the major sterol found in the prepupae regardless of the dietary sterol included in the diet.   Even when no sterol was added to the diet of the nurse bees, 24-methylenecholesterol was the major sterol identified in the prepupal sterols (26).

Another phytophagous hymenopteran, the solitary bee, *Megachile rotundata*, also appears to be unable to dealkylate and utilizes dietary sterols by pathways similar to those found in the honey bee (27). In this same study, four omnivorous species (*Dolichovespula maculata, Vespula maculifrons, Formica exsectoides, and Solenopsis invicta*), all predatory to some extent, contained high levels of cholesterol, which is a reflection of their usual dietary sterols.   The phytophagous bees *A. mellifera* and *M. rotundata* evolved from predacious ancestors (28) and have adapted to utilizing dietary sterols that include little or no cholesterol for structural purposes. Further, at least in the honey bee, C_{27} sterols may not be needed for physiological purposes since this insect produces the C_{28} ecdysteroid, makisterone A (Figure 3) from campesterol (29). Makisterone A is the major ecdysteroid found in all post-embryonic stages of the honey bee (larvae, pupae, adults) (30-32). The solitary cactus bee, *Diadasia rinconis*, was recently shown to utilize dietary sterols similarly to the honey bee, i.e. it does not dealkylate and 24-methylenecholesterol is the major sterol isolated from this bee (33). Quite surprisingly though, 20-hydroxyecdysone is the major ecdysteroid of *D. rinconis* rather than makisterone A.

An interesting aspect of sterol metabolism was discovered in studies with a leaf-cutting ant, *Atta cephalotes isthmicola*, an insect which cultivates fungi for food (34). In a thorough investigation, no cholesterol could be found in the brain or nervous tissue of this insect, providing the first evidence of a functional nervous system in an animal entirely lacking cholesterol. The major sterols of the workers and soldiers were $\Delta^{5,7}$-24-methylsterols, which is not surprising, since the

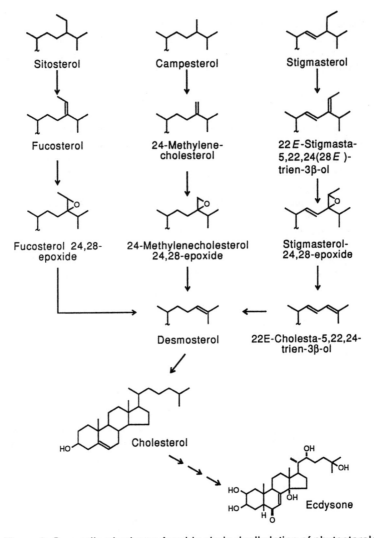

Figure 2. Generalized scheme for side chain dealkylation of phytosterols in insects

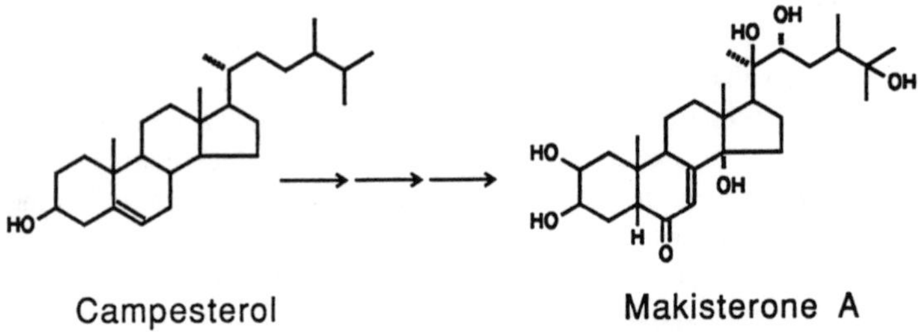

Figure 3. Biosynthesis of makisterone A

insect feeds on fungi, which contain $\Delta^{5,7}$-24-methylsterols as their major sterols. In a more recent report, pupae of another leaf-cutting ant, *Acromyrmex octospinosus*, were also found to contain mostly $\Delta^{5,7}$-24-methylsterols, but in this case a small amount of cholesterol was also present, the source of which was not determined (*35*). Although C_{28} ecdysteroids were the major ecdysteroids of this species, a lesser amount of 20-hydroxyecdysone and perhaps ecdysone were also present (*36,37*). From the evidence available, it appears that these leaf-cutting ants are similar to the honey bee in being unable to dealkylate phytosterols.

Coleoptera - Beetles

For years, it was assumed generally that phytophagous Coleoptera dealkylated and produced cholesterol from dietary phytosterols, as first demonstrated in the boll weevil, *Anthonomus grandis* (*38*). However, the confused flour beetle, *Tribolium confusum*, of the family Tenebrionidae, subsequently was shown to produce large amounts of 7-dehydrocholesterol (*ca.* 50% of the total sterols) (*39*). Regardless of the dietary sterol, over 90% of the total sterols isolated from *T. confusum* consisted of cholesterol and 7-dehydrocholesterol. These results were obtained with either cholesterol, 7-dehydrocholesterol, sitosterol or campesterol as the dietary sterols. Apparently an equilibrium is established in *T. confusum* between cholesterol and 7-dehydrocholesterol. A closely related tenebrionid, *Tribolium castaneum*, produced similar amounts of cholesterol and 7-dehydrocholesterol, but in a more distant member of this family, *Tenebrio molitor*, 7-dehydrocholesterol comprised only about 15% of the total sterols (*40*). Usually trace amounts of 7-dehydrocholesterol, an intermediate in ecdysone biosynthesis, can be found in the sterols of most insects, but there is as yet no explanation for the high levels of 7-dehydrocholesterol in these tenebrionid species.

Studies with the Mexican bean beetle, *Epilachna varivestis*, however, revealed very unusual aspects of phytosterol metabolism. Although this important crop pest dealkylates C_{28} and C_{29} phytosterols, the major products were cholestanol, other saturated sterols, and a significant amount of Δ^7-cholestenol (10-15%), rather than cholesterol (*41*). In studies with radiolabeled dietary C_{28} and C_{29} phytosterols, the Δ^5-bond was reduced before the 24-dealkylation of the side chain occurred. In more recent studies, it was determined that the side chain dealkylation proceeds similarly in *E. varivestis* to dealkylation in those species that produce cholesterol (*42*). The Mexican bean beetle is a member of the family Coccinellidae, and most species in this family are beneficial and predacious, feeding on eggs and larvae of other insects. However, the subfamily Epilachninae, to which the Mexican bean beetle belongs, includes phytophagous species in which phytophagy may have developed secondarily from predacious species. Of all the species examined with respect to sterol metabolism to date, no other species has been found to metabolize dietary phytosterols in a manner similar to the Mexican bean beetle. Comparative studies with a predacious coccinellid, *Coccinella septempunctata*, indicated that this predator utilized essentially unchanged sterols from its aphid prey (*43*, *cf.* p. 7, ref. 22). Over 97% of the sterols of *C. septempunctata* were Δ^5-sterols and cholesterol comprised over 46% of the total.

One of the first phytophagous insects found to be unable to dealkylate C_{28} and C_{29} phytosterols was the khapra beetle, *Trogoderma granarium* (*44*). Although this important stored products pest feeds on grain, it utilizes its dietary sterols unchanged, except for perhaps some selective uptake of the trace (<0.5%) amounts of cholesterol in its diet. Studies utilizing radiolabeled dietary sterols unequivocally established that the khapra beetle is unable to dealkylate the sterol side chain. *T. granarium* is a member of the family Dermestidae, which includes primarily species that feed on animal-derived materials which contain adequate amounts of cholesterol. Thus, the khapra beetle is more similar to other dermestids in its utilization of dietary sterols than it is to other stored product pests such as *T. confusum* (Tenebrionidae), even though it occupies environments similar to other stored products pests and feeds on plant-derived material containing little cholesterol. It is of interest to note that the dermestids are phylogenetically somewhat more primitive than the tenebrionids.

Lepidoptera - Moths, Butterflies

As mentioned earlier, much of the information on dealkylation and conversion of C_{28} and C_{29} phytosterols to cholesterol in insects (Figure 2) has been acquired through research with two lepidopteran species, the tobacco hornworm, *M. sexta* (*3*, and references therein) and the silkworm, *B. mori* (*5*, and references therein). Studies with *M. sexta* established that desmosterol is the terminal intermediate in the conversion of all phytosterols to cholesterol, and that fucosterol and 24-methylenecholesterol were the first intermediates in the metabolism of sitosterol and campesterol, respectively, to cholesterol (*3*). In-depth metabolic studies with *B. mori* first demonstrated the involvement of an epoxidation of the $\Delta^{24,28}$-bond of fucosterol or 24-methylenecholesterol in the dealkylation of sitosterol and campesterol (*5,45*). More recently, the metabolism of stigmasterol was elucidated in detail in another lepidopteran, *Spodoptera littoralis*, and the side chain was shown to be dealkylated via a $\Delta^{24,28}$-bond and a 24,28-epoxide as were sitosterol and campesterol (*46*). The only significant differences in the metabolism of stigmasterol are the involvement of the additional 5,22,24-triene intermediate preceding desmosterol in the pathway and reduction of the Δ^{22}-bond prior to reduction of the Δ^{24}-bond (Figure 2).

It is notable that of all Lepidoptera examined to date, none have been found to differ significantly from what is considered to be the "normal" pathways of dealkylation and production of cholesterol. This relatively highly evolved order of insects appears to have considerable uniformity with respect to the utilization and metabolism of dietary phytosterols.

Diptera - True Flies

For many years, it was believed that members of the order Diptera, as a group, were unable to dealkylate 24-alkyl sterols, based on definitive studies with the house fly, *Musca domestica* (*47,48*). These were the first studies in which a radiolabeled phytosterol such as [^{3}H]sitosterol was utilized in combination with gas-liquid chromatographic analysis. Nearly two decades later, the yellow fever mosquito, *Aedes aegypti*, was the first dipteran found capable of converting phytosterols to cholesterol (*49*). In this study, dietary radiolabeled sitosterol, campesterol, and desmosterol were fed to the larvae to assess the dealkylation capability; and feeding the 24-reductase inhibitor 25-azacholesterol (*cf.* p. 3, ref. 6) with sitosterol resulted in desmosterol accumulation, an indicator for the presence of the dealkylation mechanism (*6*). It is noteworthy that, phylogenetically, the yellow fever mosquito is a more primitive dipteran than the house fly. Recently, sterol utilization in another member of the higher Diptera, *Drosophila melanogaster*, was investigated using radiolabeled dietary sterols (sitosterol and desmosterol) in a well-defined diet; the conclusion was that this species was similar to the house fly in being unable to dealkylate and convert phytosterols to cholesterol (*50*). This clarified the results from several earlier, less rigorous studies which had left unclear whether or not this species could indeed dealkylate (*51-53*). Related research with several cactophilic *Drosophila* species (*54,55*), which are also unable to dealkylate, is in agreement with our findings with *D. melanogaster*. Complementing the elucidation of neutral sterol metabolism in *M. domestica* and *D. melanogaster*, investigations on ecdysteroid biosynthesis capabilities revealed that both species could produce makisterone A from campesterol (*56,57*). Nevertheless, C_{27} ecdysteroids are still the major ecdysteroids produced by these two species, presumably derived from trace amounts of endogenous cholesterol obtained from dietary sterols through selective uptake.

Conclusions

Of the million or so species of insects that have been identified, only a very minute percentage has been examined with respect to sterol utilization and metabolism. Certainly, of the relatively few species that have been studied, a considerable amount of diversity in sterol metabolism capabilities among insects has been discovered. This should not be too surprising when one considers the fact that insects have adapted to occupying nearly every conceivable environment during evolution and, in the course of these adaptations, some were bound to have been confronted with the need to adjust to different dietary sources of sterol. Some of the most interesting differences within orders seem to have arisen because a species or group of species has adapted secondarily to a dietary regimen very different from its ancestors even within the same family. Among the Coleoptera, this could well be the case with the Mexican bean beetle (*41*) which may derive from predacious ancestors. In the family Coccinellidae, only the subfamily to which the Mexican bean beetle belongs includes phytophagous species. The khapra beetle is another interesting example of a phytophagous species adapting to a very different environment from most other species of its family, Dermestidae. This stored products insect feeds on grain, which usually contains little or no cholesterol (*44*), whereas most dermestids feed on animals or animal products from which they obtain adequate cholesterol. However, other stored products coleopteran pests such as *T. confusum* readily carry out the dealkylation of phytosterols.

From the limited evidence available, it appears that the most primitive insects are capable of dealkylating and converting C_{28} and C_{29} phytosterols to cholesterol. The only primitive thysanuran species examined, the firebrat (6), has this capability as do all species of the next most primitive group, the hemimetabolous Orthoptera (cockroaches, locusts, grasshoppers), that have been studied. Similarly, members of the order Lepidoptera studied thus far are quite uniformly capable of dealkylating.

Until the milkweed bug was found to be unable to dealkylate C_{24}-alkyl sterols in 1977, it was assumed that all phytophagous (and omnivorous) species could dealkylate and all so-called zoophagous species that feed on animals or animal products could not. Since then, a number of other phytophagous Hemiptera have been found to lack the ability to dealkylate (18-20). However, among the more primitive Homoptera, which used to be included in the Hemiptera, at least one aphid species is definitely able to dealkylate (22). A number of phytophagous Hymenoptera, including the honey bee and certain leaf-cutting ants, are unable to produce cholesterol from phytosterols (27,33-35). However, it is interesting to note that a more primitive hymenopteran, the pine sawfly, *Neodiprion pratti*, is capable of dealkylating (23). A similar situation was noted in the discussion on the Diptera, where the higher dipteran species, *M. domestica* and *D. melanogaster*, are unable to convert phytosterols to cholesterol, whereas the more primitive *A. aegypti* is capable of this conversion (49). Thus, it does appear that, at least in certain orders, an early branching of the phylogenetic tree occurred, resulting in more primitive species that dealkylate and more advanced species that do not. This sort of relationship does not always hold true. For example, in the Coleoptera, Dermestidae are somewhat more primitive than Tenebrionidae, and the former are unable to dealkylate whereas tenebrionids can.

It is significant that, although the C_{27} ecdysteroids ecdysone and 20-hydroxyecdysone are the most commonly found molting hormones in insects, the C_{28} ecdysteroid, makisterone A, is the major molting hormone of many insects that are unable to dealkylate. In addition, the C_{29} ecdysteroid, makisterone C, has been identified as a major embryonic molting hormone of a hemipteran that lacks dealkylation capabilities. The ability to synthesize makisterone A is present even in certain insects that ordinarily have adequate cholesterol in their diets as shown for the hemipteran, *P. maculiventris*, and the higher Diptera. So this appears to be a well-conserved biochemical mechanism.

Some of this variability in sterol metabolism capabilities could well be exploited in developing new selective control methodology. As an example, it was found that certain inhibitors that block conversion of C_{28} and C_{29} phytosterols to cholesterol and adversely affect growth and development in Lepidoptera and species of other orders that dealkylate, had no effect on the honey bee, which is unable to dealkylate (58). Any technology that would limit the availability of utilizable sterol in a pest insect could provide a useful addition to our arsenal of weapons to use against insect pests.

Literature Cited

1. Ross, H. H; Ross, C. A.; Ross, J. R. P. *A Textbook of Entomology*; Fourth Edition; John Wiley & Sons: New York, 1982; 572 p.
2. In *Lipids in Evolution*; Nes, W. R.; Nes, W. D., Eds.; Plenum Press: New York, 1980; pp. 130-155 and 161-175.
3. Svoboda, J. A; Thompson, M. J; In *Comprehensive Insect Physiology, Biochemistry, and Pharmacology*; Kerkut, G. A.; Gilbert, L. I., Eds.; Pergamon Press: Oxford, 1985, Vol. 10; pp. 137-175.
4. Venderzant, E. S. In *Annual Review of Entomology*; Smith, R. F.; Mittler, T. E.; Smith, C. N., Eds.; Palo Alto, California, 1974, Vol. 19; pp. 139-160.
5. Ikekawa, N. *Experientia* **1983**, *39*, 466.
6. Svoboda, J. A.; Robbins, W. E. *Lipids* **1971**, *6*, 113.
7. Clark, A. J.; Bloch, K. *J. Biol. Chem.* **1959**, *234*, 2589.
8. Robbins, W. E.; Dutky, R. C.; Monroe, R. E.; Kaplanis, J. N. *Ann. Entomol. Soc. Am.* **1962**, *55*, 102.
9. Ritter, F. J.; Wientjens, W. H. J. M. *TNO Nieuws* **1967**, *22*, 381.
10. Allais, J. P.; Barbier, M. *Experientia* **1971**, *27*, 506.
11. Majumder, M. S. I. Ph.D. Thesis. University of Liverpool, 1978.
12. Charlet, M.; Roussel, J.-P.; Rinternecht, E.; Berchtold, J.-P.; Costet, M.-F. *J. Insect Physiol.* **1988**, *34*, 787.
13. Costet, M. F.; El Achouri, M.; Charlet, M.; Lanot, R.; Benveniste, P.; Hoffmann, J. A. *Proc. Nat. Acad. Sci. USA* **1987**, *84*, 643.
14. Corio-Costet, M.; Charlet, M.; Benveniste, P.; Hoffmann, J. *Arch. Insect Biochem. Physiol.* **1989**, *11*, 47.
15. Svoboda, J. A.; Dutky, S. R.; Robbins, W. E.; Kaplanis, J. N. *Lipids* **1977**, *12*, 318.
16. Kaplanis, J. N.; Dutky, S. R.; Robbins, W. E.; Thompson, M. J; Lindquist, E. L.; Horn, D. H. S.; Galbraith, M. N. *Science* **1975**, *190*, 681.
17. Aldrich, J. R.; Kelly, T. J.; Woods, C. W. *J. Insect Physiol.* **1982**, *28*, 857.
18. Svoboda, J. A.; Lusby, W. R.; Aldrich, J. R. *Arch. Insect Biochem. Physiol.* **1984**, *1*, 139.
19. Gibson, J. M.; Majumder, M. S. I.; Mendis, A. H. W.; Rees, H. H. *Arch. Insect Biochem. Physiol.* **1983**, *1*, 105.
20. Feldlaufer, M. F.; Herbert, E. W., Jr.; Svoboda, J. A.; Thompson, M. J. *Arch. Insect Biochem. Physiol.* **1986**, *3*, 415.
21. Feldlaufer, M. F.; Weirich, G. F.; Lusby, W. R.; Svoboda, J. A. *Arch. Insect Biochem. Physiol.* **1991**, *18*, 71.
22. Campbell, B. C.; Nes, W. D. *J. Insect Physiol.* **1983**, *29*, 149.
23. Schaefer, C. H.; Kaplanis, J. N.; Robbins, W. E. *J. Insect Physiol.* **1965**, *11*, 1013.
24. Herbert, E. W., Jr.; Svoboda, J. A.; Thompson, M. J.; Shimanuki, H. *J. Insect Physiol.* **1980**, *26*, 287.
25. Svoboda, J. A.; Thompson, M. J.; Herbert, E. W., Jr.; Shimanuki, H. *J. Insect Physiol.* **1981**, *27*, 183.

26. Svoboda, J. A.; Thompson, M. J.; Herbert, E. W., Jr.; Shimanuki, H. *J. Insect Physiol.* **1980**, *26*, 291.
27. Svoboda, J. A.; Lusby, W. R. *Arch. Insect Biochem. Physiol.* **1986**, *3*, 13.
28. Batra, S. W. T. *Sci. Am.* **1984**, *250*, 120.
29. Feldlaufer, M. F.; Herbert, E. W., Jr.; Svoboda, J. A.; Thompson, M. J. *Arch. Insect Biochem. Physiol.* **1986**, *3*, 415.
30. Feldlaufer, M. F.; Herbert, E. W., Jr.; Svoboda, J. A.; Thompson, M. J.; Lusby, W. R. *Insect Biochem.* **1985**, *15*, 597.
31. Feldlaufer, M. F.; Svoboda, J. A.; Herbert, E. W., Jr. *Experientia* **1986**, *42*, 200.
32. Rachinsky, A.; Strambi, C.; Strambi, A.; Hartfelder, K. *Gen. Comp. Endocrinol.* **1990**, *79*, 31.
33. Feldlaufer, M. F.; Buchmann, S. L.; Lusby, W. R.; Weirich, G. F.; Svoboda, J. A. *Arch. Insect Biochem. Physiol.* **1993**, *23*, 91.
34. Ritter, K. S.; Weiss, B. A.; Norrborn, A. L.; Nes, W. R. *Comp. Biochem. Physiol.* **1982**, *71B*, 345.
35. Maurer, P.; Debieu, D.; Malosse, C.; Leroux, P.; Riba, G. *Arch. Insect Biochem. Physiol.* **1992**, *20*, 13.
36. Maurer, P.; Royer, C.; Mauchamp, B.; Porcheron, P.; Debieu, D.; Riba, G. *Arch. Insect Biochem. Physiol.* **1991**, *16*, 1.
37. Maurer, P.; Girault, J.-P.; Larcheveque, M.; Lafont, R. *Arch. Insect Biochem. Physiol.* **1993**, *23*, 29.
38. Earle, N. W.; Lambremont, E. N.; Burkes, M. L.; Slatten, B. H.; Bennett, A. F. *J. Econ. Entomol.* **1967**, *60*, 291.
39. Svoboda, J. A.; Robbins, W. E.; Cohen, C. F.; Shortino, T. J. In *Insect and Mite Nutrition*; Rodriguez, J. G., Ed.; North Holland: Amsterdam, 1972; pp. 505-516.
40. Svoboda, J. A.; Lusby, W. R. *Experientia* **1994**, *50*, 72.
41. Svoboda, J. A.; Thompson, M. J.; Robbins, W. E.; Elden, T. C. *Lipids* **1975**, *10*, 524.
42. Svoboda, J. A.; Cohen, C. F.; Lusby, W. R.; Thompson, M. J. *Lipids* **1986**, *21*, 639.
43. Svoboda, J. A.; Robbins, W. E. *Experientia* **1979**, *35*, 186.
44. Svoboda, J. A.; Agarwal, N.; Robbins, W. E.; Nair, A. M. G. *Experientia* **1980**, *36*, 1029.
45. Ikekawa, N.; Fujimoto, Y.; Takasu, A.; Morisaki, M. *J. C. S. Chem. Comm.* **1980**, 709.
46. Svoboda, J. A.; Rees, H. H.; Thompson, M. J.; Hoggard, N. *Steroids* **1989**, *53*, 329.
47. Kaplanis, J. N.; Monroe, R. E.; Robbins, W. E.; Louloudes, S. J. *Ann. Ent. Soc. Am.* **1963**, *56*, 198.
48. Kaplanis, J. N.; Robbins, W. E.; Monroe, R. E.; Shortino, T. J.; Thompson, M. J. *J. Insect Physiol.* **1965**, *11*, 251.
49. Svoboda, J. A.; Thompson, M. J.; Herbert, E. W., Jr.; Shortino, T. J.; Szczepanik-Vanleeuwen, P. A. *Lipids* **1982**, *17*, 220.
50. Svoboda, J. A.; Imberski, R. B.; Lusby, W. R. *Experientia* **1989**, *45*, 983.
51. Cooke, J.; Sang, J. H. *J. Insect Physiol.* **1970**, *16*, 801.

52. Redfern, C. P. F. *Proc. Nat. Acad. Sci. USA* **1984**, *81*, 5643.
53. Parkin, C. A.; Burnet, B. J. *J. Insect Physiol.* **1986**, *32*, 463.
54. Kircher, H. W.; Rosenstein, F. U.; Fogelman, J. C. *Lipids* **1984**, *19*, 235.
55. Kircher, H. W.; Rosenstein, F. U.; Fogelman, J. C. *Lipids* **1984**, *19*, 231.
56. Feldlaufer, M. F.; Svoboda, J. A. *Insect Biochem.* **1991**, *21*, 53.
57. Pak, M. D.; Gilbert, L. I. *J. Liquid Chromatogr.* **1987**, *10*, 2591.
58. Svoboda, J. A.; Herbert, E. W., Jr.; Thompson, M. J. *Arch. Insect Biochem. Physiol.* **1987**, *6*, 1.

RECEIVED May 2, 1994

Evolution of Other Natural Products and Membranes

Chapter 8

Evolution of the Endomembrane System

D. James Morré

Department of Medicinal Chemistry, Purdue University, West
Lafayette, IN 47907

The evolutionary origins of the endomembrane
system[1] are discussed in relation to theories for the
origin of eukaryotic cells from prokaryotic ancestors
by a specific series of symbioses. Development of
ingestion mechanisms that would permit phagocytosis of
complete organisms within an enveloping membrane
derived from the host plasma membrane is considered in
relationship to organellogenesis. The modern origins
of outer chloroplast and mitochondrial membranes and
those of the nucleus are traced to plasma membrane
and, in turn, to the endomembrane system. Concepts of
membrane flow and membrane differentiation are
combined to provide a mechanism whereby eukaryotic
endomembranes originate along a chain of cell
components in subcellular developmental pathways.
This concept of endomembrane functioning may reflect
evolutionary origins and, equally important, is
expected to contribute to a better understanding of
cell structure and function in modern life forms.

A theory of evolution, frequently advanced (*1-6*) and championed by
Dr. Lynn Margulis of Boston University (*3, 5*) proposes that
eukaryotic cells arose in the late Precambrian period from
prokaryotic ancestors by a specific series of symbioses. Some of

[1]The author's views of the evolution of the endomembrane system were
developed in the early 1970's through discussion with many
scientists and colleagues, especially Professor Charles Bracker, of
the Department of Botany and Plant Pathology at Purdue University
and Dr. Hilton H. Mollenhauer, then of the Charles F. Kettering
Research Laboratories in Yellow Springs, OH. The concepts developed
were largely untestable and, therefore, timeless. The paper repro-
duced here was written in 1972 but never submitted for publication.
Except for minor editing, the principal changes were elimination of
overlapping literature citations and of the electron micrographs
illustrating the appearance of the cell components discussed.

the assumptions basis to this proposal were summarized by Margulis (5) as follows:

1) The basic dichotomy between organisms of the present-day world is between prokaryote and eukaryotes.

2) Photosynthetic eukaryotes (eukaryotic algae, green plants) and heterotrophic eukaryotes (animals, fungi, protozoa) evolved from (a) common heterotrophic (amebo-flagellate) ancestor(s).

3) The evolution of photosynthesis occurred on the ancient earth in prokaryotes, the bacteria and blue-green algae; high plants evolved abruptly from prokaryotes when the heterotrophic ancestor (2 above) acquired plastids by symbiosis.

4) Animals and most fungi evolved directly from protozoans.

5) Mitochondria were already present in the primitive eukaryotic ancestor when plastids were first acquired by symbiosis.

6) Prokaryotes evolved from primitive ancestors by single mutational steps. Eukaryotes evolved from prokaryotes by a specific series of symbioses.

In this article, I address the evolutionary origins of important cell components discussed little, it at all, in previous considerations on the origins of eukaryotic cells, i.e., the nucleus, plasma membrane and complex internal endomembrane systems which characterize modern eukaryotes. These cell components are important to normal cell functioning and serve vital roles in processes governing cell-cell interactions and the social behavior of cells, tissues, and organisms. Acquisitions of a nucleus, endomembrane system, and specialized plasma membrane emerge as essential steps along the pathway of achieving multicellularity. In fact, what we study today as differentiation of cell function, immunochemical specificity and cell-cell interaction are determined in large measure by functional characteristics of the endomembrane system which are nucleus-controlled but ultimately expressed in the surface chemistry of the plasma membrane.

With these considerations in mind, we summarize evidence to explain ontogeny of modern eukaryotic endomembrane systems and address two major evolutionary enigmas: 1) the origin of the eukaryotic nucleus and 2) the origin of the modern eukaryotic plasma membrane.

As will be developed in this report, we propose that plasma membrane was derived from endoplasmic reticulum membrane by differentiation through a system of transition elements derived ultimately from nuclear envelope. Outer membranes of both mitochondria and chloroplasts may have derived through phagocytosis from the plasma membrane of a host cell. These organelles, like modern endosymbionts which invade living cells, may require containment within the confines of a host plasma membrane to achieve stable associations. Thus we expect an origin for the outer membranes of these oxygen mediating organelles similar to that for the plasma membrane. To maintain the association between inner and outer organelle membranes and insure its continuity, it would be necessary to replicate the outer membrane derivative of the host plasma membrane in concert with inner membrane replication. These considerations, plus documented continuity between other endomembrane components and outer mitochondrial membranes (7-9), do not permit an *a priori* exclusion of the outer organelle membranes from the list of cell components derived from the endomembrane system (Table I, Figure 1). Excluded from the endomembrane system

TABLE I

COMPONENTS OF THE ENDOMEMBRANE SYSTEM

OUTER MEMBRANE OF THE NUCLEAR ENVELOPE AND ITS EXTENSIONS:

Nuclear envelope
Annulate lamellae
Rough endoplasmic reticulum

SYSTEMS OF TRANSITION ELEMENTS

Smooth endoplasmic reticulum
Primary or cytoplasmic vesicles
Golgi apparatus and Golgi apparatus equivalents
Secretory vesicles

END PRODUCTS:

Plasma membrane
Vacuole membrane (lysosomes)
Outer membranes of oxygen-mediating
organelles?
Peroxisome membranes?

Figure 1.   Diagrammatic interpretation of the endomembrane system
of eukaryotes. Only a portion of the cell and its parts are shown.
Illustrated is the central nucleus (N) with its surrounding membrane
(NM) with pores and attached polyribosomes (small arrows).
Continuous with the nuclear envelope are sheets of rough (with
ribosomes) endoplasmic reticulum (RER).   The rough endoplasmic
reticulum is continuous with smooth endoplasmic reticulum (SER).
The latter tends to have both flattened and tubular characteristics.
In association with special ER regions, called transitional regions
(large arrows) bridged by primary vesicles, are the stacked
cisternae of the Golgi apparatus (GA). Secretory vesicles (SV) bud
from the Golgi apparatus at the pole distant from the region of
association with endoplasmic reticulum.   These vesicles gain entry
to the cytoplasm, migrate to the plasma membrane (PM) at the cell
surface where they fuse to incorporate their surrounding membranes
into the plasma membrane and to release their contents to the cell's
exterior.

and its end products are the inner membrane systems of the oxygen mediating organelles. These membranes are presumed to have been gained from invading endosymbionts and are perpetuated in a manner partly under the control of the organellar genome.

Origins of the Eukaryotic Nucleus

The problem of endomembrane origins is considerably simplified by the presence of a eukaryotic nucleus. Once a cell derived an eukaryotic nucleus with a nuclear envelope, other endomembrane components (endoplasmic reticulum, Golgi apparatus, lysosomes, etc.) were likely acquired in a manner not unlike their present-day derivations (i.e., formation by processes of membrane flow and membrane differentiation (Table II) beginning with endoplasmic reticulum-like extensions of the nuclear envelope (Table I)).

The endosymbiont theory of organelle evolution imposes few restrictions on origins of the eukaryotic nucleus. The prevailing notion is that the nucleus was derived from the nucleoids present in the primitive prokaryote(s) that gained mitochondria and/or chloroplasts through serial symbiosis (3, Figure 2). Other possibilities are that the nucleus arose as a direct result of serial symbiosis or that a primitive pro-nucleus arose through joining or fusion of the nucleoid of a host cell with that of an invading endosymbiont. This latter possibility provides a double, but not necessarily homologous, set of chromosomes. Irrespective of what mechanisms we accept for the origin of the eukaryotic genome, the modern eukaryotic nucleus has acquired features absent from prokaryotic nucleoids. The presence of histones associated with nuclear DNA (1), the nuclear envelope with its two functionally distinct membranes and complex pore apparatus (10), and the production of 80 S ribosomal subunits (11) are just a few of the features that distinguish present-day eukaryotic nuclei from prokaryotic nucleoids.

The frequently voiced objection to the endosymbiont theory of organelle evolution, that up to 90% or more of the mitochondrial proteins, for example, are of nucleocytoplasmic origins, can easily be accommodated within the endosymbiont concept by assuming an origin of the nuclear genome from the nucleoid of a prokaryotic progenitor. Each would provide for common origins of the DNA of both the pro-nucleus and the oxygen-mediating organelles with much similar or identical genetic information in both. Thus, it would matter little which DNA took over what function. Components of intracellular organelles origination by symbiosis could be replaced at any time by synthetic activities directed by nuclear genes. Thus, one may view the nuclear "takeover" of mitochondria and chloroplast development as a natural consequence of the endosymbiont theory rather than a formidable deterrent to its acceptance.

Several possibilities for the origin of the nuclear envelope include its derivation from mesosomic extensions of a prokaryotic plasma membrane and a "denovo" origin. Acquisition of a primitive nucleus by endosymbiosis or by "internalization" of a primitive genome via an "invagination" mechanism are plausible variations. With endosymbiosis, a double membrane would result. With invagination, either a single or double membrane might surround the genetic material depending on the "complexity" of the invagination

TABLE II

THE ENDOMEMBRANE CONCEPT

DEVELOPMENTAL CONTINUITY
Cell components of the endomembrane system comprise a developmental and structural continuum within the cell's interior.

MEMBRANE FLOW
The process through which membranes are transferred physically from one part of the cell to another provides a mechanism for growth and extension of the endomembrane system that is consistent with its dynamic and pleomorphic properties.

MEMBRANE DIFFERENTIATION
The process by which one membrane type is progressively transformed from one type of cell component into another through changes in the composition or arrangement of its constituents, e.g., transformation of Golgi apparatus cisternae from ER-like on one face of the stack to plasma membrane-like on the opposite face of the stack. This provides a mechanism whereby different membrane types may share common origins.

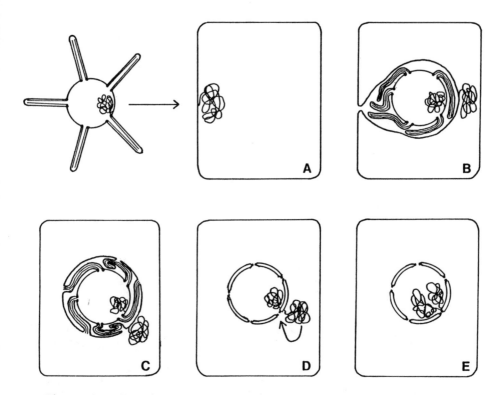

Figure 2. Hypothetical scheme to account for acquisition of a eukaryotic nucleus with double membrane, pores and duplicate sets of genetic material by endosymbiotic invasion of one procaryotic cell by a primitive flagellate. The resultant endosome would have a double membrane. The pores and pore apparatus could derive from multiple fusions of the flagellar membranes with the outer membrane of the endosome. The primitive motile apparatus of the invading flagellate could have contributed as well to the origins of the mitotic apparatus of the resulting endosymbiont.

event. Neither mechanism accounts adequately for the origin of pores in the nuclear envelope.

According to the hypothesis of Margulis (3, 5, 12), the "achromatic apparatus" of mitosis, that is the spindle, centrioles, and other nonchromatin portions of the mitotic figure, trace their origins to some primitive motile structure ancestral to flagella, cilia and other 9 + 2 structures which universally characterize eukaryotic cells. Accordingly, she regards microtubules as a phenotypic expression of the nuclear 9 + 2 genome. If one assumes (correctly or incorrectly) that the modern eukaryote genome was derived from some primitive 9 + 2 amebo-flagellae ancestor, then even more complicated schemes for the origin of the nuclear envelope are possible, i.e., the nuclear pores might represent remnants of some derivative of a primitive motile apparatus. One such scheme is summarized in Figure 2.

The origins of the nuclear envelope with its intricately-formed pore complexes pose a special problem complicated by our ignorance of how the nuclear envelope arises in contemporary cells. In many fungi and euglenoid flagellates, the nuclear envelope does not break down during nuclear division (13). Yet a complete nuclear envelope surrounds each of the daughter nuclei. In those cells where the nuclear envelope breaks down during prometaphase, it reforms in telophase, at least in part by reaggregation of membrane fragments. These fragments appear to be derived primarily from the endoplasmic reticulum (14-16) or from remnants of the dissociated nuclear membrane (17-20). Pores are present in the regenerating nuclear envelope even at its earliest stages and are morphologically indistinguishable from those of the fully-formed nucleus (21). One explanation for the early appearance of pores in regenerating nuclear envelopes is that the pores, like the nuclear envelope in some species, are never lost but stay with the condensed chromosomes while most of the rest of the envelope breaks down at prometaphase (Figure 3).

Chromosomes are attached to the nuclear membrane at multiple sites (22, 23). Some of the sites are temporary and may involve only specialized portions of the chromosome (22). They need not be fixed and may move relative to the attachment point of a sister chromatid, to a homologous chromosome or to the surrounding cytoplasm (24). In this regard, the nuclear envelope has been suggested as a primitive mechanism of chromosome transport (25). Some of the attachment sites may involve pore complexes (26). At least some of these chromatin attachments to fragments of the nuclear membrane seem to persist through mitosis and may serve as foci for the formation of the new nuclear membrane at telophase (27, Figure 3). In fact, both pore apparatus and fragments of nuclear membrane have been shown to be associated with metaphase and anaphase chromosomes by whole mount electron microscopy (27). These fragments are indistinguishable from intact nuclear membranes, have been observed from a variety of mammalian tissues, and appear intimately attached to the chromatin fibers of mitotic chromosomes (27). Barer *et al.* (14) in their studies on the origin of the nuclear membrane discuss organization of newly-formed nuclear envelope along the chromosomes until a complete double membrane is formed around the nucleus and "the fully formed membrane appears to be lifted off the surface of the chromosomes." Thus, pores and pore

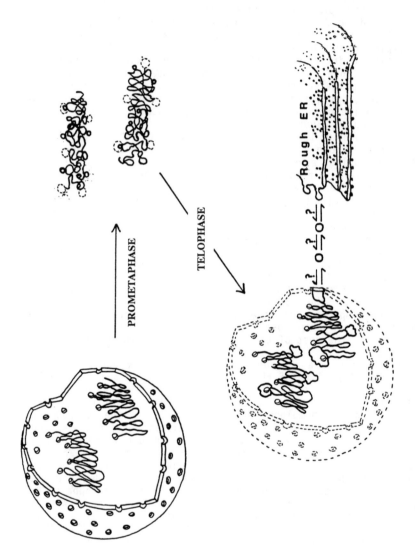

Figure 3. Diagram summarizing the events in the breakdown and reorganization of the nuclear envelope in eukaryotes.

apparatus emerge not only as potential sites of association between
nuclear envelope and chromosomes but as foci for the reformation of
a nuclear envelope during telophase.

The findings of Hsu (28) are interpreted as evidence for an
exclusive origin of annulate lamellae (cytoplasmic "double"
membranes with pores which are structurally indistinguishable from
nuclear envelope) from *inner* membrane of the nuclear envelope.
According to this interpretation, a double membrane surrounding the
nucleus with pores might have been derived from a single membrane
such as that which presently surrounds the nuclei of some
mesokaryotes (29). According to this scheme, nuclear pores arose
through association between membrane chromatin or as ribosomal
derivatives.

Nuclear pore density doubles during interphase (30) and studies
on the mode of pore information in mosquito oocytes provide new
information on their origin. In the much enlarged nucleus of the
maturing egg of *Aedes*, a karyosphore containing chromosomes is
segregated from the bulk of the nuclear material by several layers
of annulated structures formed well beneath and distant from the
existing nuclear envelope (30). The annulated structures delimiting
the karyosphore are morphologically indistinguishable from the
annulae of the nuclear envelope and annulate lamella. The annulae
arise at the peripheries of modified synaptonemal complexes and then
appear to dissociate and reassociate at the rim of the karyosphore
to form an "annulated pseudomembrane" in which annuli are connected
by fine fibers. At later stages of development, the annulated
pseudomembranes seem to "gain substance" and eventually resemble the
nuclear envelope in every detail. The model proposed by Fiil and
Moens (30) assumes that regions which appear to give rise to annulae
correspond to nuclear envelope-chromosome attachment sites (possibly
existing nuclear annulae) and that the sheets of transversal
filaments which connect the annulae are descendant from, or derived
in a manner similar to, constituents of the nuclear membrane. Thus,
pore apparatus emerge as derivatives of chromosomal or achromosomal
nuclear material whose origins may have preceded that of the nuclear
membrane.

In spite of recent progress toward elucidation of the mechanism
of nuclear envelope formation in modern eukaryotes, evolutionary
origins of the eukaryote nucleus remain obscure. Even the concept
of nuclear envelope organization around annulae derived from pre-
existing annulae offers little to help select among the various
possibilities. Yet, similarities of inner membrane of the nuclear
envelope and the procaryotic cell membrane argue for a common
parallel evolution of these two membrane types. For example, the
inner nuclear membrane is associated with DNA (1) and perhaps even
with DNA replication (21, 23) as in bacterial cells (1). There was
even a report that it contained cytochrome oxidase (20), an enzyme
associated with the plasma membrane of prokaryotes and critical to
the development of aerobic metabolism.

Similarities of inner membrane of the nuclear envelope and the
prokaryotic cell membrane argue for a common or parallel evolution
of these two membrane types. Irrespective of the mechanism, pores
and the outer membrane of the nuclear envelope have developed as a
feature unique to all eukaryotes but apparently lacking or lost from
certain mesokayotes (34, 35). The concept of nuclear envelope

organization around annulae derived from pre-existing annulae helps to explain the present day origin and continuity of complex nuclear envelopes. Derivations of pores from associations between nuclear membrane and portions of the achromatic apparatus or from chromatin emerge as possibilities for future clarification.

Formation of Endoplasmic Reticulum

The endoplasmic reticulum (ER) is a major locus of membrane biogenesis and its origin is central to the derivation of complex endomembrane systems. An explanation for the present day origins of ER in keeping with available evidence is that rough ER (ER with attached ribosomes) gives rise to rough ER, i.e., rough ER is responsible for its own formation (evidence reviewed in reference 36). According to this view, rough ER is the site of biosynthesis and/or assembly of its own membrane constituents and an important source of membrane for the formation and replacement of other endomembrane components during cellular ontogeny. Rough ER is capable of protein synthesis and synthesis of membrane lipids. We do not exclude the possibility, for example, that among endomembranes, synthesis of membrane proteins and certain phosphoglycerides such as lecithin resides predominantly in rough ER (cf. 37).

Yet, rough ER is continuous with outer membrane of the nuclear envelope (38). Although direct transfer of membrane from nuclear envelope to ER remains to be demonstrated, nuclear membranes incorporate labeled amino acids in vivo with approximately the same kinetics as rough ER (39) and have ribosomes on their cytoplasmic surface (38). They may have the capacity for membrane biosynthesis. In some cells, membranes of the Golgi apparatus appear to originate from vesicles or short tubules which are derived from specific ribosome-free sites along the outer membrane of the nuclear envelope (e.g., 40). These vesicles fuse to form flattened cisternae in a manner similar to derivation of Golgi apparatus membranes from rough ER.

Although rough ER does not appear to depend solely upon the nuclear envelope for its biogenesis, its derivation as an extension of the nuclear envelope helps to explain the evolutionary origins of ER. If this is indeed the origin of ER, then perhaps some intermediate stages have been preserved in contemporary organisms.

The hypothetical intermediate between nuclear envelope and rough ER is provided by the cytoplasmic annulate lamellae know to cytologists for many years (see 41 and 42 for reviews). Annulate lamellae are commonly found in gametes, especially oocytes (42, 43). In plants, they are found in pollen grains (43). The ultrastructure of individual lamellae is essentially identical with that of the nuclear envelope (42, 43), and most annulate lamella, like nuclear envelope and endoplasmic reticulum, have ribosomes on their cytoplasmic surfaces (42). Cisternae of annulate lamellae frequently show transitions into what is structurally typical of ER 42-47). This observation has led some to suggest that annulate lamellae are derived from ER (47, 48). Yet, most authors favor the view that annulate lamellae originate from outfoldings of the nuclear envelope (28, 46, 49) and, in turn, give rise to ER. One mode of formation frequently suggested is an organization from rows of vesicles which arise by blebbing from the outer membrane of the nuclear envelope (*41, 50*). These vesicles are then suggested to fuse and further transform into stacks of lamellae containing pores (*50*). Other mechanisms have been described where the lamellae exist from the very outset as an extension of the nuclear envelope including exclusive derivation from the inner membrane (*28*). Although details of formation of annulate lamellae are incomplete and sometimes seemingly contradictory, some relationship to nuclear envelope is clearly indicated. According to Wischnitzer (*42*), "The preponderance of evidence indicates that the major function of this transitory organelle (annulate lamellae) has to do with protein synthesis, and that in some forms ... (it) is capable of becoming transformed into a rough-surfaced endoplasmic reticulum ... ." Thus, annulate lamellae provide evolutionary continuity between nuclear envelope and ER. This evidence, plus extensively documented continuities between the outer membrane of the nuclear envelope and conventional ER, give credence to the idea that these so-called "rough" membranes (membranes with ribosomes attached) have similar (although not necessarily identical) origins and functions in membrane biogenesis. For purposes of the present discussion, we emphasize that rough ER has the capacity for membrane biosynthesis but traces it evolutionary (but not necessarily ontogenic) origins to the membranes of the nuclear envelope. We do not exclude the view of Bell et al. (*51, 52*) that early oocyte development is a time when various cell components (including rough ER) might be completely renewed in a nucleus-controlled process from extensions of the nuclear envelope.

Systems of Transition Elements

To generate the rest of the cell's endomembrane system from ER, contemporary eukaryotes apparently employ a variety of "smooth" membranes lacking ribosomes. Because these membranes are transitional in that they facilitate the conversion of one type of membrane to another, we refer to them as transition elements (Table II, Figure 4).

The most general form of a transition element is smooth endoplasmic reticulum (*38*) in direct continuity with rough ER (Figure 4). Smooth ER is a ubiquitous transitional membrane form and predominates in certain fungi (cf. *53*) and protistan forms (*54*) that lack more complicated forms of transition elements such as Golgi apparatus (Figure 5). Evidence from a variety of experiments (mostly applied to rat liver) show that smooth endoplasmic reticulum is synthesized first as rough ER or at least by rough ER (*reviewed in ref. 55*). Evidence is insufficient to decide if smooth ER derives from rough ER through loss of ribosomes or alternatively, if smooth ER arises from the outgrowth of ribosome-free regions continuous with rough ER. According to the latter, ribosomes are never attached to smooth membranes. Irrespective of mode of origin, membranes of smooth ER differ from those of rough ER in important respects. The basic unit of structure for smooth ER is a tubule (Figure 1), as opposed to a flattened cisterna for rough ER (*38*). Additionally, there may be small differences in the lipid composition (*56*) of the two membrane types. Most significant, however, is the ability of smooth ER (including smooth-surfaced

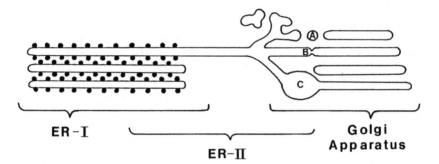

ER-I

ER-II

Golgi Apparatus

Figure 4. Types of transition elements as typified by three observed forms of continuity. *A* is via a primary vesicle which binds from one structure and migrates to and fuses with the second structure. *B* is direct membrane continuity among conjoining compartments. *C* is a variation on *B* in which adjacent flattened structures are directly continuous via a shared tubular-vesicular compartment.

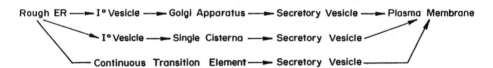

Figure 5. System of transition vesicles for delivery of endoplasmic reticulum (ER) membranes to the plasma membrane. A hierarchy of decreasing complexity is illustrated ranging from a primary vesicle coupled to a stacked Golgi apparatus to a simple continuum. At least one discontinuous, vesicular compartment if necessary to avoid the disastrous consequences of direct continuity that would allow free entry of the extracellular medium into the cell.

portions of rough ER cisternae) to give rise to small blebs or vesicles which migrate away from the reticulum and fuse to form other systems of transition elements such as Golgi apparatus cisternae (*57*). Origin of Golgi apparatus cisternae either by fusion of small ER-derived vesicles or through direct continuities with smooth ER is a widely suggested mechanism for formation of Golgi apparatus (*53*). A parallel origin from smooth (lacking ribosomes) portions of the outer membrane of the nuclear envelope (*53*) has been suggested as well. The latter observations provide additional evidence for an equivalence between outer membrane of the nuclear envelope and endoplasmic reticulum important to evolutionary considerations. An association between annulate lamellae and the forming face of dictyosomes has been reported for plants (*43*) and melanocytes (*58*) and forms the basis for the proposal of Maul (*58*) that Golgi apparatus give rise to annulate lamellae. Franke & Scherer (*59*) have recently reported the presence of "pores" in forming Golgi apparatus cisternae. Differences between dictyosomal pores and fenestrae as well as the relationship of these pores to nuclear pores remains unresolved.

The cytoplasmic vesicle or *primary vesicle* (transition vesicle) which provides structural and functional continuity between systems of transition elements also provides a link between smooth ER and more complex systems of transitional elements. According to Merriam (*19*), the cytoplasmic vesicle represents a "dynamic pool of cellular membrane systems whose common point of interconversion is the general cytoplasmic vesicle." Merriam continues, "It is of course possible that the general vesicles are homologous with the endoplasmic reticulum of other cell types where a continuity between the reticulum and other membrane systems may be apparent at a single moment in time."

Once formed, more complex systems of transition elements such as Golgi apparatus give rise to secretory vesicles which detach from their sites of origin and migrate to the plasma membrane (*53*). As the membranes of the secretory vesicle fuse with the plasma membrane, the vesicle membranes are incorporated into the plasma membrane (Figure 6). Secretory vesicles are a critical link between the plasma membrane and the rest of the endomembrane system. They are an example of a plasma membrane-forming transition element to help account for the origin of the lipoprotein constituents of the cell surface resulting from biosynthetic activities of the ER (*61*).

Isotope tracer studies show that proteins of the plasma membrane are synthesized within the cell's interior and subsequently transferred to the plasma membrane (*39 and ref. cit.*). Secretory vesicles derived from the Golgi apparatus provide a structural basis for one pathway while vesicles or direct transfer from ER provide a second pathway (Figure 6, *39*). Other pathways, including migration through the soluble cytoplasm, are also possible as is direct synthesis from plasma membrane-associated polysomes. The latter is a mechanism which might account for the origin of the plasma membrane in prokaryotes. The essential point is that endomembrane-derived structures, such as secretory vesicles, do contribute membrane constituents to the plasma membrane and in certain cell types, such as elongating pollen tubes, represent a major, if not the sole, source of new plasma membrane (*61, 62*). Thus, transitional membranes, derived from ER or nuclear envelope, become plasma membrane through conversion into secretory vesicles.

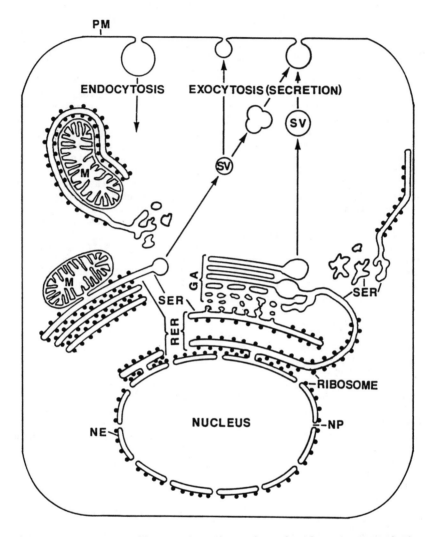

Figure 6. Diagram illustrating the modern-day functioning of the endomembrane system in secretion, exocytosis and endocytosis. Secretion and exocytosis allow delivery of products of synthesis to contiguous membrane compartments and eventually their release to the cell's exterior. Endocytosis provides a means whereby insoluble materials, large particles or even cells or cell parts, may be engulfed to gain entry into the cell's interior. Endocytosis also provides a mechanism of membrane removal from the cell surface to compensate for excess membrane delivered during secretion in non-expanding cells and for membrane replacement and renewal.

Evidence for membrane differentiation within a system of transition elements is considerable (*53, 55, 57 and ref. cit.*). Cisternae at one face of the Golgi apparatus, for example, resemble most closely ER whereas cisternae at the opposite face are more like plasma membrane. Intercalary cisternae are intermediate in appearance between these two extremes. Biochemical studies comparing ER, Golgi apparatus, and plasma membrane from rat liver provide one of the important sources of information upon which the endomembrane concept is based. These studies show that the intermediate character of transition elements is also reflected in the enzymatic activities, lipid composition, biosynthetic rate, turnover and progressive modification of the protein and lipids (e.g., glycosylation) of the membranes (*39, 53, 55, 63*). Additionally, some enzyme activities will be lost while others will be gained either by addition or deletion of enzyme molecules or by activation of enzyme molecules already associated with the membranes. Not only do the biochemical studies reflect the transitional nature of Golgi apparatus and other transitional elements but, equally important, they indicate the nature of some of the biochemical changes required to effect the transformation of ER membranes into plasma membrane. We view the process as a gradual one with changes occurring at the ER, the transition elements, and at the plasma membrane itself.

Membrane Endproducts

By involving different systems of transition elements and mechanisms of membrane flow and differentiation, it is possible to account for the origins of membranes of lysosomes, peroxisomes membranes, tonoplast membranes which surround vacuoles and other intracytoplasmic membranes not specifically associated with the inner membrane systems of the oxygen-mediating organelles, as well as the outer membrane systems of chloroplasts and mitochondria (Figure 5). Some of these latter structures, including the plasma membrane and specific secretions may be viewed as endproducts of endomembrane activities (Table I). Among the secretory endproducts are cell walls and surface coats.

According to the scheme of Figure 6, the membrane endproducts of the endomembrane system have ontogenetic origins similar to those for plasma membrane. Thus, parallel or similar evolutionary origins of endomembrane endproducts is also indicated. The origin of outer organelle membranes from host plasma membrane through symbiotic invasion has already been mentioned as a possibility. In animal cells, the vacuolar apparatus is largely derived through phagocytosis and pinocytosis from plasma membrane (*64*). In plants, the vacuole membrane is frequently likened to an "internal plasma membrane" (*65*). Even though direct derivation of vacuole membrane from plasma membrane may occur rarely if at all in plants, the two membranes appear capable of fusing in some species. In fungal cells, vacuole membrane become plasma membranes during cleavage (*66-68*). Thus, the principal derivation of plasma membrane or plasma membrane-like endproducts appears to be through biogenetic activities of the nuclear envelope - rough ER - transition element export route or endomembrane system.

The Eukaryotic Plasma Membrane: An Evolutionary Enigma

We explain present-day origins of the eukaryotic plasma membrane as a process of flow and differentiation involving biosynthetic activities of endomembrane components derived ultimately from outer membrane of the nuclear envelope.   Yet, according to the endosymbiotic theory (1-6, 69), the eukaryotic plasma membrane traces its origins to the plasma membrane of the original mycoplasma-like or amebo-flagellate progenitor that gave rise through progressive symbiotic invasions to the present-day eukaryotic cell type. Accordingly, it might be just as reasonable to assume that the endomembrane systems of present-day eukaryotes trace their origins to mesosomic invaginations of the plasma membrane of an invading symbiont rather than to derivation from nuclear envelope. Complex internal membranes are possible in modern prokayrotes (see 70) and differentiation of mitochondrial inner membranes to resemble rough ER (including "80S-type" ribosomes of similar size to cytoplasmic ribosomes) has been observed (11). Additionally, plasma membranes of present-day prokaryotes contain cytochrome oxidase and other oxygen-mediating components similar to those of the internal membranes of mitochondria and chloroplasts. Even the inner membrane of the nuclear envelope seems to contain a vestige of such activities while these enzymes appear to be lacking from the eukaryotic plasma membrane (71). Additionally, sterols are present in low concentration or are absent from the prokaryotic plasma membrane (70) and eukaryotic ER but are concentrated in the eukaryotic plasma membrane (71). As pointed out by Raff and Mahler (70), ER, with it concentration of microsomal electron transport components and cytochromes, more closely resembles the prokaryotic plasma membrane than does the present-day eukaryotic plasma membrane.
How then might this enigma be resolved? One less obvious possibility is that the plasma membrane of eukaryotes is of relatively recent evolutionary origin. The structure corresponding to the original plasma membrane of the amoeboid progenitor may have been lost in some evolutionary event in much the same manner the plasma membrane of the "mother cell" is discarded during bacterial spore formation (68).   Such a process is also repeated in ascomycetes where ascospores are formed within asci by a cytokinetic process called free cell formation and then are released through rupture of the original ascus plasma membrane (68).
In spite of the considerable uncertainty, the eukaryotic plasma membrane may very well be of recent evolutionary origins in view of its marked differences from plasma membranes of modern prokaryotes. Its present form is so closely tied to activities of the endomembrane system that evolution of the endomembrane system in all likelihood preceded a modern derivation of the plasma membrane. The reader is reminded of these possibilities in Figure 7 which summarizes the overall concept and indicates that modern eukaryotic plasma membranes need not trace their origin directly to the plasma membrane of the progenitor cell.

Cell Walls and Surface Coats

The cell wall, a nearly universal structure in prokaryotes might

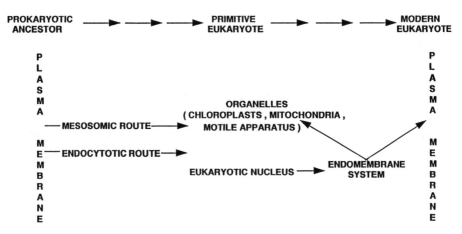

Figure 7. Summary diagram illustrating the postulated pathways for origins of the endomembrane system and the modern derivation of the eukaryotic plasma membrane via synthetic activities of the endomembrane system.

have presented a barrier to endosymbiosis as mentioned by Stanier (*69*) but not an unsurmountable barrier as evidence by *Rhizobium* endosymbiosis of higher plants (*72*). In any event, prokaryotic cell walls are compositionally distinct from eukaryotic cells walls and surface coats. A peptidoglycan is characteristic of all prokaryotes and absent from eukaryotes. Other differences also exist (*69*). A logical extension of the summary diagram of Figure 6 is re-acquisition of the ability to form external surface constituents by eukaryotic plasma membranes so that, of the cell components under consideration, surface coats would be the most recently evolved. A recent derivation of plasma membrane would explain the extreme biochemical diversity of eukaryotic surface coats and why common structural patterns (especially comparing diverse algal groups, fungi, insects, higher plants and vertebrates) are the exception rather than the rule. Evolution of surface coats during the late Precambrian would also help to explain the rarity of Precambrian fossils. It is the durable surface coats of plants and animals that are most often fossilized and most Precambrian eukaryotic organisms are assumed to have been soft-bodies, (i.e., lacking surface coats, exoskeletons or shells).

Concluding Comments

A precise ordering of events in the evolution of the eukaryotic cell is not possible. It is difficult to exclude even a fully independent origin of one or each of the individual cell components with eventual amalgamation into an interdependent system. What I present is a hypothesis to explain how plasma membranes and other endomembrane components are renewed during cell division and maintained at other times by processes of membrane biogenesis, flow, and differentiation along clearly defined subcellular developmental pathways (Figure 6). The evolutionary implications of the endomembrane concept must remain secondary to the primary considerations which focus on similarities and relationships among endomembrane components and provide a functional and developmental integration of a variety of cellular membranes into a single dynamic continuum operating along major pathways within the cell. The endomembrane concept is also significant in that it facilitates comprehension of cellular complexity by students of cytology and cell (and molecular) biology and contributes to a fundamental understanding of the workings of living cells. A major secretory route, the important roles of the endomembrane system involving synthesis, turnover, and differentiation of membranes relate to normal and in turn, abnormal processes of cell division, growth and differentiation, and to the acquisition of specific characteristics of the cell surface important to homeostasis and transformation.

Literature Cited

1.   Nass, S. *Int. Rev. Cytol.* 1969, *25*, 55-129.
2.   Raven, P. H. *Science* 1970, *169*, 641-646.
3.   Margulis, L. In *Origin of Eukaryotic Cells*; Yale University Press, New Haven, CT, 1970.
4.   Schnepf, E. and Brown, R. M. In *Results and Problems in Cell Differentiation*; Reinert J. and Ursprung, H., Eds.; Springer-Verlag, Heidelberg, 1971; pp. 299-322.
5.   Margulis, L. *American Scientist* 1971, *59*, 230-235.
6.   Sager, R. *Cytoplasmic Genes and Organelles*; Academic Press, New York, NY, 1972.
7.   Bracker, C. E. and Grove, S. N. *Protoplasma* 1971, *73*, 15-35.
8.   Franke, W. W. and Kartenbeck, J. *Protoplasma* 1971, *73*, 35-41.
9.   Morré, D. J., Merritt, W. D. and Lembi, C. A. *Protoplasma* 1971, *73*, 43-49.
10. Franke W. W. *Z. Zellforsch.* 1970, *105*, 405-429.
11. Stutz, E. and Noll, H. *Proc. Natl. Acad. Sci. USA* 1967, *57*, 774-781.
12. Sagan L. *J. Theoret. Biol.* 1967, *14*, 225-274.
13. Leedale, G. F. *Ann. New York Acad. Sci.* 1970, *175*, 429-453.
14. Barer, R., Joseph, S. and Meek, G. A. *Exptl. Cell Res.* 1959, *18*, 179-182.
15. Porter, K. R. and Machado, R. D. *J. Biochem. Biophys. Cytol.* 1969, *7*, 167-180.
16. Robbins, E. and Gonatas, N. K. *J. Cell Biol.* 1964, *21*, 429-463.
17. Bajer, A. and Mole-Bajer, J. *Chromosoma* 1968, *27*, 448-473.
18. Davies, H. G. and Toose, J. J. *J. Cell Sci.* 1966 *1*, 331-350.
19. Merriam, R. W. *Expt. Cell Res.* 1961, *22*, 93-107.
20. Roth. K. E. and Daniels, E. W. *J. Cell Biol.* 1962, *12*, 57-78.
21. Mizumo, N. S., Stoops, C. E. and Peiffer, R. L. *J. Mol. Biol.* 1971, *59*, 517-525.
22. Moens, P. B. *Chromosoma* 1969, *28*, 1-25.
23. Comings, D. E. and Kakefuda, T. *J. Mol. Biol.* 1968, *33*, 225-229.
24. Hollande, M. A. and Valentin, J. *C. R. Acad. Sci. (Paris)* 1968, *266D*, 367-370.
25. Pickett-Heaps, J. D. *Cytobios.* 1969, *3*, 257-280.

26. Maul, G. G. *J. Cell Biol.* 1970, *47*, 132a.
27. Comings, D. E. and Okada, T. A. *Expt. Cell Res.* 1970, *63*, 62-68.
28. Hsu, W. W. *Z. Zellforsch.* 1967, *82*, 376-390.
29. Dodge, J. D. *Protoplasma* 1971, *73*, 145-157.
30. Fiil, A. and Moens, P. B. *Chromosoma* 1973, *41*, 37-62.
31. Ryter, A. *Bact. Rev.* 1968, *32*, 39-54.
32. Berezney, R. *J. Cell Biol.* 1972, *55*, 18a.
33. Jarasch, E.-O. and Franke, W. W. *J. Biol. Chem.* 1974, *249*, 7245-7254.
34. Dodge, J. D. *The Fine Structure of Algal Cells*; Academic Press, New York, NY, 1973.
35. Dodge, J. D. and Crawford, R. M. *J. Phycol.* 1970, *6*, 137-149.
36. Morré, D. J. *Ann. Rev. Pl. Physiol.* 1975, *26*, 441-481.
37. Jungalwala, F. B. and Dawson, R. M. C. *Biochem. J.* 1970, *117*, 481-490.
38. Porter, K. R. In *The Cell*; Brachet, J. and Mirsky, A. E., Eds.; Academic Press, New York, NY, 1961; pp. 621-675..
39. Franke, W. W., Morré, D. J., Deumling, B., Cheetham, R. D., Kartenbeck, J., Jarasch, E.-D. and Zentgraf, H.-W. *Z. Naturforsch.* 1971, *26b*, 1031-1039.
40. Bracker, C. E., Grove, S. N., Heintz, C. E. and Morré, D. J. *Cytobiol.* 1971, *4*, 1-8.
41. Kessel, R. G. *J. Ultrastruct. Res.* 1968, *24* (Suppl. 10), 1-82.
42. Wischnitzer, S. *Int. Rev. Cytol.* 1970, *27*, 66-100.
43. Scheer, U. and Franke, W. W. *Planta* 1973, *107*, 145-159.
44. Babbage, P. C. and King P. E. *Z. Zellforsch.* 1970, *107*, 15-22.
45. Rebhun, L. *J. Ultrastruct. Res.* 1961, *5*, 208-225.
46. Bal, A. K., Jubinville, F., Cousineau, G. H. and Inoué, S. *J. Ultrastruct. Res.* 1968, *25*, 15-28.
47. Franke, W. W. and Scheer, U. *Cytobiol.* 1971, *4*, 317-329.
48. Hoage, T. R. and Kessel, R. G. *J. Ultrastruct. Res.* 1968, *24*, 6-32.
49. Gwynn, I., Barton, R. and Jones, P. C. T. *Z. Zellforsch.* 1971, *112*, 390-395.
50. Kessel, R. G. *J. Cell Biol.* 1963, *19*, 391-414.
51. Bell, P. R., Frey-Wyssling, A. and Mühlethaler, K. *J. Ultrastruct. Res.* 1966, *15*, 108-121.
52. Bell, P. R. and Mühlethaler, K. *J. Cell Biol.* 1964, *20*, 235-248.
53. Morré, D. J., Mollenhauer, H. H. and Bracker, C. E. In *Results and Problems in Cell Differentiation*; Reinert, J. and Ursprung, H., Eds.; II. Origin and Continuity of Cell Organelles, Springer-Verlag, Berlin, 1971; pp. 82-126.
54. Franke, W. W., Eckert, W. A. and Krien S. *Z. Zellforsch* 1971, *119*, 577-604.
55. Morré, D. J. and Mollenhauer, H. H. In *Dynamics of Plant Ultrastructure*; Robards, A. W., Ed.; McGraw-Hill, New York, NY, 1974; pp. 84-137.
56. Glaumann, H. and Dallner, G. *J. Lipid Res.* 1968, *9*, 720-729.
57. Grove, S. N., Bracker, C. E. and Morré, D. J. *Amer. J. Bot.*, 1970, *57*, 245-266.
58. Maul, G. *J. Ultrastruct. Res.* 1970, *24*, 6-32.
59. Franke, W. w. and Scheer, U. *J. Ultrastruct. Res.* 1972, *40*, 132-144.

60. Morré, D. J., Keenan, T. W. and Mollenhauer, H. H. In *Adv. Cytopharmacol.*; Clementi, F. and Ceccarelli, B., Eds.; Raven Press, New York, NY, 1971, 1; pp. 159-182.
61. Morré, D. J. and VanDerWoude, W. J. In *Molecules Regulating Growth and Development;* Hay, E. D., King, T. J. and Papaconstantinou, J., Eds.; Academic Press, New York, NY, 1974; pp. 81-111.
62. VanDerWoude, W. J., Morré, D. J. and Bracker, C. E. *J. Cell Sci.* 1971, *8*, 331-351.
63. Morré, D. J., Franke, W. W., Deumling, B., Nyquist, S. E., and Ovtracht, L. In *Biomembranes*; Manson, L. A., Ed.; Plenum Press, New York, NY, 1971; Vol. 2; pp. 95-104.
64. Essner, E. and Novikoff, A. B. *J. Cell Biol.* 1962, *15*, 289-312.
65. Buvat, R. In *Results and Problems in Cell Differentiation*; Reinert, J. and Ursprung, H., Eds.; II Origin and Continuity of Cell Organelles; Springer-Verlag, Berlin, 1971; pp. 127-157.
66. Gay, J. L. and Greenwood, A. D. In *The Fungus Spore*; Madelin, M. F., Ed.; Butterworth, London, 1966; pp. 67-84.
67. Hohl, H. and Hamamoto, S. *Am. J. Bot.* 1967, *54*, 1131-1139.
68. Bracker, C. E. *Ann. Rev. Phytopath.* 1967, *5*, 343-374.
69. Stanier, R.Y. In *Organization and Control in Eurkaryotic Cells*; Charles, H. P. and Knight, B. C. J. G., Eds.; Cambridge University Press, Cambridge, England, 1970; pp. 1-38.
70. Raff, R. A. and Mahler, H. R. *Science* 1972, *177*, 575-582.
71. Benedetti, E. L. and Emmelot, P. In *Ultrastructure in Biological Systems*; Dalton, A. and Haguenau, F., Eds.; The Membranes, Academic Press, New York, 1968; pp. 33-120.
72. Nutman, P. S. In *Symbiotic Associations*; Nutman, P. S. and Mosse, B., Eds.; Williams and Wilkins, New York, NY, 1963; pp. 51-63.

RECEIVED May 2, 1994

Chapter 9

Evolution of Structure and Function of Fatty Acids and Their Metabolites

James L. Kerwin

Department of Botany, AJ–30, University of Washington, Seattle, WA 98195

As integral membrane components of all phyla except archaebacteria, fatty acids are vital for growth and development. Archaebacteria are unique in having membrane lipids based on diphytanyl glycerol diethers and tetraethers rather than fatty acid - linked glycerol moieties found in eubacteria and eukaryotes. A primary strategy used by eubacteria to regulate membrane fluidity is the incorporation of branched chain and/or cyclopropyl moieties. Monounsaturated fatty acids are common in this group, but polyunsaturated fatty acids are rarely found in nonphotosynthetic members of these phyla. The proliferation of organelles in eukaryotes occurred concomitantly with the elaboration of polyunsaturated fatty acids. Synthesis of complex systems of endomembranes may be facilitated by bulk or localized areas of enhanced fluidity, necessary for budding and fusion, which can be imparted by increased fatty acid unsaturation. Oxygenated fatty acids are synthesized by all major phyla except archaebacteria using one of three major pathways: cytochromes P450, lipoxygenases, and cyclooxygenases. These oxygenated derivatives often have regulatory rather than structural functions.

Fatty acids and their metabolites are involved in the regulation of the growth and morphogensis of all living organisms. The diverse functions of these lipids include their structural role in cell membranes (1, 2) , cell walls and exoskeletons (3, 4); control of gene expression (5-10); a source of cellular energy (11,12); intra- and intercellular signal transduction (13-17); regulation of brain function (18, 19) and calcium metabolism (20-22); determinants of host specificity exhibited by microorganisms (23,24); modulation of enzyme activity (25, 26) including covalent attachment (27, 28); and mediation of inflammatory and other immune responses to disease, wounding or microbial infection (29-32).

All organisms share basic developmental processes such as membrane synthesis, cell division, replication of nucleic acids, and ribosome-mediated protein synthesis.

0097–6156/94/0562–0163$11.42/0

Many of these processes are affected in part by lipids. Since organisms as diverse as archaebacteria and mammals have very different lipid compositions, there must be a variety of structural features which can direct the same developmental events. This review will summarize the structural features of fatty acids and their metabolites that are characteristic of major taxonomic groups. A brief synopsis of the comparative enzymology of fatty acid chain elongation is presented, leading to a more detailed discussion of the phylogenetic distribution of fatty acid desaturases. The final section covers the production and function of oxygenated fatty acids among different taxonomic groups. When possible, the structure of fatty acids and their metabolites are related to different strategies used by organisms to direct specific developmental events.

Nomenclature and Structure

Several schemes for designating fatty acid structures have been developed and are in common use in the literature (Figure 1). Most compounds have trivial names, often bestowed before their complete structure was known. These names will be used after first providing the corresponding systematic name. Carbon chain length is based on the length of the longest hydrocarbon chain containing the carboxyl group. A widely accepted shorthand designation involves the use of two numbers separated by a colon, followed by ω and a third number. The first number corresponds to the chain length, and the second to the number of ethylenic bonds, and the third to the number of carbon atoms from the ethylenic carbon furthest from the carboxyl moitey to and including the terminal methyl group (33). For instance 20:4 *omega* 6 describes all-*cis*-5,8,11,14-eicosatetraenoic acid (arachidonic acid). This nomenclature assumes the presence of *cis*-methylene-interrupted ethylenic bonds most commonly encountered in fatty acids of natural origin. The ω (or, alternately, *n-*) numbering system is somewhat confusing since its reference point is the terminal methyl group, while the standard system of chemical nomenclature would designate the carboxyl carbon as the 1-carbon (34, 35). Another common system of nomenclature common in the literatutre uses $\alpha, \beta, \gamma, \omega$, and $\omega-1$ to denote the C-2, C-3, C-4, terminal, and penultimate carbons, respectively. These designations will be used sparingly in this review. Fatty acids with methyl groups at the last and penultimate positions are referred to as *iso-* and *anteiso-* acids, while other substituents are designated by the number of the carbon on which they occur, starting at the carboxyl terminus. Markley (36) has reviewed fatty acid classification and nomenclature. Nuances of nomenclature for eicosanoids and other oxygenase products, especially prostaglandins and other cyclic fatty acid metabolites, are presented in detail by Smith (37).

The majority of naturally occurring fatty acids are 14 to 22 carbons in length, with varying degrees of *cis-* methylene-interrupted unsaturation; however, structural variations of this standard pattern can include: chain lengths from C-2 to C-80; methyl and to a lesser extent ethyl branching; *cis-* and/or *trans-* conjugated unsaturation; acetylenic compounds; mono-, di- and trihydroxyl substitution; addition of keto and epoxy groups; various cyclic products including cyclopropenoid, cyclopentenoid and furanoid moieties; substitution with sulfur or

A.

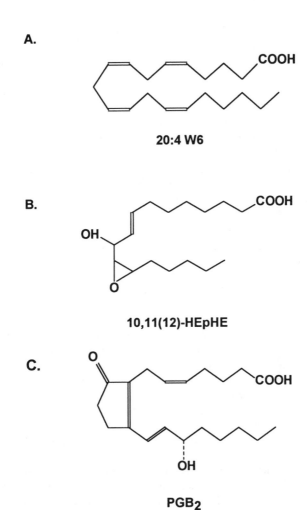

20:4 W6

B.

10,11(12)-HEpHE

C.

PGB₂

Figure 1. Fatty acid nomenclature - A. Arachidonic acid; 20:4ω6; 5-*cis*, 8-*cis*, 11-*cis*, 14-*cis*-eicosatetraenoic acid.   B. 10, 11(12)-HEpHE; 10-hydroxy-11(12)-epoxy-8-*trans*-heptadecanoic acid.   C. Prostaglandin B₂ (PGB₂); 15α-hydroxy-9-ketoprosta-5-*cis*-8(12), 13-*trans*-trienoic acid.

fluorine; and combinations of one or more of these features (35, 38-39). Although over 600 naturally-occurring fatty acids have been described and many more than that synthesized (40), fewer than two dozen make up the bulk of these lipids produced by organisms.

Less than 1% of the fatty acid in most organisms is present as nonesterified fatty acid; therefore, any consideration of alternative strategies by organisms to complete specific developmental events or biochemical processes, especially those mediated by membranes, cannot completely ignore the form in which fatty acids are found in the cell. Fatty acids are frequently esterified through O-acyl, O-alkyl, O-alkenyl or N-acyl linkages, to form diverse classes of lipids: acylglycerols; phosphatidylglycerols; ceramides, including sphingo- and glycosphingolipids; phosphonolipids; sulfatides; glycosylglycerides; acylornithines; steryl esters; mycolic acids and other wax or wax-like esters; and proteolipids (3, 34, 41-44). Limited discussion of these lipids is included when necessary to provide a suitable context for chemical structure - biological activity relationships.

Physical Properties Affecting Function

Free fatty acids are slightly water-soluble compounds whose effect on biological systems will be determined by the nature of their hydrocarbon moieties; the propensity of the carboxyl group to release a proton (K_a) in the case of free fatty acids; or the nature of the groups to which they are esterified. At pH 7 ca. 99% of free fatty acids are present as carboxylate anions, largely due to resonance energy associated with hybridization of the $O=C-O^- \leftrightarrow O^--C=O$ double bond (45), which greatly enhances their water solubility. Fatty acids, regardless of chain length, have pK_as near 4.8 [K_a of 1.3×10^{-5} (46)], which, although weak in comparison with mineral acids, ranks them as one of the most acidic classes of organic substances.

Chain length and degree of unsaturation, especially the latter, are the major characteristics of the hydrocarbon portion of fatty acids that will affect their biological effects following incorporation into membranes. In some groups of organisms, e.g. most bacteria, branched chain or cyclopropyl fatty acids are assumed to perform functions comparable to unsaturated compounds in other organisms (47, 48).

The orthodox view is that greater fatty acid unsaturation or the mixing of long and short fatty acyl chains in one molecule (usually phospholipid) is usually associated with a decreased phase transition temperature and, therefore, increased membrane fluidity (49-53). Membrane phase transition temperatures can be roughly correlated with fatty acid melting points. Increasing the chain length of both fatty acids in a phospholipid will add 15 - 20° C to a bilayer melting point, while introduction of a double bond can decrease this temperature by over 20° C (54, 55); however, the position of the double bond (56, 57), the nature of the headgroup (58) and the degree of unsaturation (59) all affect the melting point, and there is no simple linear relationship between the number of double bonds and physical properties.

Until recently, it has been assumed that the the high melting point of a distearoyl or other homologous saturated fatty acid pair esterified to a phospholipid is due primarily to the rigid extended *trans* conformation of these molecules, and that the introduction of even a single *cis* double bond interrupts this close, stable packing of the hydrocarbon chains (26, 45, 60, 61). This interpretation has been challenged by Applegate and Glomset (62-64) in a series of molecular modeling studies using an MM2 molecular mechanics program (65). Using all-*cis*-4,7,10,13,16, 19-docosahexaenoic acid, two conformations were identified which had straight chain axes resulting from methylene carbon alignment. The most interesting conformation had the carbons of six double bonds projecting from the methylene axis in two perpendicular planes, resulting in an angle iron-shaped molecule. Packed arrays of hexaenes in the presence or absence of saturated hydrocarbons in this confromation could generate very close intermolecular packing. Comparable packing could be modelled for 1,2-diacylglycerols, with the planar surfaces of adjacent docosahexaenoic acids brought into contact back-to-back, and saturated moieties packed into an interplanar V groove. Subsequent modeling showed that while *sn*-1-stearoyl-2-oleoyl diacylglycerol can adopt only irregular shapes as shown in previous studies (66, 67), *sn*-1-stearoyl diacylglycerols with polyenoic *sn*-2 fatty acids can assume regular shapes, which are dependent on the position of the double bonds. Experiments to test these predictions, which contradict current ideas on increased membrane fluidity with increasing fatty-acyl unsaturation, are in progress (J. Glomset, personal communication).

   A further complication with biological membranes is the probable existence of laterally segregated lipid domains (68-71). When combined with the ability of mixtures of lipids to adopt different phases [micellar, liquid crystalline, hexagonal (72, 73)], the interpretation of lipid-mediated biological processes based on membrane or even organelle fatty acid composition must be done with caution.

   Data on molar volumes, surface tensions, viscosities, thermal characteristics and related physical properties for fatty acids and their esters in the crystal and liquid states, which may be useful in interpreting experimental results on natural membranes, have been compiled (74, 75).

Distribution of Fatty Acids Among Major Taxonomic Groups

A number of phylogenetic schemes have been developed, many based on the division of organisms into five kingdoms: Monera, Protista, Fungi, Plantae, and Animalia (76-78). Others have suggested that there should be as many as nine kingdoms (79). Recently Woese et al. (80), on the basis of ribosomal RNAs and other molecular features, have proposed that living organisms be divided into three main domains: the Bacteria, the Archaea and the Eucarya, each comprised of two or more kingdoms. All of these schemes have been challenged (81, 82) using a variety of arguments. Rather than arguing for one major classification scheme over another, this review will use the synopsis edited by Parker (83) and the two empire - eight kingdom scheme outlined by Cavalier-Smith (84) as the bases for taxonomic divisions and phylogeny simply because the two references provide the most complete listing currently available of the various taxa.

   In addition to a previous monograph on lipid evolution (85), articles discussing related topics include lipid roles in prebiotic structures (86), prebiotic fatty acid synthesis (87), and a possible primitive fatty acid biosynthetic pathway based on glycoaldehyde (88). There are numerous reviews on specific classes of fatty acids, and the structures of fatty acids characteristic of groups of organisms. For detailed lists of fatty acids found in specific organisms or groups of organisms, the reader is referred to these reviews, many of which are listed in Table I (89-281). Table I summarizes structural features of fatty acids characteristic of major phyla. It is not meant to be a complete review, but rather a list of the variety of fatty acids synthesized by groups of living organisms. Details on some specific groups below the phylum level is included when unique classes of lipid are found in those groups.

Biosynthesis of Saturated Fatty Acids

There have been many reviews of saturated fatty acid synthesis (34-36, 282-291). A brief survey of pertinent aspects of comparative enzymology based on those references is presented here. The two enzyme systems which operate sequentially in fatty acid synthesis are acetyl-CoA carboxylase (ACC) and fatty acid synthetase (FAS). Straight-chain fatty acid synthesis operates using acetyl-CoA as the chain initiator, while branched-chain compounds are synthesized using isobutyryl-, isovaleryl- or 2-methylbutyryl-CoA esters. In either case malonyl-CaA acts as the unit for fatty acid chain extension and NADPH (reduced nicotinamide adenine dinucleotide phosphate) as the hydrogen donor. The repeated condensation of malonyl-CoA with the appropriate chain initiator by FAS ultimately produces the appropriate end-product, usually 16:0, palmitic acid, in the great majority of organisms (292, 293).

There are two main types of FAS: Type I FAS occurring in animals, fungi, protozoa and some bacteria, and type II characteristic of bacteria and green plants including algae. Type I FAS are stable, high molecular weight (410,000 to 2×10^6 daltons) multifunctional complexes. The enzyme from mammals and birds consists of two subunits, each of which can catalyze all steps in fatty acid synthesis except for condensation (β-keto-acyl synthetase). Dimerization restores this activity.

In type II FAS, each reaction has to be catalyzed by a monofunctional polypeptide. In type II enzyme systems, 4'-phosphopantheine, which is involved in binding malonyl from malonyl-CoA, is attached to a 10,000 dalton apoprotein. In type I FAS, this cofactor is an integral part of a >180,000 dalton subunit. Another difference is that type II systems have a holoprotein known as acyl carrier protein (ACP) which functions like CoA in animals and fungi, and all FAS reactions are carried out on ACP-bound substrates.

Potential sources of differences in fatty acid compostion among different groups of organisms, therefore, are the use of different substrates as the chain extenders (i.e. branched vs. straight-chain compounds) and metabolic control of fatty acid chain termination. These latter are complex and regulated by a variety of nutritional, developmental and environmental factors. Chain termination mechanisms can involve β-ketoacyl-ACP synthetase specificity (from 2-16 carbons); palmitoyl-ACP: β-ketostearoyl-ACP synthetase (16-18 carbons); stearoyl-(oleoyl)-CoA:β-ketoeicosanoyl-CoA synthetase (18-20 carbons); FAS systems catalyzing condensation of acyl-CoA and malonyl-CoA (20-n carbons); or, for 2-carbon

Table I. Fatty acids characteristic of major groups of organisms

Taxa[1]	Fatty Acids	References

Viruses
 saturated 11:0 - 24:0
 unsaturated 16:1ω7,9; 18:1ω7,9; 18:2ω6;18:3ω3; 20:1ω9,11;
   20:2, 20:3, 20:4, 20:5 all ω6; 22:1ω9; 22:4, 22:5,
   22:6ω3, 6 or 9; 24:1ω9; 24:4ω6
 other 14:0 -OH; 17:0 methylene   89-96

Empire Bacteria
 Kingdom Eubacteria
  Subkingdom Negibacteria
   Rhodobacteria, Chlorobacteria, Phycobacteria (Cyanobacteria),
    Deinobacteria
    saturated 14:0-20:0 (odd and even C)
    unsaturated 14:1-18:1 (odd and even); 18:2; 18:3ω3, ω6, 19:1
    other *iso*-15:0, 16:0, 17:0, 17:1; *anteiso*-15:0, 17:0;
      cyclopropyl   97-108
   Subkingdom Spirochaetae
    saturated 12:0-18:0
    unsaturated 18:1-18:3
    other *iso*-14:0-18:0   109-111
  Subkingdom Posibacteria
   Gram positives, Mycoplasmas, Myxobacteria
    saturated 13:0-19:0 (odd and even C)
    unsaturated 16:1ω5,-7, -9; 18:1ω7,-9; 18:2ω6; 18:3ω3, 20:5ω3
     other *iso*-15:0 & 17:0 *anteiso*-15:0 & 17:0; various
      hydroxy acids, esp. 2-OH 12:0, cyclohexl
        112-132
   Actinomycetes
    saturated 8:0-32:0 (even); 11:0-23:0 (odd)
    unsaturated 14:1ω5; 15:1ω6; 16:1ω7; 17:1ω8; 18:1ω9;
     19:1ω10; 20:1 - 26:1 (even) ω9; phleic acids;
    others mycolic acids: 28 to 90-carbon *iso*-alkyl hydroxy acids;
     various *iso*- & *anteiso*; 33 to 56-carbon
     (di)cyclopropyl ± mono-unsaturated; long chain
     methoxy   133-143

Kingdom Archaea
 Archaebacteria
  other 80-95% diphytanyldiglycerol diether/tetraether lipids
       144-154

Continued on next page

Table I. Continued

Empire Eukaryota
Superkingdom Archezoa
  Microsporidia, Metamonada
    host-mediated   155

Superkingdom Metakaryota
Kingdom Protozoa
Subkingdom Euglenozoa
  saturated   16:0-22:0
  unsaturated 16:1; 18:1; 16:2ω3, ω4, ω7; 16:3ω3, ω5;
              16:4ω1, ω3; 18:2, 18:3 ω3   156-159
Subkingdom Parabasalia
  Trichmonadea
  saturated   16:0-18:0
  unsaturated   18:1ω9, 18:2ω6, 20:4ω6   160-161
Subkingdom Sarcomastigota
  Proterozoa, Mycetozoa, Dinoflagellata, Ciliophora
  saturated   12:0-20:0
  unsaturated 16:1ω7; 18:1ω9; 18:2, 18:3, 20:2, 20:3, 20:4ω6 , 22:5
              & 22:6ω6
  other   *iso*-15:0-19:0 (odd)   162-165
Subkingdom Mesozoa
  No data
Kingdom Chromista
Subkingdom Cryptista
  saturated   14:0-22:0
  unsaturated 16:1-16:4, 18:1-18:4, 20:4ω6,
              20:5ω3, 22:6ω6   166-168
Subkingdom Chromophyta
  saturated   14:0-22:0
  unsaturated 16:1-16:4, 18:1ω9, 18:2ω6, 18:3 & 18:4ω3, 20:1,
              20:4ω6, 20:5ω3, 22:5ω3 & ω6
                              166, 169-177

  Phaeophyceae
  saturated   14:0-24:0 (even)
  unsaturated 16:1, 18:1ω9; *trans* 16:1ω3; 18:2, 2 0:4ω6;
              18:3, 18:4, 20:5ω3
  other   hydroxy-methyl(N,N,N-trimethyl) alanyl   144-148
Kingdom Fungi
  Archemycota, Ascomycota, Basidiomycota
  saturated   10:0-24:0
  unsaturated 14:1 n-5; 16:1, 18:1ω7 &9; 18:2ω6, 18:3ω3,
              20:1ω11; 20:2, 20:3, 20:4ω6; 20:3, 20:5ω33, 22:1,
              24:1ω9
  other   mono-, di- & trihydroxy; keto-hydroxy; epoxy;
              various *iso*- & *anteiso*   35, 178-181

Table I. Continued

Kingdom Plantae
Subkingdom Biliphyta
Rhodophyceae

saturated	12:0-20:0
unsaturated	16:1ω5 & ω7, *trans*ω13; 16:4ω3; 18:1ω7 & 9;
	18:2, 18:3, 18:4, 20:5ω3; 18:2, 18:3, 20:4ω6
other	*iso*-22- and 24-carbon (un)satd. 182-187

Subkingdom Viridiplantae
Chlorophyta

saturated	14:0-20:0ω
unsaturated	16:1-16:4, all isomers; 18:19; 18:2, 18:3,
	20:2-20:4,ω6; 18:3, 20:3, 20:4, 20:5ω3, 22:6ω
other	*trans*-16:1 & 18:1; 23 to 32-carbon, saturated
	and monounsaturated 143, 182, 188-192

Bryophyta

saturated	10:0 - 30:0 (even); 15:0, 17:0
unsaturated	18:2, 18:3ω6; 16:3, 18:3, 18:4, 20:4ω6, 20:5ω3
other	acetylenic; cyclopentenonyl 193-197

Pteridophyta

saturated	14:0-18:0
unsaturated	16:1; 18:1; 16:2; 18:2, 20:4ω6; 16:3, 18:3, 20:5ω3
	198

Spermatophyta

saturated	14:0-20:0 (even); 22:0-32:0 (rare)
unsaturated	16:1ω7; 18:1ω9; 18:2ω6; 18:3ω3; 18:3ω6 (rare);
	22:1ω9
other	hydroxy; epoxy; hydroxy-epoxy; keto; furanoid;
	acetylenic; cyclo- propenoid; cyclopentenoid;
	conjugated; polyhydroxyalkanoates 199-210

Kingdom Animalia
Subkingdom Radiata
Porifera

saturated	14:0-30:0
unsaturated	16:1-26:1ω7, ω9 & various other (ω5 to ω17);
	18:2 - 26:2ω6 (even); 18:3-26:3 ω3 (even); 18:4,
	20:4, 20:5, 22:5, 22:6 n-3
others	aldehydes; methyl esters; pyrroles; crasserides
	(cyclitols); *trans*-long chain dienoates; various *iso*-,
	anteiso- 211-219

Cnidaria

saturated	16:0-26:0 (even)
unsaturated	16:1-26:1 (even) ω7, ω9; 18:2, 20:4, 22:4, 22:5
	all ω6; 18:3, 18:4, 20:5, 22:5, 22:6 all n-3
other	7-methyl-7-hexadecanoate, pyrroles, chlorohydrins
	211, 220-223

Continued on next page

Table I. Continued

Subkingdom Bilateria
Mollusca
saturated   14:0-20:0
unsaturated   16:1-22:1 (even), ω7, ω9, ω11, ω13, ω15;
                 18:4-22:4, ω6; 18:3, 20:3, 20:5, 22:5, 22:6, ω3
other   conjugated dienes; 4,8,12-trimethyl-decanoate
                 (from phytol)   211, 224- 231
Arthropoda
Subphylum Crustaceae
saturated   14:0-20:1 (even);13:0-19:0 (odd)
unsaturated   14:1-20:1, ω5; 16:1-22:1, ω7, ω9; 18:1-22:1, ω11;
                 16:2, ω4, ω6, ω7; 18:3, 18:4, 20:4, 20:5, 22:5,
                 22:6 all ω3; 18:2. 18:3, 20:2, 20:4, 22:4, 22:5, all ω6
other   tri- and tetramethyl derivatives; *iso-* and *anteiso*-13:0-
                 18:0   211, 232-242
Subphylum Insecta
saturated   14:0-20:0 (even);13:0-19:0 (odd)
unsaturated   14:1, 15:1 ω5; 14:1, 16:1 ω7; 16:1, 17:1, 18:1,
                 20:1 ω9, ω11; 18:2, 18:3, 20:2, 20:3, 20:4 ω6;
                 18:3, 20:3, 20:5 ω6
other   *iso-, anteiso*-13:0-16:0; acid monoesters;
                 hydroxyacids; 12-54 carbon wax esters;
                 oxo- and dioc-acids   243-255
Echinodermata
saturated   14:0-26:0 (even); 15:0; 17:0
unsaturated   14:1 ω5; 16:1ω5, ω7 & ω9, ω11; 18:1 ω7, ω9
                 & ω13; 20:1 ω7, ω9, ω11& ω15; 20:2, 22:2 ω9
                 & ω11; 18:3, 20:3, 20:4, ω6; 18:3, 18:4, 20:3,
                 20:5, 22:5, 22:6, all ω3
other   conjugated dienes; *iso*-17:0   211, 256-262
Chordata
saturated   12:0-24:0 (even); 11:0-19:0 (odd)
unsaturated   16:1, 18:1 ω7; 18:1, 19:1, 20:1ω9; 16:2, 16:3ω4;
                 16:2, 18:2, 18:3, 20:2, 20:3, 20:4, 22:4, 22:5 all ω6;
                 16:3, 16:4, 18:3, 18:4, 20:4, 20:5, 22:5, 22:6 all ω3
other   various *iso-* & *anteiso*; phytanic; 24-32-carbon ±
                 unsaturated; di- & trimethyl   263-281

---

[1] Taxonomic groups presented with slight modifications from Table 2, reference 84.

through medium chain lengths, termination can be regulated by the specificity of thioesterase, acyltransferase, β-ketoacyl synthetase, β-oxidation, or compartmentalization of enzymatic activities within various cell types.

Fatty Acid Desaturation

Fatty acid desaturation, a second major source of variation in the phylogenetic distribution of fatty acids, has been reviewed in detail (294-302). Some bacteria have a unique anaerobic system for production of monounsaturated fatty acids. This mechanism is involved in elongation of medium-chain length *cis*-3-unsaturated fatty acyl intermediates, and functions via β,γ-dehydration of β-OH intermediates. It should be noted that this process cannot generate methylene-interrupted polyunsaturated fatty acids.

The more common mechanism of fatty acid desaturation, found in all eukaryotes and some bacteria, involves the oxygen-mediated introduction of a *cis*-double bond into preformed fatty acyl chains, with oxygen acting as the terminal electron acceptor. Animals can desaturate fatty acids at positions 4, 5, 6, 8, and 9 carbons from the carboxyl terminus, but cannot introduce double bonds between a preexisting double bond and the methyl terminus. Plants, however, can desaturate e.g. $18:1\omega9 \Rightarrow 18:2\omega6 \Rightarrow 18:3\omega3$. Though these desaturation characteristics are generally true for the various phyla, there is no absolute phylogenetic consistency. *Ochromonas* and some protists apparently have both types of desaturase systems; the cockroach *Periplaneta* synthesizes $18:2\omega6$; and the livers of hen, goat and pig can desaturate $18:1\omega6 \Rightarrow 18:2\omega6$.

As additional organisms are analyzed, it is likely that more exceptions to the general characteristics outlined above will emerge. For example, it has recently been shown that the non-photosynthetic bacterium *Shewanella putrefaciens* can produce polyunsaturated fatty acids using either the aerobic or anaerobic pathway of desaturation (303).

Phylogeny and Fatty Acid Occurrence

An excellent review by Irwin (164) is the last comprehensive review of the phylogenetic distribution of fatty acids of which I am aware. Figure 2, based on the updated phylogenetic scheme of Cavalier-Smith used in Table I (84), summarizes our current understanding of the development of fatty acid composition among major groups of organisms. Several major trends emerge from this synopsis, and will be discussed in subsequent sections.

Viruses. The phylogenetic position of viruses is uncertain, and they are not even considered as living organisms by many researchers. Because of this, and their inability to synthesize fatty acids, viruses will now be discussed separately from other groups of organisms. Although lipids are probably a constituent of most if not all virion envelopes, they have largely been ignored (304), despite evidence that the composition of the lipid matrix effects virulence, infectivity and replication (95, 96, 305-308). Viral lipid composition often reflects that of the host plasma membrane,

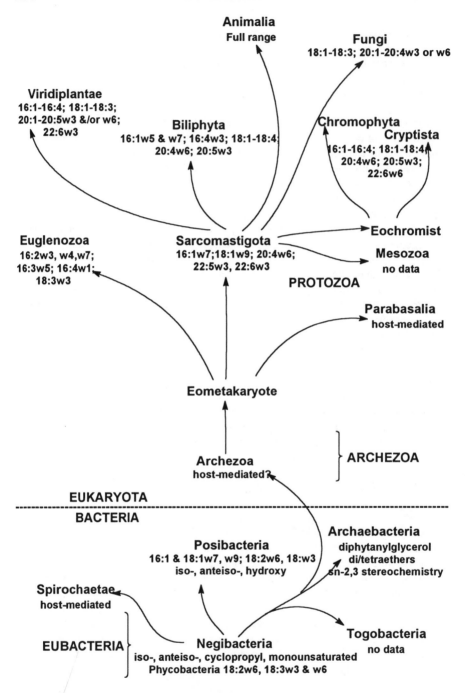

Figure 2. Summary of the phylogenetic distribution of fatty acids.

from which viruses bud (94); however, at least some viruses apparently selectively incorporate host lipids (95, 96, 309, 310). Selective incorporation appears to be regulated in part by the structure of viral coat proteins, which require specific lipids for proper assembly. Some viruses even require N- , O- or S-acylation of proteins, usually with 14:0 (myristic) or 16:0 (palmitic) acids, for structural assembly, conformation and stability (311, 312).

More recently it has been shown that flockhouse virus, a positive-strand RNA virus, requires specific fatty acyl chains in glycerophospholipids for *in vitro* replication. Maximum activity was found in compounds with 18:2 and 18:3 moieties (313). There is indirect evidence of a requirement for membranes/lipid vesicles for RNA synthesis by other viruses (314-316).

Some viruses affect the metabolism of arachidonic acid (20:4ω6), including the production of prostaglandins and other eicosanoids involved in mammalian immune responses to microbial infection (318-321). A recent report supports the presence of at least one gene in orthopoxviruses that redirects host eicosanoid metabolism, and these metabolites are an obligate requirement for viral replication (322). Fatty acid oxygenase products are discussed in more detail below.

Since so little data is available for viral fatty acids, it is not possible at this time to describe trends in e.g. selective incorporation of lipids by different groups of these organisms, or to relate lipid composition of insect vs. plant vs. bacterial pathogens to their host range, development or phylogeny. Initial work on the effects of fatty acids on virion assembly and nuclei acid synthesis suggest that research in this neglected area will yield useful and surprising information on viral biology.

Archaebacteria. Archaebacteria are unique in that their membrane lipids are based solely or primarily on diphytanylglycerol diethers (archaeol) or diphytanylglycerol tetraethers (caldarchaeol), as opposed to the fatty acid ester - linked glycerol moieties found in eubacteria and eukaryotes (148). Methanogenic archaebacteria are unique in possessing phospholipid and glycolipid derivatives of these di- and tetraethers. Another unique feature of this group is the *sn*-2,3-diphytanyl stereochemistry, rather than *sn*-1,2 configuration found in other groups of organisms. It is also likely that many members of this group, especially thermophilic species, have a single rather than double-layer plasma membrane (323, 324). Although archaebacteria thrive in a variety of very extreme environments (149, 325, 326), their polar lipids are completely saturated, there is no alkyl chain length variability, and the phytanyl chains are consistently branched. A variety of strategies have been adopted by these organisms to regulate membrane fluidity, including: addition of pentacyclic rings to shorten the length of tetraethers; change the ratio of diether/tetraether and the addition of macrocyclic diethers to decrease the freedom of motion; the incorporation of highly sulfated glycolipids by extreme halophiles; the probable adoption of of U-shaped hydrocarbon chain conformations by some tetraether moieties; and the presence of both diether bilayers and tetraether bipolar monolayers in the membranes of some species (144-154). These isoprenoid chain polymers promote a highly fluid state, that nonetheless is thermostable and highly resistant to free passage of salts and higher MW molecules (327, 328). This type of molecular architecture is probably an adaptation to the extreme environments in which these organisms are commonly found.

Eubacteria. As more species from diverse habitats are examined, many conclusions regarding fatty acid composition among major groups of bacteria have had to be discarded; however, some generalizations remain valid. All groups have straight-chain saturated and/or monounsaturated 15-to 18-carbon fatty acids. Aerobic posibacteria usually have *iso-* and *anteiso-* 15- to 19-carbon compounds, but branched-chain fatty acids can be found in a wide range of taxa. In some *cocci* species, for example, 100% of the fatty acids can be branched. In negibacteria, characterized by the presence of lipopolysacchadire in their outer membranes, straight-chain and hydroxy fatty acids ae common, as are cyclopropyl derivatives. These latter can also be found, though less common, in gram positive species such as *Clostridium* and *Lactobacillus*. Mycobacteria ae easily differentiated from other major taxa by their synthesis of 2-alkyl-2-OH long chain (24-90 carbons) fatty acids known as mycolic acids (329-331). Acidophilic-thermophilic species often synthesize ω-cyclohexyl fatty acids, and cyanobacteria and marine bacteria are characterized by relatively high levels of polyunsaturated fatty acids.

The use of fatty acids for chemotaxonomy at the strain, species, genus, or higher taxon level is well-established for many groups of bacteria (331-334), although patterns at the species or strain level usually vary in relative composition rather than the in the presence or absence of specific fatty acids.

Transition to Eukaryota. A primary feature of the transition from a prokaryotic to a eukaryotic system of organization in biological organisms was the acquisition of discrete organelles delimited by membranes. It is highly suggestive that development of an extensive endomembrane system occurred with the rapid proliferation of polyunsaturated fatty acids. The phylogenetic scheme used in this review is certainly not universally accepted; however, whatever scheme is followed, and whatever group is accepted as the most primitive of eukarotic taxa, it is evident that a diverse array of polyunsaturated fatty acids was synthesized by the earliest of these organisms. If the Sarcomastigota is accepted as one of the most primitive groups of eukaryotes, then many of the unsaturated fatty acids found in fishes and mammals, from 18:2 through 22:6, were present very early in the development of these organisms. This was followed by either a further elaboration of fatty acid synthetic capabilities, as was the case for the Animalia, or by specialization and simplification, as occurred in most higher plants and fungi.

The proliferation of polyunsaturated fatty acids in eukaryotic organisms should have a physiological basis. It must also be assumed that these physiological functions could not be sufficiently regulated by e.g. branched chain or cyclopropyl fatty acids, which can regulate membrane fluidity as well as unsaturated fatty acids in many instances. One possibility is the emergence of fatty acid oxygenase products as pivotal regulatory molecules, which are formed primarily from polyunsaturated fatty acids, as discussed in the final section of this review. The evolution of an endomembrane system capable of intracellular transport by coordinated processes of budding and fusion are additional physiological processes in which polyunsaturated fatty acids probably play a pivotal role.

Membrane fusion (which broadly includes the processes of karyogamy, plasmogamy, budding, exocytosis, endocytosis and phagocytosis) is a very complex event which, although extensively studied, is little understood. Along with specific proteins, calcium and other cations, and proteins, there have been numerous studies using model and biological systems which have implicated phospholipid headgroups and fatty acyl composition in membrane fusion (335-344). Despite the lack of precise knowledge of biochemical and biophysical events involved in fusion, it is generally assumed that membranes must be relatively fluid in order to fuse, i.e. near or above their transition temperature, which is defined as the temperature or transition from a gel to the more fluid liquid crystalline state. Using model membrane systems, this temperature is generally higher for phospholipids with two saturated acyl moieties, and progressively reduced as the degree of unsaturation of constituent fatty acids increases, although even this assumption is now suspect (62-64).

It has been argued that the development of phagotrophy may have been the major driving force in the evolution of the eukaryotic cell (345-347). The option of using fatty acid unsaturation as one method of optimizing this and related processes requiring fluid membranes must have been sufficiently advantageous that it was exploited by the earliest eukaryotes. Membrane fluidity directly or indirectly affects most cellular processes (348-350), and the ability to regulate fatty acid desaturation allowed eukaryotes to radiate into divergent environmental niches.

Fatty Acid Function and Structure in Eukaryota. Examination of Figure 2 shows that there is no strong correlation at the phylum level with fatty acid composition. There are certainly trends among major taxa as outlined previously, but few dogmatic beliefs have survived the continuing examination of the fatty acid composition of additional groups of organisms. Many of the similarities in fatty acid composition of unrelated taxa can be attributed to parallel or convergent evolution, driven by the need to accomplish a given physiological function. One of the most obvious examples of this is the use of very long chain fatty acids (in combination with other classes of lipid) to maintain water balance by insects, birds, mammals, green plants, sponges and mycobacteria (143, 351-355). Another example is the use of specific fatty acids by a variety of eukaryotes as well as bacteria such as *Bacillus subtilis* to regulate protein kinase C, an enzyme with a pivotal role in signal transduction (356, 357). Many if not all taxa modify fatty acids using oxygenase enzymes to produce hydroxy and related derivatives, which often serve as regulatory molecules in development and defense responses. This topic will now be covered in the last part of this review.

Phylogenetic Distribution of Fatty Acid Oxygenases. Hydroxy fatty acids and related oxygenated compounds are widespread among eubacteria and eukaryotes. There are three main pathways of synthesis of these products (Figures 3-5): cytochrome P450 (epoxygenase) which is a microsomal (eukaryotes) or cytosolic (prokaryotes) monooxygenase that requires cytochrome C, NADPH/NADP+ and

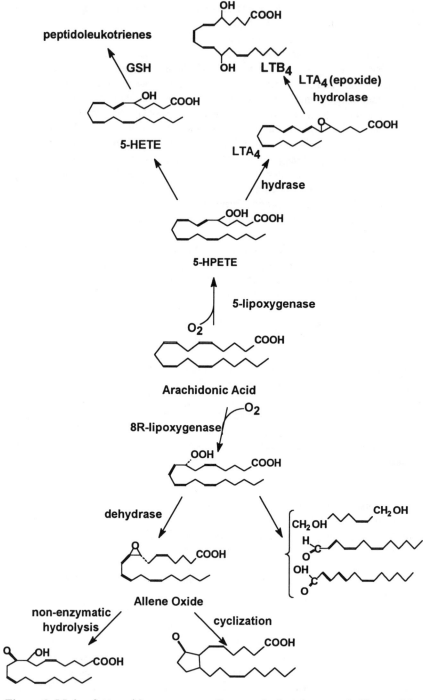

Figure 3. Major fatty acid oxygenase pathways - leukotrienes and allene oxide. Adapted from (37, 451-456).

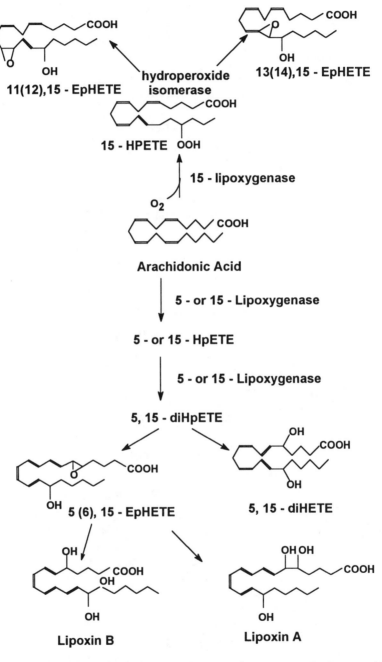

Figure 4. Major fatty acid oxygenase pathways - lipoxygenase, hydroperoxide isomerase and lipoxins. Adapted from (37).

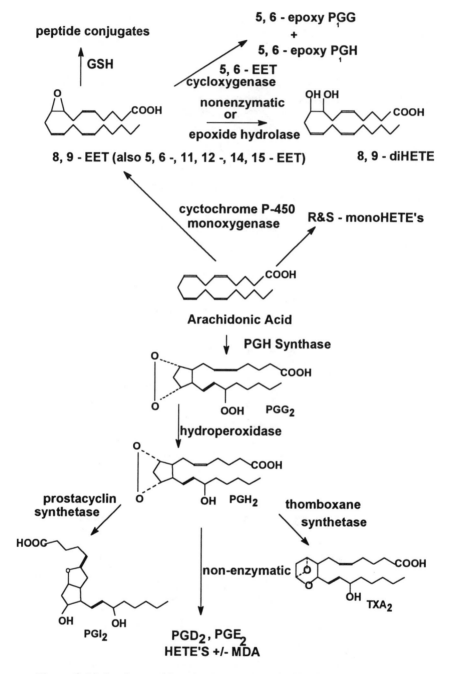

Figure 5. Major fatty acid oxygenase pathways - cytochrome P450 and cyclooxygenase. Adapted from (37).

O_2; cyclooxygenase, an O_2-requiring microsomal dioxygenase; and lipoxygenase, which are cytosolic enzymes, some of which need Ca^{2+}, and also requiring O_2. Cytochrome P450 enzymes have varying degrees of substrate specificity, but many can oxidize a variety of lipophilic compounds, and exhibit little enantioselectivity or regioselectivity. The diooxygenase enzymes usually only oxidize polyunsaturated fatty acids with a 1,4-*cis*-pentadiene structure (358), and have a high degree of positional specificity when fatty acids such as 20:4ω6 are used as substrates. Table II presents a brief survey of the phylogenetic distribution of fatty acid oxygenase enzymes.

Enteric bacteria growing in an anaerobic or microaerobic environment can synthesize hydroxy acids by hydration of double bonds. *Pseudomonas*, for example, can synthesize 10D-hydroxy-18:0 from 18:1ω9, and 10D-hydroxy-18:1ω6 (282, 359, 360). These bacteria also have epoxide hydrolase activity, and can form e.g. 9, 10-dihydroxy 18:0 from 9,10-epoxy 18:0. Many eubacteria have an O_2-dependent monooxygenase (cytochrome P450) that can catalyze hydroxylation and epoxidation of fatty acids and hydrocarbons: *n*-alkanes are metabolized to primary alcohols, fatty acids to ω-OH compounds, and unsaturated hydrocarbons to epoxides (282).

Cytochrome P450 enzymes are heme-containing proteins that are probably present in all organisms (361), and dozens of different enzymes can be found in a single species. It is likely that cytochromes P450 are the most ancient of the fatty acid oxygenase enzymes, and this superfamily of oxygenases has probably existed for more than 3.5 billion years (408, 440). Cytochromes P450 have been extensively studied not only because they act on a variety of endogenous substrates such as steroids, alkanes, fatty acids and hydroxy fatty acids, but because they are often pivotal in the oxidative, peroxidative and reductive metabolism of a wide variety of xenobiotic compounds (441-444). It has been proposed that these enzymes are involved in regulation of gene transcription, affecting homeostasis, growth, and differentiation (445, 446), by regulating steady-state levels of fatty acids and other endogenous ligands.

This group of enzymes will probably become even more widely investigated due to the relatively recent discovery of the importance of epoxygenase-generated bioactive compounds in mammals (409, 414, 447-450). A second equally exciting development was the discovery by Brash and colleagues that a kind of dehydrase documented in corals, starfish, green plants, and probably red algae is a cytochrome P450 (451- 455). These enzymes convert hydroperoxy fatty acids produced by lipoxygenases to an allene oxide, which, in plants, can then be converted by allene oxide cyclase, to metabolic precursors of jasmonic acids (456). These recent studies clarified the biosynthesis of cyclopentenone fatty acids discovered earlier by Zimmerman and others in a variety of plants (457-459). Elucidation of this biosynthetic pathway explains not only the generation of jasmonates in plants, which are phytohormones which regulate many physiological processes (456, 460-464), but also explains the non-cyclooxygenase pathway for synthesis of prostaglandins by sea whip corals including *Plexaura homomalla*, which were the biological sources for pioneering studies on the structural elucidation of cyclic eicosanoids (465-467).

Table II. Phylogenetic distribution of fatty acid oxygenase enzymes.

Taxa	Oxygenase Enzymes	References
Viruses		
	Mediate host eicosanoid metabolism	318-322
Bacteria		
	Cytochrome P450	48, 282, 362-366
Cyanobacteria		
	Lipoxygenase	367, 368
Eukaryota		
Protozoa		
	Cytochrome P450	369-372
	Lipoxygenase?	373
Chromista		
	Lipoxygenase	374-377
	Cytochrome P450	
Fungi		
	Cytochrome P450	378-383
	Lipoxygenase	384-387
Plantae		
Biliphyta		
	Lipoxygenase	388-392
Viridiplantae		
	Lipoxygenase	393-398
	Cytochrome P450	399-401
Animalia		
Radiata		
	Cytochrome P450	402
	Lipoxygenase	403-407
Bilateria		
	Cytochrome P450	408-416
	Lipoxygenase	417-429
	Cyclooxygenase	430-439

Given the apparent antiquity of this group of enzymes, their widespread occurrence, their importance in the development and survival of organisms, and the relatively large data base available on gene and amino acid sequences, cytochromes P450 offer one of the best opportunities for relating phylogeny and fatty acid metabolism. Evidence continues to accumulate that the p450 superfamily has and continues to evolve due to a variety of selective pressures including secondary plant metabolites (468-470). It is now possible to alter the substrate specificity of cytochrome P450 using standard techniques in molecular biology, in some instances by a single amino acid substitution (471-474). Increased understanding of structural features that determine substrate binding and catalytic properties of these enzymes, coupled with research on enzymes from other groups of organisms, can potentially clarify the phylogenetic relationships of major phyla.

Lipoxygenases have been known in higher plants for over 60 years (475, 476) and more recently were identified in animals (477, 478) and algae (388-392, 479, 480). Much of the interest in the hydroxy fatty acids produced via the lipoxygenase pathway has been generated because of their role in the inflammatory responses of mammals (29-32) and fish (427, 428, 481, 482). A role for these products has also recently been postulated as part of plant defense responses against microbial invasion (483-488). This is just one class of dioxygenase, enzymes which catalyze the introduction of an intact O_2 into a variety of substrates (489).

The major lipoxygenases act at C-5, C-12 and C-15 (from the carboxyl terminus) of polyunsaturated fatty acids, but positional specificity of oxygenation can be variable (423). There is evidence that direct oxidation of one class of compound by lipoxygenase can lead to carbon-centered free radical-catalyzed cooxygenation of a second class (490), and that transcellular metabolism (491-495) or intracellular metabolism by different classes of oxygenases also occurs. Specific oxygenation at C-8 of arachidonic acid has also been induced in mammals (496) and is constitutive in some corals and starfish (497, 498). These factors complicate interpretation of phylogenetic relationships based on the presence of specific fatty acid metabolites.

Nucleotide sequences and deduced amino acid sequences have been used to investigate the phylogeny of cyclooxygenase and lipoxygenase enzymes (499, 500) and related enzymes such as 5-lipoxygenase-activating protein (501). Toh and colleagues have shown that: (1) Cyclooxygenase, which initiates formation of prostaglandins (PGs), is distantly related to peroxidases, and developed independently from lipoxygenases; (2) 12-lipoxygenases evolved by gene duplication at least twice; (3) Thromboxane synthase (PGI synthase) is closely related to the cytochrome P450 III family, while cytochromes P450 involved in ω-oxidation are related to the p450 IV group. In addition, an epoxide hydrolase specific for leukotriene A_4 (LTA$_4$ hydrolase) shows weak homology with amino peptidases. Although the data base for these enzymes is not as extensive as that for cytochrome P450, a phylogenetic trees have been generated for animal and plant lipoxygenases; PGD synthases and a family of secretory proteins known as lipocalins; aldo-keto reductases which include PGF synthases; and LTA$_4$ hydrolases and aminopeptidases (499, 500).

At this time it appears that cyclooxygenases are found exclusively in the animal kingdom (502-503) including birds (504), insects (432-435), reptiles (438, 505) and fish (428, 506). The recent discovery of a second inducible isozyme class of cyclooxygenase (507-510) with ca. 60% amino acid homology with the intensively investigated constitutive isozyme has added an additional level of complexity to comparative analyses. There are dozens of naturally occurring PGs and PG metabolites. Further cataloguing of the diversity of these compounds in animal phyla should aid in understanding phylogeny of major animal taxa.

Conclusions

This review has presented a survey of fatty acids and their oxygenase metabolites for major phyla. General trends have been developed, but without using computer-assisted analyses or other statistical treatment. The value of statistical analyses is well established for establishing phylogenetic relationships (499, 500, 511) on all levels of classification from subspecies to phylum. As research continues on the presence and structure of unusual fatty acids (512-515), enzyme site-specific mutation and substrate specificity (516, 517) and 3-dimensional structure of enzymes involved in fatty acid synthesis and metabolism (518), these more sophisticated analyses will be useful in confirming and modifying existing concepts of phylogeny.

Acknowledgments

Writing of this review was supported in part by a grant from the National Institutes of Health (5 RO1 AI22993).

Literature Cited

1. Gennis, R.B. Biomembranes - Molecular Structure and Function; Springer-Verlag: New York, 1989, pp. 1-35.
2. Rouser, G. Membrane composition, structure, and function. In Membrane Fluidity in Biology; R.C. Aloia, Editor; Academic Press: New York, 1983, pp. 235-289.
3. Kolattukudy, P.E. (Editor). Chemistry and Biochemistry of Natural Waxes; Elsevier : Amsterdam, 1976, pp 1-459.
4. Elias, P.M.; Brown, B.E.; Ziboh, V.A. J. Invest. Dermatol. 1980, 74, 230-233.
5. Amri, E.; Bertrand, B.; Ailhaud, G.; and Grimaldi, P.A. J. Lipid Res. 1991, 32, 1449-1456.
6. Grimaldi, P.A.; Knobel, S.M.; Whitesell, R.R.; Abumrad, N.A. Proc. Natl. Acad. Sci. USA 1992, 89, 10930-10934.
7. Chirala, S.S. Proc. Natl. Acad. Sci. USA 1992, 89, 10232-10236.
8. Tebbey, P.W.; Buttke, T.M. Biochim. Biophys. Acta 1992, 1171, 27-34.
9. Clarke, S.D.; Armstrong, M.K.; Jump, D.B. J. Nutr. 1990, 120, 225-231.
10. Das, U.N. Prostaglandins Leukot. Essent. Fatty Acids 1991, 42, 241-244.
11. Neely, J.R.; Morgan, H.E. 1974. Annu. Rev. Physiol. 1974, 36, 413-459.
12. van der Vusse, G.J.; Glatz, J.F.C.; Stam, H.C.G.; Reneman, R.S. Physiol. Rev. 1992, 72, 881-940.

13. Rainsford, K.D. Pharmacol. Res. 1992, 25, 335-346.
14. Glasgow, W.C.; Afshari, C.A.; Barrett, J.C.; Eling, T.E. J. Biol. Chem. 1992, 267, 10771-10779.
15. Haliday, E.M.; Ramesha, C.S.; Ringold, G. EMBO J 1991, 10, 109-115.
16. Strauch, M.A.; de Mendoza, D; Hoch, J.A. Molec. Microbiol. 1992, 6, 2909-2917.
17. Nunez, E.A. Prostaglandins Leukot. Essent. Fatty Acids 1993, 48, 1-4.
18. Devane, W.A.; Hanus, L; Breuer, A; Pertwee, R.G.; Stevenson, L.A.; Griffin G.; Gibson, D; Mandelbaum, A.; Etinger, A.; Mechoulam, R. Science 1992, 258, 1946-1949.
19. Horrobin, D.F. Prostaglandins Leukot. Essent. Fatty Acids 1992, 46, 71-77.
20. Vacher, P.; McKenzie, J; Dufy, B. Am. J. Physiol. 1992, 263, E903-E912.
21. Kimoto, M.; Javors, M.A.; Ekholm J.; Siafaka-Kapadai, A.; Hanahan, D.J. Arch. Biochem.Biophys. 1992, 298, 471-479.
22. Malcolm, K.C.; Fitzpatrick, F.A. J. Biol. Chem. 1992, 267, 19854-19858.
23. Spaink, H.P.; Sheeley, D.M.; van Brussel, A.A.N.; Glushka, J.; York, W.S.; Tak, T.; Geirger, O.; Kennedy, E.P.; Reinhold, V.N.; Lugtenberg, B.J.J. Nature 1991, 354, 125-130.
24. Kerwin, J.L. Can. J. Microbiol. 1984, 30, 158-161.
25. Khan, W.A.; Blobe, G.C.; Hannum, Y.A. J. Biol. Chem. 1992, 267, 3605-3612,
26. Stubbs, C.D.; Smith, A.D. Biochim. Biophys. Acta 1984, 779, 89-137.
27. Towler, D.A.; Gordon, J.I.; Adams, S.P.; Glaser, L. Annu. Rev. Biochem. 1988, 57, 69-99.
28. Gordon, J.I.; Duronio, R.J.; Rudnick, D.A.; Adams, S.P.; Gokel, G.W. J. Biol. Chem. 1991, 266, 8647-8650.
29. Sardesai, V.M. J. Nutr. Biochem. 1992, 3, 562-579.
30. Fukushima, M. Eicosanoids 1990, 3, 189-199.
31. Galliard, T. In Biochemistry of Wounded Plant Tissues; G. Kahl, Editor; Walter de Gruyte & Company: Berlin, 1978, pp. 155-201.
32. Chakrin, L.W.; Bailey, D.M. (Editors); The Leukotrienes - Chemistry and Biology; Academic Press: New York, 1984, pp. 1-308.
33. Holman, R.T. Prog. Chem. Fats Other Lipids, 1966, 9, 3-.
34. Gunstone, F.D. In Comprehensive Organic Chemistry - The Synthesis and Reactions of Organic Compounds - Volume 5 Biological Compounds; E. Haslam, Editor; Pergamon Press: Oxford, 1979, pp. 587-632.
35. Weete, J.D. Lipid Biochemistry of Fungi and Other Organisms; Plenum Press: New York, 1980, pp. 1-388.
36. Markley, K.S. In Fatty Acids - Their Chemistry, Properties, Production and Uses - Second Edition; K.S.Markley, Editor; Interscience Publishers: New York, 1961, Part I, pp. 23-249.
37. Smith, D.L. In Handbook of Eicosanoids: Prostaglandins and Related Lipids, Volume I . Chemical and Biochemical Aspects Part A; A.L. Willis, Editor; CRC Press: Boca Raton, 1987, pp. 47-83.
38. Badami, R.C.; Patil, K.B. Prog. Lipid Res. 1981, 19, 119-153.
39. Rezanka, T. 1989, 28, 147-187.
40. Ogliaruso, M.A.; Wolfe, J.F. Synthesis of Carboxylic Acids, Esters and their Derivatives; John Wiley and Sons: Chichester, 1991, pp. 1-684.

41. Hawthorne, J.N.; Ansell, G.B. (Editors); Phospholipids; Elsevier Biomedical Press: Amsterdam, 1982, pp. 1-484.
42. Asselineau, J. The Bacterial Lipids; Holden-Day: San Francisco, 1966, pp. 1-372.
43. Stumpf, P.K.; Conn, E.E. (Editors); Biochemistry of Plants Volume 4. Lipids; Academic Press, New York, 1980, p. 1-693.
44. Razin, S.; Rottem, S. (Editors); Membrane Lipids of Prokaryotes; Academic Press: New York, Curr. Topics Membr. Transport 1982, 17, 1-383.
45. Mead, J.F.; Alfin-Slater, R.B.; Howton, D.R.; Popjak, G. Lipids - Chemistry, Biochemistry, and Nutrition; Plenum Press: New York, pp. 23-48.
46. Garvin, J.E.; Karnovsky, M.L. J. Biol. Chem. 1956, 221, 211-222.
47. Kaneda, T. Microbiol. Rev. 1991, 55, 288-302.
48. Fulco, A.J. Prog. Lipid Res. 1983, 22, 133-160.
49. Small, D.M. J. Lipid Res. 1984, 25, 1490-1500.
50. Melchior, D.L. In Membrane Lipids of Prokaryotes; S. Razin, S. Rottem, Editors; Academic Press: New York, Curr. Topics Membr. Transport 1982, 17, 263-316.
51. Mabrey, S.; Strutevant, J.M. Proc. Natl. Acad. Sci. USA 1976, 73, 3862-3866.
52. Kimelberg, H.K. In Physical Methods on Biological Mambranes and Their Model Systems; F.Conti, W.E. Blumberg, J. de Gier, F. Pocchiara, Editors; Plenum Press: New York, 1985, pp. 261-276.
53. Hazel, J.; Prosser, C.L. Physiol. Rev.1974, 54, 620-667.
54. Ladbrooke, M.D.; Chapman, D. Chem. Phys. Lipids 1969, 3, 304-367.
55. Kunau, W.-H. Angewandte Chem. 1976, 15, 61-122.
56.Christie, W.W.; Holman, R.T. Lipids 1967, 1, 407-423.
57. Barton, P.G.; Gunstone, F.D. J. Biol. Chem. 1975, 4470-4476.
58. Hauser, H.; Phillips, M.C. Prog. Surf. Membr. Sci. 1979, 13, 297-404.
59. Coolbear, K.P.; Berde, C.B.; Keough, K.M.W. Biochem. 1983, 22, 1466-1473.
60. Cullis, P.R.; deKruijff, B. Biochim. Biophys. Acta 1979, 559, 399-420.
61. Small, D.M. J. Lipid Res. 1984, 25, 1490-1500.
62. Applegate, K.R.; Glomset, J.A. J. Lipid Res. 1986, 27, 658-680.
63. Applegate, K.R.; Glomset, J.A. J. Lipid Res. 1991, 32, 1635-1644.
64. Applegate, K.R.; Glomset, J.A. J. Lipid Res. 1991, 32, 1645-1655.
65. Burkert, U.; Allinger, N.L. Molecular Mechanics. American Chemical Society Monograph No. 177; American Chemical Society: Washington, DC, 1982, pp. 1-339.
66. Seelig, A.; Seelig, J. Biochem. 1977, 16, 45-50.
67. Stubbs, C.; Kouyama, T.; Kinosita, K.; Ikegami. A. Biochem. 1981, 20, 2800-2810.
68. Rodgers, W; Glaser, M. Proc. Natl. Acad. Sci. 1991, 88, 1364-1368.
69. Davenprot, L.; Knutson, J.R.; Brand, L. In Subcellular Biochemistry; J.R. Harris, A.H. Etenadi, Editors; Plenum Press: New York, 1989, Volume 14, pp. 145-188.
70. Nicolson, G.L. Current Topics Dvptl. Biol. 1979, 13, 305-338.
71. Krisovitch, S.M.; Regen, S.L. J. Amer. Chem. Soc. 1992, 114, 9828-9835.

72. Hui, S.W. Curr. Topics Membr. Transp. 1987, 29, 29-70.
73. Lafleur, M.; Bloom, M.; Cullis, P.R. Biochem. Cell Biol. 1990, 68, 1-8.
74. Singleton, W.S. *In* Fatty Acids - Their Chemistry, Properties, Production, and Uses - Second Edition; K.S. Markley, Editor; Interscience Publishers: New York, 1960, Part 1, pp. 499-678.
75. Lutton, E.S. *In* Fatty Acids - Their Chemistry, Properties,, Production and Uses - Second Edition; K.S. Markley, Editor; Interscience Publishers: New York, 1967, Part 4, pp. 2583- 2641.
76. Margulis, L; Schwartz, K.V. Five Kingdoms - An Illustrated Guide to the Phyla of Life on Earth; W.H. Freeman and Company: San Francisco, 1982, pp. 1-338.
77. Whittaker, R.H. Q. Rev. Biol. 1959, 34, 210-226.
78. Whittaker, R.H.; Margulis, L. Biosystems 1978, 10, 3-18.
79. Cavalier-Smith, T. Biosystems 1981, 14, 461-481.
80. Woese, C.R.; Kandler, O; Wheelis, M.L. Proc. Natl. Acad. Sci USA 1990, 87, 4576-4579.
81. Cammarano, P.; Palm, P.; Creti, P; Ceccarelli, E.; Sanagelantoni, A.M.; Tiboni, O. J. Molec. Evol. 1992, 34, 396-405.
82. Forterre, P.; Benachenhou-Lahfa, N.; Confalonieri, F.; Duguet, M.; Elie, C.; Labedan, B. BioSystems 1993, 28, 15-32.
83. Parker, S.P. (Editor in Chief) Synopsis and Classification of Living Organisms; McGraw-Hill: New York, 1982, Volume 1, pp. 1-1166, Volume 2, pp. 1-1232.
84. Cavalier-Smith, T. *In* Evolution of Life - Fossils, Molecules, and Culture; S. Osawa, T. Honjo, Editors; Springer-Verlag: Tokyo, 1991, pp. 271-304.
85. Nes, W.R.; Nes, W.D. Lipids in Evolution; Plenum Press: New York, 1980, pp. 1-244.
86. Deamer, D.W.; Oro, J. Biosystems 1980, 12, 167-175.
87. Nooner, D.W.; Oro, J. *In* Hydrocarbon Synthesis from Carbon Monoxide and Hydrogen; E.L. Kugler, F.W. Steffgen, Editors; American Chemical Society Monograph No. 178; American Chemical Society: Washington, DC, 1979, pp.159-171.
88. Weber, A.L. J. Mol. Evol. 1991, 32, 93-100.
89. Blough, H.A.; Tiffany, J.M. Adv. Lipid Res. 1973, 11, 267-339.
90. Blough, H.A.; Lawson, D.E.M. Virology 1968, 36, 286-292.
91. Blough, H.A.; Tiffany, J.M. Curr. Topics Microbiol. Immunol. 1975, 70, 1-30.
92. Kruse, C.A.; Wisnieski, B.J.; Popjak, G. Biochim. Biophys. Acta 1984, 797, 40-50.
93. Schlesinger, H.R.; Wells, H.J.; Hummeler, K. J. Virol. 1973, 12, 1028-1030.
94. Lenard, J.; Compans, R.W. Biochim. Biophys. Acta 1974, 344, 51-94.
95. Barnes, J.A.; Pehowich, D.J.; Allen, T.M. J. Lipid Res. 1987, 28, 130-137.
96. Klenk, H.D.; Choppin, P.W. Virology 1970, 40, 939-947.
97. Nevenzel, J.C. *In* Marine Biogenic Lipids, Fats, and Oils, Volume 2; R.G. Ackman, Editor; CRC Press: Boca Raton, Florida, pp. 3-48.
98. Holton, R.W.; Blecker, H.H.; Stevens, T.S. Science, 1968, 160, 545-547.
99. Nichols, B.W.; Wood, B.J.B. Lipids, 1967, 3, 46-50.
100. Boon, J.J.; DeLeeuw, J.W. *In* The Cyanobacteria; P. Fay, C. Van Baalen, Editors; Elsevier Science: Amsterdam, 1987, pp. 472-534.

101. Kenyon, C.N.; Rippka, R.; Stanier, R.Y. Arch. Mikrobiol. 1972, 83, 216-236.
102. Kenyon, C.N. *In* The Photosynthetic Bacteria; R.K. Clayton, W.R. Sistrom, Editors; Plenum: New York, 1978, pp. 281-313.
103. Kenyon, C.N.; Gray, A.M. J. Bacteriol.1974, 120, 131-138.
104. Knudsen, E.; Jantzen, E.; Bryn, K.; Ormerod, J.G.; Sirevag, R. Arch. Microbiol. 1982, 132, 149-154.
105. Imhoff, J.F. *In* Green Photosynthetic Bacteria; Olson, J.G. Ormerod, J. Amesz, E. Stackebrandt, H.G. Truper, Editors; Plenum: New York, 1988, pp. 223-232.
106. Anderson, R.; Huang, Y. J. Bacteriol. 1992, 174, 7168-7173.
107. Nalik, H.P.; Muller, K.-D.; Ansorg, R. J. Med. Microbiol. 1992, 36, 371-376.
108. Ditandy, T.; Imhoff, J.F. J. Gen. Microbiol. 1993, 139, 111-117.
109. Kaneda, T. Microbiol. Rev. 1991, 55, 288-302.
110. Livermore, B.P.; Johnson, R.C. J. Bacteriol. 1974, 120, 1268-1273.
111. Meyer, H.; Meyer, F. 1971, Biochim. Biophys. Acta 1971, 231, 93-106.
112. Fautz, E.; Rosenfelder, G.; Grotjahn, L. J. Bacteriol. 1979, 140, 852-858.
113. Nichols, P.D.; Stulp, B.K.; Jones, J.G.; White, D.C. Arch. Microbiol. 1986, 146, 1-6.
114. Poen, E.; Aufderheide, M.; Diekmann, H.; Kroppenstedt, R.H. Arch. Microbiol. 1984, 137, 295-301.
115. Wirsen, C.O.; Jannasch, H.W.; Wakeham, S.G.; Canuel, E.A. Current Microbiol. 1987, 14, 319-322.
116. Intriago, P.; Floodgate, G.D. J. Gen. Microbiol. 1991, 137, 1503-1509.
117. Oyaizu, H.; Komagata, K. J. Gen. Appl. Microbiol. 1981, 27, 57-107.
118. Yabuuchi, E.; Moss, C.W. FEMS Microbiol. Lett. 1982, 13, 89-91.
119. Yamanaka, S.; Rudo, R.; Kawaguchi, A.; Komagata, K. J. Gen. Appl. Microbiol. 1988, 34, 57-66.
120. Holt, S.C.; Forcier, G.; Takacs, B.J. Infect. Immun. 1979, 26, 298-304.
121. Intriago, P; Floodgate, G.D. J. Gen. Microbiol. 1991, 137, 1503-1509.
122. Johns, R.B.; Perry, G.J. Arch. Microbiol. 1977, 114, 267-271.
123. Nichols, P.D.; Shaw, P.M.; Jones, J.G.; White, D.C. Arch. Microbiol. 1986, 146, 1-6.
124. Suzuki, K.; Komagata, K. Int. J. Syst. Bacteriol. 1983, 33, 188-200.
125. Collins, M.D.; Goodfellow, M.; Minnikin, D.E. J. Gen. Microbiol. 1982, 128, 2503-2509.
126. Kaneda, T. Can. J. Microbiol. 1983, 29, 364-368.
127. Suzuki, K.; Saito, K.; Kawaguchi, A.; Okuda, S.; Komagata, K. J. Appl. Microbiol. 1981, 27, 261-266.
128. Verhulst, V.; van Hespen, H.; Symons, F.; Eyssen, H. J. Gen. Microbiol. 1987, 133, 275-282.
129. Tadayon, R.A.; Carroll, K.K. Lipids 1971, 6, 820-825.
130. Collins, M.D.; Jones, D. FEMS Microbiol. Lett. 1983, 18, 131-134.
131. Goodfellow, M.; Collins, M.D.; Minnikin, D.E. J. Appl. Bacteriol. 1980, 48, 269- 276.
132. Collins, M.D. FEMS Microbiol. Lett. 1982, 13, 295-297.
133. Takayama, K.; Qureshi, N. *In* The Mycobacteria - A Sourcebook, Part A; G.P. Kubica, L.G. Wayne, Editors; Marcel Dekker: New York, 1984, pp. 315-344.

134.Asselineau, C.; Asselineau, J. *In* The Mycobacteria - A Sourcebook, Part A; G.P. Kubica, L.G. Wayne, Editors; Marcel Dekker: New York, 1984, pp. 345-359.

135.Campbell, I.M.; Naworal, J. J. Lipid Res. 1969, 10, 593-598.

136.Gautier, N.; Lopez Marin, L.M.; Laneelle, M.-A.; Daffe, M. FEMS Microbiol. Lett. 1992, 98, 81-88.

137.Brennan, P.J. *In* Microbial Lipids; C. Ratledge, C. Wilkinson, Editors; Academic Press: New York, 1988, pp. 203-298.

138.Hung, J.G.; Walker, R.W. Lipids 1970, 5, 720-722.

139.Asselineau, C.P.; Montrozier, H.L.; Prome, J.C. Eur. J. Biochem. 1969, 28, 102-109.

140.Takayama, K.; Qureshi, N. Lipids 1978, 13, 575-579.

141.Takayama, K.; Qureshi, N.; Jordi, H.C.; Schnoes, H.K. *In* Biological/Biomedical Applications of Liquid Chromatography; G.L. Hawk, Editor; Marcel Dekker: New York, 1979, Volume 10, pp. 91-101 & Volume 12, pp. 375-394.

142.Lederer, E. *In* The Mycobacteria - A Sourcebook, Part A; G.P. Kubica, L.G. Wayne, Editors; Marcel Dekker: New York, 1984, pp. 361-378.

143.Rezanka, T. Prog. Lipid Res. 1989, 28, 147-187.

144. Harwood, J.L.; Russell, N.J. Lipids in Plants and Microbes; George Allen and Unwin: London, 1984, pp. 1-162.

145. Sprott, G.D. J. Bioenerget. Biomembr., 1992, 24, 555-566.

146. Kates, M. Prog. Chem. Fats Other Lipids 1978, 15, 301-342.

147. Langworthy, T.A. *In* Biochemistry of Thermophily; S.M. Friedman, Editor; Academic Press: New York, 1978, pp. 11-30.

148. DeRosa, M.; Gambacorta, A. Prog. Lipid Res. 1988, 27, 153-175.

149.Langworthy, T.A. *In* The Bacteria. The Treatise on Structure and Function; C.R.Woese, R.S. Wolfe, Editors; Academic Press: New York, 1985, Volume VIII, pp. 459-497.

150. Langworthy, T.A.; Pond, J.L. *In* Thermophiles: General, Molecular, and Applied Microbiology; T.D. Brock, Editor); Wiley-Interscience: New York, pp. 459-497.

151. Kates, M. *In* Biological Membranes: Abberations in Membrane Structure and Function; M.L. Karnovsky, A.M. Leaf, A.C. Bolis, Editors; Alan R. Liss: New York, 1988, pp. 357-384.

152. Kates, M. *In* Handbook of Lipid Research; M. Kates, Editor; Plenum Press: New York, 1990, pp. 1-122.

153.Stern, J.; Freisleben, H.-J.; Janku, S; Ring, K. Biochim. Biophys. Acta 1992, 1128, 227-236.

154.DeRosa, M.; Tricone, A.; Nicolaus, B.; Gambacorta, A. *In* Life Under Extreme Conditions; G. di Prisco, Editor; Springer-Verlag: Berlin, 1991, pp. 61-87.

155.Blair, R.J.; Weller, P.F. Mol. Biochem. Parasitol. 1987, 25, 11-18.

156.Korn, E.D. J. Lipid Res. 1964, 5, 352-362.

157.Erwin, J.; Bloch, K. Biochem. Z. 1963, 338, 496-511.

158.Korn, E.D. Biochem. Biophys. Res. Commun. 1964, 14, 352-362.

159.Constantopoulos, G.; Bloch, K. J. Biol. Chem. 1967, 242, 3538-3542.

160.Holz, G.G.; Lindmark, D.G.; Beach, D.H.; Neale, K.A.; Singh, B.N. Acta Univ. Carol., Biol.1986, 30, 299-311.
161.Lindmark, D.G.; Beach, D.H.; Singh, B.N.; Holz, G.G. Biochem. Protozool. 1991, 329-335.
162.Erwin, J.; Bloch, K. 1963. J. Biol. Chem. 1963, 238, 1618-1624.
163.Meyer, H.; Holz, G.G. J. Biol. Chem. 1963, 241, 5000-5007.
164.Erwin, J. A. In Lipids and Biomembranes of Eukaryotic Microorganisms; J.A. Erwin, Editor; Academic Press: New York, 1973, pp. 41-143.
165.Hallegraeff, G.M.; Nichols, P.D.; Volkman, J.K.; Blackburn, S.I.; Everitt, D.A. Phycol.1991, 591-599.
166.Pohl, P.; Zurheide, F. In Marine Algae in Pharmaceutical Science; H.A. Hoppe, T. Levring, Y. Tanaka, Editors; Walter de Gruyter: Berlin, 1979, pp. 473-523.
167.Harwood, J.L.; Pettitt, T.P.; Jones, A.L. In Biochemistry of the Algae and Cyanobacteria; L.J. Rogers, J.R. Gallon, Editors,; Clarendon Press: Oxford, pp. 49-67.
168.Holz, Z.Z., Jr. In Biochemistry and Physiology of Protozoa, 2nd Edition; M. Levandowsky, S.H. Hunter, L. Provasoli, Editors; Academic Press: New York, Vol. 4, pp. 301-331.
169.Ackman, R.G.; Tocher, A.S. J. Fish. Res. Board Canada 1968, 22, 1107-1120.
170.Chuecas, L.; Riley, J.P. J. Mar. Biol. 1969, 49, 97-116.
171.Wantanabe, T.; Ackman, R.G. J Fish. Res. Board Canada 1974, 31, 401-409.
172.Creamer, J.C.; Bostock, R.M. Physiol. Mol. Plant Pathol. 1986, 28, 215-255.
173.Kerwin, J.L.; Duddles, N.D. J. Bacteriol. 1989, 171, 3831-3839.
174.Jones, A.L.; Harwood, J.L. Phytochem. 1992, 31, 3397-3403.
175.Hiscock, S. A Field Guide to the British Brown Seaweeds, Field Studies; 1972, 5, 1-44.
176.Smith, K.L.; Harwood, J.L. Phytochem. 1984, 23, 2469-2473.
177.Jamieson, G.; Reid, H. Phytochem. 1972, 11 1423-1432..
178.Wassef, M.K. Adv. Lipid Res. 1977, 15, 159-232.
179.Mau, J.-L.; Beelman, R.B.; Ziegler, G.R. Phytochem. 1992, 31, 4059-4064.
180.Kerwin, J.L. In Ecology and Metabolism of Plant Lipids; G. Fuller, W.D. Nes, Editors; American Chemical Society: Washington, D.C., 1987, pp. 329-342.
181.Stredanska, S.; Sajbidor, J. Folia Microbiol. 1992, 37, 357-359.
182.Kayama, M.; Araki, S.; Sato, S. In Marine Biogenic Lipids, Fats, and Oils; R.G. Ackman, Editor; CRC Press: Boca Raton, 1989, pp. 3-48.
183.Johns, R.B.; Nichols, P.D.; Perry, G.J. Phytochem. 1979, 18, 799-802.
184.Pohl, P.; Wagner, H.; Passig, T. Phytochem. 1968, 7, 1565-1572.
185.Ackman, R.G.; Tocher, C.S. J. FIsh. Res. Bd. Canada 1968, 25, 1603-1620.
186.Nichols, B.W.; Appleby, R.S. Phytochem. 1969, 8, 1907-1915.
187.Radunz, A. Hoppe-Seyler's Z. Physiol. Chem. 1968, 349, 1091-1094.
188.Schneider, H.; Gelpi, E.; Bennet, E.O.; Oro, J. Phytochem. 1970, 9, 613-617.
189.Rezanka, T.; Podojil, M. Lipids, 1984, 19, 472-473.
190.Pohl, P.; Passig, T.; Wagner, H. Phytochem. 1971, 10, 1505-1513.
191.Douglas, A.G.; Douraghi,-Zadeh, K.; Eglinton, G. Phytochem. 1969, 8, 285-293.
192.Rezanka, T.; Podojil, M. J. Chromatogr. 1989, 463, 397-408.

193.Ichikawa, T.; Namikawa, M.; Yamada, K.; Sakai, K.; Kondo, K. Tetrahedron Lett. 1983, 24, 3337-3340.
194.Gellerman, J.L.; Schlenk, H. Experientia 1964, 20, 426-xxx.
195.Karunen, P. *In* Bryophytes - Their Chemistry and Chemical Taxonomy; H.D. Zinsmeister, R. Mues, Editors; Clarendon Press: Oxford, 1990, pp. 121-141.
196.Kohn, G.; Demmerle, S.; Vandekedhove, O.; Beutelmann, P. Phytochemistry 1987, 26, 2101.
197.Kohn, G.; Vierengal, A.; Vandekerhove, O.; Hartmann, E.; Phytochemistry 1987, 26, 2271.
198.Nichols, B.W.; Stubbs, J.M.; James, A.T. *In* The Biochemistry of Chloroplasts; T.W. Goodwin, Editor; Academic Press: New York, 1966, pp. 677.
199.Badami, R.C.; Patil, K.B. Prog. Lipid Res. 1981, 19, 119-153.
200.Plattner, R.D.; Kleiman, R. Phytochem. 1977, 16, 255-256.
201.Raju, P.K.; Reiser, R. Lipids 1966, 1, 10-15.
202.Smith, C.R., Jr.; Wilson, T.L.; Bates, R.B.; Scholfield, C.R. J. Org. Chem. 1962, 27, 3112-3117.
203.Vickery, J.R. J. Amer. Oil Chem. Soc.1980, 57, 87-91.
204.Stumpf, P.K. *In* The Biochemistry of Plants - Volume 9 Lipids: Structure and Function; P.K. Stumpf, Editor; Academic Press: New York, pp. 121-136.
205.Hilditch, T.P.; Williams, P.N. The Chemical Constitution of Natural Fats, IVth Edition; Chapmann and Hall: London, 1964.
206.Holman, R.T. Prog. Chem. Fats Other Lipids 1978, 16, 9-29.
207.Tulloch, A.P. *In* Chemistry and Biochemistry of Natural Waxes; P.E. Kolattukudy, Editor; Elsevier: Amsterdam, 1976, pp. 236-287.
208.Kolattukudy, P.E.; Croteau, R.; Buckner, J.S. *In* Chemistry and Biochemistry of Natural Waxes; P.E. Kolattukudy, Editor; Elsevier: Amsterdam, pp. 289-347.
209.Poirier, Y.; Dennis, D.; Klomparens, K.; Nawrath, C.; Somerville, C. FEMS Microbiol. Rev.1992, 103, 237-246.
210.Gibbs, R.D. Chemotaxonomy of Flowering Plants - Volume 1; McGill-Queen's University Press: Montreal, 1974, pp. 493-523.
211.Joseph, J.D. Prog. Lipid Res. 1982, 21, 109-153.
212.Joseph, J.D. Prog. Lipid Res. 1979, 18, 1-30.
213.Litchfield, C.; Greenberg, A.J.; Noto, G.; Morales, R.W. Lipids 1976, 11, 567-570.
214.Morales, R.W.; Greenberg, A.J.; Noto, G.; Litchfield, C. J. Am. Oil Chem. Soc. 1977, 54, 145A.
215.Walkup, R.D.; Jamieson, G.C.; Ratcliff, M.R.; Djerassi, C. Lipids 1981, 16, 631-646.
216.Cimino, G.; deStefano, S.; Minale, L. Experientia 1975, 31, 1387-1389.
217.Costantino, V.; Fattorusso, E.; Mangoni, A. J. Org. Chem. 1993, 58, 186-191.
218.Carballeira, N.M.; Shalabi, F. Lipids 1990, 25, 835-840.
219.Carballeira, N.M.; Reyes, E.D. Lipids 1990, 25, 69-71.
220.Joseph, J.D. *In* Marine Biogenic Lipids, Fats, and Oils, Volume 2; R.G. Ackman, Editor; CRC Press: Boca Raton, Florida, pp. 49-143.
221.Cimino, G.; deStefano, S.; Minale, L. Experientia 1975, 31, 1387-1389.

222.Stillway, L.W. Comp. Biochem. Physiol. Series B 1976, 53, 535.
223.White, R.H.; Hager, L.P. Biochem. 1977, 16, 4944-4948.
224.Johns, R.B.; Nickols, P.D.; Perry, G.J. Comp. Biochem. Physiol. Series B 1980, 65, 207-214.
225.deMoreno, J.E.A.; Pollero, R.J.; Moreno, V.J.; Brenner, R.R. J. Exp. Mar. Biol. Ecol. 1980, 48, 263.
226.Ackman, R.G. Hooper, S.N.; Ke, P.J. Comp. Biochem. Physiol. Series B 1971, 39, 579-587.
227.Paradis, M.; Ackman, R.G. Lipids 1977, 12, 170-176.
228.Ben-Mlih, F.; Marty, J.-C.; Fiala-Medioni, A. J. Lipid Res. 1992, 33, 1797-1806.
229.Joseph, J.D. Prog. Lipid Res. 1982, 22, 109-153.
230.Voogt, P.A. In The Molluscs Volume I; Academic Press: New York, 1983, pp. 329-370.
231.Carballeira, N.M.; Anastacio, E. J. Nat. Prod. 1992, 12, 1783-1786.
232.Ackman, R.G.; Hooper, S.N. Lipids 1970, 5, 417-421.
233.Lee, R.F. Comp. Biochem. Physiol. Series B 1975, 51, 263-272.
234.Morris, R.J. Comp. Biochem. Physiol. Series B 1971, 40, 275-281.
235.Moore, J.W. Estuarine Coastal Mar. Sci. 1976, 4, 215-224.
236.Bottino, N.R. Mar. Biol. 1974, 27, 197.
237.Pearce, R.E.; Stillway, L.W. Lipids, 1977, 12, 544-549.
238.Joseph, J.D.; Fender, D.S. J. Amer. Oil Chem. Soc. 1977, 54, 145A.
239.Pascal, J.-C.; Ackman, R.G. Chem. Phys. Lipids 1976, 16, 219.
240.Sargent, J.R.; Lee, R.F.; Nevenzel, J.C. In Chemistry and Biochemistry of Natural Waxes; P.E. Kolattukudy, Editor; Elsevier: Amsterdam, 1976, pp. 50-91.
241.Farkas, T.; Storebakken, T.; Bhosle, N.B. Lipids 1988, 23, 619-622.
242.Bottino, N.R.; Gennity, J.; Lilly, M.L.; Simons, E.; Finne, G. Aquaculture 1980, 19, 139-148.
243.Jackson, L.L.; Blomquist, G.J. In Chemistry and Biochemistry of Natural Waxes; P.E. Kolattukudy, Editor; Elsevier: Amsterdam, 1976, pp. 201-233.
244.Tulloch, A.P. Lipids 1970, 5, 1-12.
245.Tamaki, Y. Lipids 1968, 3, 186-189.
246.Chibnall, A.C.; Latner, A.L.; Williams, E.F.; Ayre, C.A. Biochem. J. 1984, 28, 313.
247.Tulloch, A.P. Chem. Physics Lipids 1971, 6, 235-265.
248.Blomquist, G.J.; Dillwith, J.W. In Comprehensive Insect Biochemistry and Pharmacology Volume 3 - Integument, Respiration and Circulation; G.A. Kerkut, L.I. Gilbert, Editors; Pergamon Press: Oxford, 1985, pp. 117-154.
249.Uscian, J.M.; Miller, J.S.; Howard, R.W.; Stanley-Samuelson, D.W. Comp. Biochem. Physiol. Series B 1992, 103, 833-838.
250.Armold, M.; Blomquist, G.J.; Jackson, L.L. Comp. Biochem. Physiol. 1969, 31, 685-692.
251.Jackson, L.L.; Baker, G.L. Lipids 1970, 5, 239-246.
252.Hanson, B.J.; Cummins, K.W.; Cargil, A.S.; Lowry, R.R. Comp. Biochem. Physiol. Series B 1985, 80, 257-276.

253. Stanley-Samuelsson, D.W.; Dadd, R.H. Insect Biochem. 1983, 13, 549-558.
254. Beenakkers, A.M.T.; Van der Horst, D.J.; Van Marrewijk, W.J.A. Prog. Lipid Res. 1985, 24, 9-67.
255. Blomquist, G.J.; Nelson, D.R.; de Renobales, M. Arch. Insect Biochem. Physiol. 1987, 6, 227-265.
256. Allen, W.V. J. Mar. Biol. Assoc., U.K. 1968, 48, 521-533.
257. Takagi, T.; Eaton, C.A.; Ackman, R.G. Can. J. Fish. Aquat. Sci. 1980, 37, 195-202.
258. Rodegker, W.; Nevenzel, J.C. Comp. Biochem. Physiol. Series B 1964, 11, 53-60.
259. Falk-Petersen, I.B.; Sargent, J.R. Mar. Biol. 1982, 69, 291-298.
260. Sargent, J.R.; Falk-Petersen, I.B.; Calder, A.G. Mar. Biol. 1983, 72, 257-xxx.
261. Mita, M.; Ueta, N. Comp. Biochem. Physiol. Series B 1992, 102, 15-18.
262. Mita, M.; Ueta, N. Comp. Biochem. Physiol. Series B 1989, 92, 319-322.
263. Morris, R.J.; Culkin, F. In Marine Biogenic Lipids, Fats, and Oils, Volume II; R.G. Ackman, Editor; CRC Press: Boca Raton, 1989, pp. 145-178.
264. Ackman, R.G.; Lamothe, F. In Marine Biogenic Lipids, Fats, and Oils, Volume II; R.G. Ackman, Editor; CRC Press: Boca Raton, 1989, pp. 179-381.
265. Henderson, R.J.; Tocher, D.R. Prog. Lipid Res. 1987, 26, 281-347.
266. Ackman, R.G. In Advances in FIsh Science and Technology; J.J. Connell, Editor; Fishing News Books Ltd.: Farnham, Surrey, U.K., 1980, pp. 86-103.
267. Sargent, J.R. In Biochemical and Biophysical Perspectives in Marine Biology, Volume 3; D.C. Malins, J.R. Sargent, Editors; Academic Press: New York, 1976, pp. 149-212.
268. Faulkner, D.J. Natural Prod. Reports 1991, 8, 97-147.
269. Faulkner, D.J. Natural Prod. Reports 1992, 9, 323-364.
270. Al-Hassan, J.M.; Afzal, M; Ali, M.; Thomson, M.; Fatima, T.; Fayad, S.; Criddle, R.S. Comp. Biochem. Physiol. Series B 1986, 85, 41-47.
271. Body, D.R. J. Sci. Food. Agric. 1983, 34, 388-392.
272. Sargent, J.R.; Gatten, R.R.; Henderson, R.J. Pure Appl. Chem. 1981, 53, 967-871.
273. Sargent, J.R.; Lee, R.F.; Nevenzel, J.C. In Chemistry and Biochemistry of Natural Waxes; P.E. Kolattakudy, Editor; Elsevier: Amsterdam, 1976, pp. 50-91.
274. Greene, D.H.S.; Selivonchick, D.P. Prog. Lipid Res. 1987, 26, 53-85.
275. Clarke, A. In Marine Biogenic Lipids, Fats, and Oils, Volume II; R.G. Ackman, Editor; CRC Press: Boca Raton, 1989, pp. 383-398.
276. Jacob, J. In Chemistry and Biochemistry of Natural Waxes; P.E. Kolattukudy, Editor; Elsevier: Amsterdam, 1976, pp. 94-146.
277. Buckner, J.S.; Kolattukudy, P.E. In Chemistry and Biochemistry of Natural Waxes; P.E. Kolattukudy, Editor; Elsevier: Amsterdam, 1976, pp. 148-200.
278. Rosenthal, M.D. Prog. Lipid Res. 1987, 26, 87-124.
279. Sastry, P.S. Prog. Lipid Res. 1985, 24, 69-176.
280. Ziboh, V.A.; Chapkin, R.S. Prog. Lipid Res. 1988, 27, 81-105.
281. Downing, D.T. In Chemistry and Biochemistry of Natural Waxes; P.E. Kolattukudy, Editor; Elsevier: Amsterdam, 1976, pp. 18-48.

282.Fulco, A.J. Prog. Lipid Res. 1983, 22, 133-160.
283.Bloch, K. Adv. Enzymol. Related Areas Molec. Biol. 1977, 45, 1-84.
284.Vagelos, P.R. In Biochemistry of Lipids; T.W. Goodwin, Editor; Butterworths: London, 1974, pp. 99-104.
285.Volpe, J.J.; Vagelos, P.R. Annu. Rev. Biochem. 1973, 42, 21-60.
286.Wakil, S.J.; Stoops, J.K.; Joshi, V.C. Annu. Rev. Biochem. 1983, 52, 537-579.
287.Wakil, S.J.; Stoops, J.K. The Enzymes 1983, 16, 3-61.
288.Harwood, J.L. Annu. Rev. Plant Physiol. Mol. Biol. 1988, 39, 101-138.
289.Harwood, J.L. Prog. Lipid Res. 1979, 18, 55-86.
290.Stumpf, P.K. In The Biochemistry of Plants; P.K. Stumpf, E.E. Conn, Editors; Academic Press: New York, 1980, pp.177-203.
291.Stumpf, P.K. In Fatty Acid Metabolism and its Regulation; S. Numa, Editor; Elsevier: Amsterdam, 1984, pp. 155-199.
292.Wakil, S.J. Annu. Rev. Biochem. 1963, 32, 369-406.
293.Stumpf, P.K. Annu. Rev. Biochem. 1969, 38, 159-207.
294.Brenner, R.R. Mol. Cell. Biiochem. 1974, 3, 41-52.
295.Sprecher, H. In New Trends in Nutrition, Lipid Research, and Cardiovascular Diseases; N.G. Bazan, R. Paoletti, J.M. Iacono, Editors; Alan R. Liss: New York, 1981, pp.15.
296.Ayala, S.; Gaspar, G.; Brenner, R.R.; Peluffo, R.O.; Kunau, W. J. Lipid Res. 1973, 14, 296-305.
297.Bloch, K. Acc. Chem. Res. 1969, 2, 193-202.
298.Kinsella, J.E.; Broughton, K.S.; Whelan, J.W. J. Nutr. Biochem. 1990, 1, 123-141.
299.Brenner, R.R. In Function and Biosynthesis of Lipids; N.G. Bazan, R.R. Brenner, N.M. Giusto, Editors; Plenum Press: New York; pp. 85-101.
300.Walker, P.; Woodbine, M. In The Filamentous Fungi Volume 2 Biosynthesis and Metabolism; J.E. Smith, D.R. Berry, Editors; John Wiley and Sons: New York, 1976, pp. 137-158.
301.Ackman, R.G. In CRC Handbook of CHromatography - Lipids - Volume 1; H.K. Mangold, G. Zweig, J. Sherma, Editors; CRC Press: Boca Raton, 1984, pp. 95-240.
302.Holloway, P.W. In The Enzymes, Volume XVI - Lipid Enzymology; P.D. Boyer, Editor; Academic Press: New York, 1983, pp. 63-83.
303.Nichols, D.S.; Nichols, P.D.; McMeekin, T.A. FEMS Microbiol. Lett. 1992, 98, 117-122.
304.Shastri-Bhalla, K.; Funk, C.J.; Consigli, R.A. J. Invertebr. Pathol. 1993, 61, 69-74.
305.St.Geme, J.W.; Martin, H.L.; Davis, C.W.C.; Mead, J.F. Pediatr. Res. 1977, 11, 174-177.
306.Yamamoto, T.; Tanada, Y. J. Invertebr. Pathol. 1977, 30, 279-281.
307.Blough, H.A.; Tiffany, J.M. Curr. Topics Microbiol. Immunol. 1975, 70, 1-30.
308.Bramhall, J.; Wisnieski, B. In Virus Receptors, Part 2; K. Lonberg-Holm, L. Philipson, Editors; Chapman and Hall: London, 1981, pp. 141-153.
309.Moore, N.F.; Moore, J.C.; Kelley, D.C. Microbiologica 1984, 7, 267-272.
310.Kruse, C.A.; Wisnieski, B.J.; Popjak, G. Buichim. Biophys. Acta 1984, 797, 40-50.

311.Moscufo, N.; Chow, M. J. Virol. 1992, 66, 6849-6857.

312.Towler, D.A.; Gordon, J.I.; Adams, S.P.; Glaser, L. Annu. Rev. Biochem. 1988, 57, 69-99.

313.Wu, S.-X.; Ahlquist, P.; Kaesberg, P. Proc. Natl. Acad. Sci. USA 1992, 89, 11136-11140.

314.Bienz, K.; Egger, D.; Pfister, T.; Troxler, M. J. Virol. 1992, 66, 2740-2747.

315.Takeda, N.; Kuhn, R.J.; Yang, C.-F.; Takegami, T.; Wimmer, E. J. Virol. 1986, 60, 43-53.

316.Molla, A.; Paul, A.V.; Wimmer, E. Science 1991, 254, 1647-1651.

317.Abubakar, S.; Boldogh, I.; Albrecht, T. Arch. Virol. 1990, 113, 255-266.

318.Fitzpatrick, F.A.; Stringfellow, D.A. J. Immunol. 1980, 125, 431-437.

319.Palumbo, G.J.; Pickup, D.J.; Frederickson, T.N.; McIntyre, L.J.; Buller, R.M.L. Virology 1989, 172, 262-273.

320.Strayer, D.S.; Korber, K.; Dombrowski, J. J. Immunol. 1988, 140, 2051-2059.

321.Villani, A.; Cirino, N.M.; Baldi, E.; Kester, M.; McFadden, E.R.; Panuska, J.R. J. Biol.Chem. 1991, 266, 5472-5479.

322.Palumbo, G.J.; Glasgow, W.C.; Buller, R.M.L. Proc. Natl. Acad. Sci. USA 1993, 90, 2020-2024.

323.Gliozzi, A.; Rolandi, R.; DeRosa, M.; Gambacorta, A. J. Membr. Biol. 1983, 75, 45-56.

324.DeRosa, M.; Gambacorta, A. J. Mol. Biol. 1985, 182, 131-149.

325.DeRosa, M.; Gambacorta, A.; Gliozzi, A. Microbiol. Rev. 1986, 50, 70-80.

326.Brock, T.D.; Brock, K.M.; Belley, R.T.; Weiss, R.L. Arch. Microbiol. 1972, 84, 54-68.

327.Yamauchi, K.; Sakamoto, Y.; Moriya, A.; Yamada, K.; Hosokawa, T.; Higuchi, T.; Kinoshita, M. J. Am. Chem. Soc. 1990, 112, 3188-3191.

328.Yamauchi, K.; Doi, K.; Yoshida, Y.; Kinoshita, M. Biochim. Biophys. Acta 1993, 1146, 178-182.

329.Minnikin, D.E. *In* The Biology of Mycobacteria; C. Ratledge, J.L. Stanford, Editors; Academic Press: London, 1982, Volume 1, pp. 95-184.

330.Brennan, P.J. *In* Microbial Lipids; C. Ratledge, C. Wilkinson, Editors; Academic Press: London, 1988, pp. 203-298.

331.Portaels, F.; Dawson, D.J.; Larsson, L.; Rigouts, L. J. Clin. Microbiol. 1993, 31, 26-30.

332.Luquin, M.; Ausina, V.; Lopez Calahorra, F.; Belda, F.; Garcia, B.; Celma, C.; Prats, G. J.Clin. Microbiol. 1991, 29, 120-130.

333.Tornabene, T.G. Meth. Microbiol. 1985, 18, 209-234.

334.Tuner, K.; Baron, E.J.; Summanen, P.; Finegold, S.M. J. Clin. Microbiol. 1992, 30, 3225-3229.

335.Leonards, K.S.; Ohki, S. Biochem. 1984, 23, 2718-2725.

336.Lucy, J.A. *In* Structure of Biological Membranes; Abrahamsson, Pascher, Editors; Plenum Press: New York, 1976, pp. 275-291.

337.Lucy, J.A. *In* Biological Membranes; D. Chapman, Editor; Academic Press: London, 1982, pp. 367-415.

338.Poste, G.; Nicholson, G.L. (Editors) Membrane Fusion; Elsevier/North Holland: Amsterdam, 1978, Cell Surface Reviews, Volume 5.

339.Roos, D.S.; Choppin, P.W. J. Cell Biol. 1985, 101, 1578-1590.

340. Blumenthal, R. Curr. Top. Membr. Transp. 1987, 29, 203-254.
341. Plattner, H. *In* Membrane Fusion; J. Wilschut, D. Hoekstra, Editors; Marcel Dekker: New York, 1991, pp. 571-598.
342. Lindblom, G.; Rilfors, L. Biochim. Biophys. Acta 1989, 988, 221-256.
343. Verkleij, A.J. Biochim. Biophys. Acta 1984, 779, 43-63.
344. Bentz, J.; Ellens, J. Colloids Surfaces 1986, 30, 65-112.
345. Stanier, R.Y. Symp. Soc. Gen. Microbiol. 1970, 20, 1-38.
346. Cavalier-Smith, T. Nature 1975, 256, 463-468.
347. Cavalier-Smith, T. Ann. N.Y. Acad. Sci. 1987, 503, 17-54.
348. McElhaney, R.N. Curr. Top. Membr. Trans. 1982, 17, 317-380.
349. Lenaz, G.; Parenti Castelli, G. *In* Structure and Properties of Cell Membranes Volume 1 - A Survey of Molecular Aspects of Membrane Structure and Function; G. Benga, Editor; CRC Press: Boca Raton, 1985, pp. 93-136.
350. Spector, A.A.; Yorek, M.A. J. Lipid Res. 1985, 26, 1015-1035.
351. Blomquist, G.J.; Nelson, D.R.; de Renobales, M. Arch. insect Biochem. Physiol. 1987, 6, 227-265.
352. Tiffany, J.M. Adv. Lipid Res. 1987, 22, 1-62.
353. Ziboh, V.A.; Chapkin, R.S. Prog. Lipid Res. 1988, 27, 81-105.
354. Hadley, N.F. The Adaptive Role of Lipids in Biological Systems; John Wiley: New York, 1985, pp. 102-148.
355. Kolattukudy, P.E. (Editor) Chemistry and Biochemistry of Natural Waxes; Elsevier: Amsterdam, 1976.
356. Strauch, M.A.; de Mendoza, D.; Hoch, J.A. Mol. Microbiol. 1992, 6, 2909-2917.
357. Bell, R.M.; Burns, D.J. J. Biol. Chem. 1991, 266, 4661-4664.
358. Fitzpatrick, F.A.; Murphy, R.C. Pharmacol. Rev. 1989, 40, 229-241.
359. Schroepfer, G.; Bloch, K. J. Biol. Chem. 1965, 237,54-63.
360. Schroepfer, G.J. Jr.; Niehaus, W.G. Jr.; McCloskey, J.A. J. Biol. Chem. 1970, 245, 3798-3801.
361. Helson, D.R.; Strobel, H.W. Mol. Biol. Evol. 1987, 4, 572-593.
362. Peterson,J.A.; Kusunose, M.; Kusunose, E.; Coon, M.J. J. Biol. Chem. 1967, 242, 4334-4340.
363. Reuttinger, R.T.; Olson, S.T.; Boyer, R.F.; Coon, M.J. Biochem. Biophys. Res. Commun.1974, 57, 1011-1017.
364. Asselineau, J. The Bacterial Lipids; Hermann/Holden-Day: San Francisco, 1966, pp. 1-287.
365. Sligar, S.G.; Murray, R.I. *In* Cytochrome P450 Structure, Mechanism, and Biochemistry; Plenum: New York, 1986, pp. 429-503.
366. Sariaslani, F.S.; Omer, C.A. CRC Crit. Rev. Plant Sci. 1992, 11, 1-16.
367. Andrianarison, R.-H.; Beneytout, J.-L.; Tixier, M. Plant Physiol. 1989, 91, 1280-1287.
368. Beneytout, J.-L.; Andrianarison, R.-H.; Rakotoarisoa, Z.; Tixier, M. Plant Physiol. 1989, 91, 367-372.
369. Bundy, G.L. Adv.Prostaglandin, Thromboxane, Leukotriene Res. 1985, 14, 229-262.
370. Chekhonadskikh, T.V.; Polyakova, N.E.; Pankova, T.G.; Salganki, R.I. Biochem. 1987, 52, 327-331.

371.Gupta, S.; Varma, V.; Bose, S.K.; Tekwani, B.L.; Shukla, O.P. Med. Sci. Res. 1987, 15, 1379-1380.
372.Pandey, V.C.; Saxena, N.; Bose, S.K.; Dutta, G.P.; Shukla, O.P.; Ghatak, S. IRCS Med. Sci. 1986, 14, 346-347.
373.Hokama, Y.; Yokochi, L.; Abad, M.A.; Shigemura, L.; Kimura, L.H.; Okano, C.; Chou, S.C. Res. Commun. CHem. Pathol. Pharmacol. 1982, 38, 169-172.
374.Hamberg, M.; Herman, C.A.; Herman, R.P. Biochim. Biophys. Acta 1986, 877, 447-457.
375.Hamberg, M.; Herman, R.P.; Jacobsson, U. Biochim. Biophys. Acta 1986, 879, 410-418.
376.Hamberg, M. Lipids 1989, 24, 249-255.
377.Herman, R.P.; Luchini, M.M. Exp. Mycol. 1989, 13, 372-379.
378.Morris, D.R.; Hager, L.P. J. Biol. Chem. 1966, 241, 1763-1768.
379.Hallenberg, P.F.; Hager, L.P. Meth. Enzymol. 1978, 52, 521-529.
380.Jayanthi, C.R.; Madyastha, P.; Madyastha, K.M. Biochem. Biophys. Res. Commun. 1982, 106, 1262-1268.
381.Shoun, H.; Sudo, Y.; Seto, Y.; Beppu, T. J. Biochem. 1983, 94, 1219-1229.
382.Shoun, H.; Sudo, Y.; Beppu, T. J. Biochem. 1985, 97, 755-763.
383.Schunck, W.-H.; Gross, B.; Wiedmann, B.; Mauersberger, S.; Koepke, K.; Kiessling, U.; Strauss, M.; Gaestel, M.; Mueller, H.-G. Biochem. Biophys. Res. Commun. 1989, 161, 843-850.
384.Schecter, G.; Grossman, S. Int. J. Biochem. 1983, 15, 1295-1304.
385.Matsuda, Y.; Satoh, T.; Beppu, T.; Arima, K. Agr. Biol. Chem. 1976, 40, 963-976.
386.Jensen, E.C.; Ogg, C.; Nickerson, K.W. Appl. Environ. Microbiol. 1992, 58, 2505-2508.
387.Lyudnikova, T.A.; Rodopulo, A.K.; Egorov, I.A. Appl. Biochem. Microbiol. 1984, 20, 142-144.
388.Gerwick, W.H.; Moghaddam, M.; Hamberg, M. Arch. Biochem. Biophys. 1991, 290, 436- 444.
389.Maghaddam, M.F.; Gerwick, W.H. Phytochem. 1990, 29, 2457-2459.
390.Moghaddam, M.F.; Gerwick, W.H.; Ballantine, D.L. J. Biol. Chem. 1990, 265, 6126-6130.
391.Burgess, J.R.; de la Rosa, R.I.; Jacobs, R.S.; Butler, A. Lipids 1991, 26, 162-165.
392.Hamberg, M. Biochem. Biophys. Acta 1992, 188, 1220-1227.
393.Veldink, G.A.; Vliegenthart, J.F.G.; Boldingh, J. Prog. Chem. Fats other Lipids 1977, 15, 131-166.
394.Gardner, H.W. *In* Autooxidation in Food and Biological Systems; M.G. Simac, M. Karel, Editors; Plenum Press: New York, 1980, pp. 447-504.
395.Vick, B.A.; Zimmerman, D.C. *In* The Biochemistry of Plants -Volume 9; P.K. Stumpf, Editor; Academic Press: Orlando, Florids, 1987, pp. 53-90.
396.Siedow, J.N. Annu. Rev. Plant Physiol. Plant Mol. Biol. 1991, 42, 145-188.
397.Hildebrand, D.F. Physiol. Plant. 1989, 76, 249-253.
398.Mack, A.J.; Peterman, T.K.; Siedow, J.N. Isozymes: Curr. Top. Biol. Med. Res. 1987, 13,127-154.
399.Bozak, K.R.; Yu, H.; Sirevag, R.; Christoffersen, R.E. Proc. Natl. Acad. Sci. USA 1990, 87, 3904-3908.

400.Gabriac, B.; Werck-Reichhart, D.; Teutsch, H.; Durst, F. Arch. Biochem. Biophys. 1991, 288, 302-309.
401.Vetter, H.-P.; Mangold, U.; Schroder, G.; Marner, F.-J.; Werck-Reichhart, D.; Schroder, J. Plant Physiol. 1992, 100, 998-1007.
402.Livingstone, D.R. Trans. Biochem. Soc. 1990, 18, 15-19.
403.Bayer, F.M.; Weinheimer, A.J. (Editors) Prostaglandins in Plexaura homomalla: Biology, Utilization, and Ecology of a Major Medical Marine Resource; University of Miama Press: Miami, 1974.
404.Gerhart, D.J. Biochem. Syst. Ecol. 1986, 14, 417-421.
405.Corey, E.J.; d'Alarcao, M.; Matsuda, S.P.T.; Lansbury, Jr., P.T. J. Am. Chem. Soc. 1987, 109, 289-290.
406.Bundy, G.L. Adv. Prostagland., Thromboxane, Leukotriene Res. 1985, 14, 229-262.
407.Schneider, W.P.; Bundy, G.L.; Lincoln, F.H.; Daniela, E.G.; Pike, J.E. J. Am. Chem. Soc. 1977, 99, 1222-1232.
408.Nelson, D.R.; Kamataki, T.; Waxman, D.J.; Guengerich, F.P.; Estabrook, R.W.; Feyereisen, R.; Gonzalez, F.J.; Coon, M.J.; Gunsalus, I.C.; Gotoh, O.; Okuda, K.; Nebert, D.W. DNA Cell Biol. 1993, 12, 1-51.
409.Fitzpatrick, F.A.; Murphy, R.C. Pharmacol. Rev. 1989, 40, 229-241.
410.Matsubara, S.; Yamamoto, S.; Sogawa, K.; Yokotani, N.; Fuji-Kuriyama, Y.; Haniu, M.; Shively, J.E.; Gotoh, O.; Kusunose, E.; Kusunose, M. J. Biol. Chem. 1987, 262, 13366-13376.
411.Capdevila, J.H.; Wei, S.; Yan, J.; Karara, A.; Jacobson, H.R.; Falck, J.R.; Guengerich, F.P.; DuBois, R.N. J. Biol. Chem. 1992, 267, 21720-21726.
412.VanRollins, M.; Frade, P.E.; Carretero, O.A. Biochim. Biophys. Acta 1988, 966, 133-149.
413.VanRollins, M. Lipids, 1990, 25, 481-490.
414.McGiff, J.C. Annu. Rev. Pharmacol. Toxicol. 1991, 31, 339-369.
415.Laethem, R.M.; Koop, D.R. Mol. Pharmacol. 1992, 42, 958-963.
416.Bradfield, J.Y.; Lee, Y.-H.; Keeley, L.L. Proc. Natl. Acad. Sci. USA 1991, 88, 4558-4562.
417.Spector, A.A.; Gordon, J.A.; Moore, S.A. Prog. Lipid Res. 1988, 27, 271-323.
416.Needleman, P.; Turk, J.; Jakschik, B.A.; Morrison, A.R.; Leikowith, J.B. Annu. Rev.Biochem. 1986, 55, 69-102.
417.Sardesai, V.M. J. Nutr. Biochem. 1992, 3, 562-579.
418.Schewe, T.; Rapoport, S.M.; Kuhn, H. Adv. Enzymol. 1986, 58, 191-271.
419.Salafsky, B.; Wang, Y.-S.; Kevin, M.B.; Hill, H.; Fusco, A.C. J. Parasitol. 1984, 70, 584-591.
420.Salafsky, B.; Fusco, A.C. Exp. Parasitol. 1985, 60, 73-81.
421.Leid, R.W.; McConnell, L.A. Clin. Immunol. Immunopathol. 1983, 28, 67-76.
422.Meijer, L.; Maclouf, J.; Bryant, R.W. Dev. Biol. 1986, 114, 22-33.
423.Yamamoto, S. Biochim. Biophys. Acta 1992, 1128, 117-131.
424.Greene, D.H.S.; Selivonchick, D.P. Prog. Lipid Res. 1987, 26,53-85.
425.Henderson, R.J.; Tocher, D.R. Prog. Lipid Res. 1987, 26, 281-347.
426.Brash, A.R.; Hughes, M.A.; Hawkins, D.J.; Boeglin, W.E.; Song, W.-C.; Meijer, L. J. Biol.Chem. 1991, 266, 22926-22931.
427.German, J.B.; Berger, R. Lipids, 1990, 25, 849-853.

428. Al-Hassan, J.M.; Afzal, M.; Ali, M.; Thomson, M.; Fatima, T.; Fayad, S.; Criddle, R.S. Comp. Biochem. Physiol. 1986, 85B, 41-47.
429. Ragab, A.; Bitsch, C.; Thomas, J.M.F.; Bitsch, J.; Chap, H. Insect Biochem. 1987, 17, 863-870.
430. Veldink, G.A.; Vliegenthart, J.F.G. *In* Studies in Natural Products Chemistry, Volume 9; Atta-ur-Rahman, Editor; Elsevier: Amsterdam, 1991, pp. 559-589.
431. Smith, W.L.; Marnett, L.J. Biochim. Biophys. Acta 1991, 1083, 1-17.
432. Wakayama, E.J.; Dillwith, J.W.; Blomquist, G.J. Insect Biochem. 1986, 16, 895-902.
433. Lange, A.B. Insect Biochem. 1984, 14, 551-556.
434. Stanley-Samuelson, D.W.; Loher, W. Ann. Entomol. Soc. Amer. 1986, 79, 841-853.
435. Brady, U.E. Insect Biochem. 1983, 13, 443-451.
436. Andersen, N.H.; Hartzell, C.J.; De, B. Adv. Prostagland. Thromboxane Leukotriene Res. 1985, 14, 1-43.
437. Guilette, L.E., Jr. *In* Progress in Comparative Endocrinology; A. Epple, C. Scanes, M. Stetson, Editors; Wiley-Liss: New York, 1990, pp. 603.
438. Gobbetti, A.; Zerani, M.; DiFiore, M.M.; Botte, V. Prostaglandins 1993, 45, 159-166.
439. Fukushima, M. Eicosanoids, 1990, 3, 189-199.
440. Nelson, D.R.; Strobel, H.W. Mol. Biol. Evol. 1987, 4, 572-593.
441. Sariaslani, F.S. Adv. Appl. Microbiol. 1991, 36, 133-177.
442. Black, S.D.; Coon, M.J. Adv. Enzymol. 1987, 60, 35-87.
443. Guengerich, F.P. Prog. Drug Metab. 1987, 10, 1-54.
444. Ortiz de Montellano, P.R. (Editor) Cytochrome P450 - Structure, Mechanism, and Biochemistry; Plenum: New York, 1986.
445. Nebert, D.W. Mol. Endocrinol. 1991, 5, 1203-1214.
446. Nebert, D.W.; Gonzalez, F.J. Annu. Rev. Biochem. 1987, 56, 945-993.
447. Capdevila, J.H.; Falck, J.R.; Estabrook, R.W. FASEB J. 1992, 6, 731-736.
448. Karara, A.; Makita, K; Jacobson, H.R.; Flack, J.R.; Guengerich, F.P.; DuBois, R.N.; Capdevila, J.H. J. Biol. Chem. 1993, 268, 13565-13570.
449. Hu, S.; Kim, H.S. Eur. J. Pharm. 1993, 230, 215-221.
450. Kikuta, Y.; Kusnose, E.; Endo, K.; Yanmamoto, S.; Sogawa, K.; Fujii-Kuriyama, Y.; Kusunose, M. J. Biol. Chem. 1993, 268, 9376-9380.
451. Song, W.-C.; Brash, A.R. Science 1991, 253, 781-784.
452. Brash, R.R.; Baertschi, S.W.; Harris, T.M. J. Biol.Chem. 1990, 265, 6705-6712.
453. Brash, A.R.; Hughes, M.A.; Hawkins, D.J.; Boeglin, W.E.; Song, W.-C.; Meijer, L. J. Biol.Chem. 1991, 266, 22926-22931.
454. Song, W.-C.; Baertschi, S.W.; Boeglin, W.E.; Harris, T.M.; Brash, A.R. J. Biol. Chem. 1993, 268, 6293-6298.
455. Rabionovitch-Chable, H.; Cook-Moreau, J.; Brweton, J.-C.; Rigaud, M. Biochem. Biophys. Res. Commun. 1992, 188, 858-864.
456. Hamberg, M.; Gardner, H.W. Biochim. Biophys. Acta 1992, 1165, 1-18.
457. Zimmerman, D.C.; Feng, P. Lipids 1978, 13, 313-316.
458. Zimmerman, D.C.; Vick, B.A. Plant Physiol. 1970, 46, 445-453.
459. Vick, B.A.; Zimmerman, D.C. Plant Physiol. 1982, 69, 1103-1108.

460. Grimes, H.D.; Koetje, D.S.; Franceschi, V.R. Plant Physiol. 1992, 100, 433-443.

461. Ricknauer, M.; Bottin, A.; Esquerre-Tugaye, M.-T. Plant Physiol. Biochem. 1992, 30, 579-584.

462. Yamagishi, K.; Mitsomuri, C.; Takahashi, K.; Fujino, K.; Koda, Y.; Kikuts, Y. Plant Mol. Biol. 1993, 21, 539-541.

463. Nojiri, H.; Yamane, H.; Seto, H.; Yamaguchi, I.; Murofushi, N.; Yoshihara, T.; Shibaoka, H. Plant Cell Physiol. 1992, 33, 1225-1231.

464. Croft, K.P.C.; Juttner, F.; Slusarenko, A.J. Plant Physiol. 1993, 101, 13-24.

465. Corey, E.J.; Washburn, W.N.; Chen, J.C. J. Am. Chem. Soc. 1973, 95, 2054-2055.

466. Corey, E.J.; Ensley, H.E.; Hamberg, M.; Samuelsson, B. J. Chem. Soc. Chem. Commun. 1975, 277-278.

467. Corey, E.J.; d'Alarco, M.; Matsuda, S.P.T.; Lansbury, P.T., Jr.; Yamada, Y. J. Am. Chem.Soc. 1987, 109, 289-290.

468. Gonzalez, F.J. Pharmacol. Rev. 1988, 40, 243-288.

469. Cohen, M.B.; Schuler, M.A.; Berenbaum, M.R. Proc. Natl. Acad. Sci. USA 1992, 89, 10920-10924.

470. Gonzalez, F.J.; Nebert, D.W. Trends Genet. 1990, 6, 182-186.

471. Halpert, J.R.; He, Y.-a. J. Biol. Chem. 1993, 268, 4453-4457.

472. Hsu, M.-H.; Griffin, K.J.; Wang, Y.; Kemper, B.; Johnson, E.F. J. Biol. Chem. 1993, 268, 6939-6944.

473. Lindberg, R.L.P.; Negishi, M. Nature, 1989, 339, 632-634.

474. Kronbach, T.; Kemper, B.; Johnson, E.F. Biochem. 1991, 30, 6097-6102.

475. Hauge, S.M.; Aitkenhead, W. J. Biol. Chem. 1931, 93, 657-665.

476. Theorell, H.; Holman, R.T.; Akeson, A. Acta Chem. Scand. 1947, 1, 571-576.

477. Hamberg, M.; Samuelsson, B. J. Biol. Chem. 1967, 242, 5329-5335.

478. Hamberg, M.; Samuelsson, B, Proc. Natl. Acad. Sci. USA 1967, 71, 3400-3404.

479. Bernart, M.W.; Whatley, G.G.; Gerwick, W.H. J. Nat. Prod. 1993, 56, 245-259.

480. Gerwick, W.H.; Bernart, M.W. In Advances in Marine Biotechnology: Pharmaceutical and Bioactive Natural Products; O.R. Zaborsky, D.H. Attaway, D.H., Editors; Plenum Press: New York, 1992, pp. 101-152.

481. German, J.B.; Kinsella, J.E. Biochim. Biophys. Acta 1986, 879, 378-387.

482. Al-Hassan, J.M.; Thomson, M.; Ali, M.; Fayad, S.; Elkhawad, A.; Thulesius, O.; Criddle, R.S. Toxicon 1986, 24, 1009-1014.

483. Galliard, T. In Biochemistry of Wounded Plant Tissues; G. Kahl, Editor; Walter de Gruyter: Berlin, 1978, pp. 155-201.

484. Kato, T.; Yamaguchi, Y.; Namai, T.; Hirukawa, T. Biosci. Biotech. Biochem. 1993, 57, 283-287.

485. Namai, T.; Kato, T.; Yamaguchi, Y.; Hirukama, T. Biosci. Biotech. Biochem. 1993, 57, 611-613.

486. Ohta, H.; Shida, K.; Peng, Y.-L., Furusawa, I. Plant Physiol. 1991, 97, 94-98.

487. Bostock, R.M.; Stermer, B.A. Annu. Rev. Phytopathol. 1989, 27, 343-371.

488. Bostock, R.M.; Kuc, J.A.; Laine, R.A. Science 1982, 212, 67-69.

489. Jefford, C.W.; Cadby, P.A. Prog. Chem. Organic Natural Prod. 1981, 40, 191-265.

490.Lund, E.; Diczfalusy, U.; Bjorkhem, I. J. Biol. Chem.1992, 267, 12462-12467.
491.Serhan, C.N. J. Bioenerg. Biomembr. 1991, 23, 105-122.
492.Carroll, M.A.; Balazy, M.; Margiotta, P.; Falck, J.R.; McGiff, J.C. J. Biol. Chem. 1993, 268, 12260-12266.
493.Serhan, C.N. J. Lab. Clin. Med. 1993, 121, 372-374.
494.Peters-Golden, M.; Feyssa, A. Am. J. Physiol. 1993, 264, L438-L447.
495.Marcus, A.J. Prog. Hemost. Thromb. 1993, 8, 127-142.
496.Hughes, M.A.; Brash, A.R. Biochim. Biophys. Acta 1991, 1081, 347-354.
497.Bundy, G.L.; Nidy, E.G.; Epps, D.E.; Mizsak, S.A.; Wnuk, R.J. J.Biol. Chem. 1986, 261, 747-751.
498.Corey, E.J.; Matsuda, S.P.T.; Nagata, R.M.; Cleaver, M.B. Tetrahedron Lett. 1988, 29, 2555-2558.
499.Toh, H.; Yokoyama, C.; Tanabe, T.; Yoshimoto, T.; Yamamoto, S. Prostaglandins 1992, 44, 291-315.
500.Toh, H.; Urade, Y.; Tanabe, T. Mediators Inflam. 1992, 1 223-233.
501.Vickers, P.J.; O-Neill, G.P.; Mancini, J.A.; Charleson, S.; Abramovitz, M. Mol. Pharm. 1992, 42, 1014-1019.
502.Smith, W.L.; Marnett, L.J. Biochim. Biophys. Acta 1991, 1083, 1-17.
503.Smith, W.L. Biochem. J. 1989, 259, 315-324.
504.Johnson, R.W.; Curtis, S.E.; Dantzer, R.; Kelley, K.W. Physiol. Behavior 1993, 53, 127-131.
505.Dubois, D.H.; Guillette, L.J., Jr. J. Exp. Zool. 1992, 264, 253-260.
506.Wade, M.G.; Van der Kraak, G. J. Exp. Zool. 1993, 266, 108-115.
507.Smith, W.L.; DeWitt, D.L. J. Biol. Chem. 1993, 268, 6610-6614.
508.Meade, E.A.Fletcher, B.S.; Kujubu, D.A.; Perrin, D.M.; Herschman, H.R. J. Biol. Chem. 1991, 267, 4338-4344.
509.Xie, W.; Chipman, J.G.; Robertson, D.L.; Erikson, R.L.; SImmons, D.L. Proc. Natl. Acad. Sci. USA 1991, 88, 2692-2696.
510.Sirois, J.; Richards, J.S. J. Biol. Chem. 1992, 267, 6382-6388.
511.Yang, P.; Vauterin, L.; Vancanneyt, M.; Swings, J.; Kersters, K. System. Appl. Microbiol. 1993, 16, 47-71.
512.Miersch, O.; Bruckner, B.; Schmidt, J.; Sembdner, G. Phytochem. 1992, 31, 3835-3837.
513.Fang, J.; Comet, P.A.; Brooks, J.M.; Wade, T.L. Comp. Biochem. Physiol. 1993, 104B, 287-291.
514.Jareonkitmongkol, S.; Shimizu, S.; Yamada, H. Biochim. Biophys. Acta 1993, 1167, 137-141.
515.Moore, B.S.; Poralla, K.; Floss, H.G. J. Am. Chem. Soc. 1993, 115, 5267-5274.
516.Watanabe, T.; Haeggstrom, J.Z. Biochem. Biophys. Res. Commun. 1993, 192, 1023-1029.
517.Smith, D.A.; Jones, B.C. Biochem. Pharmacol. 1992, 44, 2089-2098.
518.Boyington, J.C.; Gaffney, B.J.; Amzel, L.M. Science 1993, 260, 1482-1486.

RECEIVED May 2, 1994

Chapter 10

Evolution of Lignan and Neolignan Biochemical Pathways

Norman G. Lewis and Laurence B. Davin

Institute of Biological Chemistry, Washington State University,
Pullman, WA 99164-6340

Lignans and neolignans are present in pteridophytes, gymnosperms and angiosperms, all of which are putatively derived from simple algal precursors. This progressive evolution from an algal precursor to terrestrial vascular plants required elaboration of the phenylpropanoid pathway. Some lignans and neolignans participate in lignin synthesis and hence have important roles in determining physicochemical and mechanical properties of cell walls, whereas others act as antioxidants, biocides (fungicides, bactericides and antiviral agents) and perhaps even as cytokinins.

Based on structural patterns known to date, the simplest lignans/neolignans are in the pteridophytes. As evolution proceeded to gymnosperms and angiosperms, this was accompanied by a progressive increase in the structural complexity of the lignans/neolignans, which apparently peaked in the Magnoliiflorae. Growing evidence links these changes to altered (or improved) defense functions.

There are three modes of coupling mechanisms involved in lignan/neolignan and lignin syntheses in pteridophytes and woody gymnosperms/angiosperms, each having evolutionary significance: The first involves monolignol oxidations catalyzed by H_2O_2-dependent peroxidase(s). These initially catalyze formation of (neo)lignans which undergo further conversion to give lignins. Woody plants also contain O_2-requiring laccases which catalyze the coupling of coniferyl/sinapyl alcohols to give racemic (neo)lignans, and which are presumed to be subsequently converted into lignins via cooperative involvement with the peroxidases previously mentioned. By contrast, optically active lignans are formed by stereoselective coupling engendered by hitherto uncharacterized O_2-requiring oxidases.

0097–6156/94/0562–0202$12.50/0
© 1994 American Chemical Society

There are few examples of lignans/neolignans in the non-woody monocotyledons, except for the formation of dihydroxytruxillic and truxinic acids in certain superorders. These metabolites are apparently formed via photochemical dimerization rather than by enzymatically engendered phenolic coupling i.e., the stereoselective and non-stereoselective coupling enzymes tentatively appear to be absent in the monocotyledons. This decreased reliance upon a full extension of the phenylpropanoid pathway presumably prevents the formation of woody tissue. This, in turn, helps to explain why monocotyledons are generally more readily amenable to biodegradation.

Many plant biotechnological strategies are directed towards introducing fundamental changes in the properties of (woody) plant cell walls and their biopolymers. Alternatively, other approaches attempt to alter levels of specific constituents such as those imparting defense functions. The purposes of such manipulations are manifold: To optimize wood properties for lumber and paper production, to produce new biopolymers which are expected to be valuable as replacements for petroleum-derived polymers, to alter nutritional qualities of foodstuffs, to increase the production of pharmacologically important compounds, and to either introduce or enhance novel defense functions in plants (i.e., antioxidants, biocidal properties and the like). Such approaches, in fact, represent exploitation of ongoing evolutionary processes by calculated human intervention.

Of the biochemical processes occurring within vascular plants, the phenylpropanoid pathway (leading to lignans, neolignans and lignins) is arguably the most important for the successful land colonization by plants. Indeed, it is the changes in specific emphases of this pathway that lead to the fantastic variations in cell wall properties (e.g., woody versus non-woody tissues), as well as the different capabilities in defense functions which affect long-term health, viability and durability of all plant forms. The study of the evolution of this pathway uncovers the hitherto unrecognized importance of a number of phenylpropanoid coupling enzymes, the systematic exploitation of which will be harnessed to achieve the most compelling biotechnological goals. A discussion of our understanding of how the pathway evolved follows.

The points of emergence of specific plant biochemical pathways and their correlation with various stages of evolution have been the focus of many scientific studies. The approaches typically used involve correlating chemotaxonomical findings (i.e., metabolite occurrence) with perceived evolutionary relationships, and/or the analysis of ancient or fossilized plant material remains. The latter approach is limited since the metabolic products of interest (particularly low molecular weight compounds) have frequently been irretrievably lost during the passage of time. Consequently, in many instances, pathways can only be indirectly inferred, e.g., by the appearance of distinct morphological traits which suggest the presence of a particular biochemical pathway. An example would be the observation of something akin to water/nutrient conducting vessels in fossilized plant tissue suggestive of lignified cell walls.

With these cautionary remarks in mind, a discussion of the possible evolution of the lignan/neolignan pathway is presented. The approach taken *by necessity* addresses its emergence relative to that of the *overall* phenylpropanoid pathway from which it is derived. But, in order to place it in the correct perspective, a brief description of the current theories put forward for land colonization by plants is first required.

It is generally accepted that life began in an aquatic environment with the formation of unicellular organisms from which all other forms evolved (for examples see ref. *1* and *2*). How this occurred still remains a subject of intensive debate and uncertainty. Of such organisms, the green algal group, Charophyceae (which includes *Coleochaete, Nitella* and *Spirogyra*), has been proposed to be the ancestor of land plants. This assertion is based largely upon both ultrastructural considerations *(1)* and the relatively close homology of Group II introns in the $tRNA^{Ala}$ and $tRNA^{Ile}$ genes of chloroplast DNAs in algal and vascular plants *(2)*.

Unfortunately, the existing fossil record does not yet reveal (in any convincing manner) how evolutionary progression from algal to terrestrial vascular forms occurred *(3)*. Clearly, multicellular organisms needed to evolve possessing certain essential characteristics i.e., high water potential for active metabolism, mechanisms to provide an effective means of obtaining and transporting water and nutrients, the ability to minimize effects of sudden temperature and humidity changes (e.g., rapid desiccating environments) and cells able to withstand compressive forces. Plants also had to be able to withstand new challenges e.g., to UV radiation, wind, rain, pathogen attack, etc. This required the development of plants with altered cell wall characteristics, as well as an ability to form novel metabolic products to help counter these new challenges.

Niklas *(4)* proposed that the early bryophytes and tracheophytes separately evolved from an ancestral green algal precursor able to survive on land (= hemitracheophytes). He rationalized that they developed elongated cells (i.e., hemitracheids) through which water conduction occurred and, thus, possessed primitive features from which the present day hydroids (water-conducting cells in mosses) and tracheids evolved. The earliest indication (to our knowledge) that plants contained cell wall lignins came from nitrobenzene oxidation results of early Silurian plant fossils (i.e., *Eohostimella*) which apparently gave traces of *p*-hydroxybenzaldehyde and vanillin *(4, 5)*; these fossilized remains were proposed to be forerunners of early gymnosperms. From such limited data, it was concluded that lignin formation coincided with the appearance of vascular tissues in the late Silurian/early Devonian time frame *(4, 5)*.

Such interpretations must be approached with great caution: As Taylor *(3)* logically points out, the fossil record (from both structural and morphological perspectives) is far too incomplete to draw any firm conclusions. Indeed, the earliest land plant forms may have appeared in the Ordovician period and Taylor has proposed that *Cooksonia* plants should be given consideration as the forerunners of early vascular plant forms. But this too is open to question since many of the specimens examined apparently lack any primitive vascular system *(6)*. Given this state of confusion, it is clearly not yet possible to ascribe the appearance of lignin in developing plant forms to a particular time frame or plant type.

Regardless of how evolutionary progressions occurred, various specialized features were needed to secure some of the first advantages for successful land adaptation. This required either the development of new biochemical pathways, such as those leading to the phenylpropanoids and/or the selective amplification of pre-existing pathways.

Phenylpropanoid metabolism is initiated via deamination of the amino acids, Phe **1** and (in some instances) Tyr **2** *(7)*, these conversions being catalyzed by the enzymes, phenylalanine ammonia-lyase (PAL) and tyrosine ammonia-lyase (TAL), respectively. With one apparent exception, *Dunaliella (8)*, this pathway is absent in algae. Nevertheless, the essential absence of lignins (and related phenylpropanoids) in algae strongly implies that only those acquiring the pathway were able to make the transition to a terrestrial environment.

Thus, the acquisition of the phenylpropanoid pathway (in one form or another) can be rationalized as essential for land adaptation by plants. It is, however, unknown how and when the pathway first appeared and in what form. Lignins, the major metabolites, are required for *secondary* cell wall thickening processes, leading to formation of the elements suitable for water and nutrient conduction, structural support and wound-healing layers. Lignins are apparently absent from primary walls *(9)*. Instead these often contain cell-wall bound hydroxycinnamic acids (e.g., *p*-coumaric **4** and ferulic **9** acids), the couplings of which are putatively required to maintain wall integrity *(10, 11)*. But since hydroxycinnamic acid formation biochemically precedes synthesis of the lignin monomers, *p*-coumaryl **7**, coniferyl **12** and sinapyl **17** alcohols (see Figure 1), this type of reinforcing mechanism (if operative) may have *preceded* lignification. A similar line of reasoning can be made for suberin: it is a phenylpropanoid-acetate pathway polymeric product with important roles in maintaining water-diffusion resistant barriers and in wound-healing. Being partly hydroxycinnamate-derived (Bernards *et al.*, *J. Biol. Chem.*, in press), it may also have *preceded* lignification.

To confuse matters further, several dimeric phenylpropanoid (neo)lignans have been implicated in lignin synthesis *(12)* (see later). Others display potent antiviral *(13)*, antifungal *(14)*, bactericidal *(15)*, insecticidal *(16)* and putative cytokinin *(17, 18)* properties. Therefore, it is difficult at first inspection to determine whether phenylpropanoid metabolism initially evolved for water/nutrient conduction, structural support or defense. (Figure 1 shows the overall biochemical pathway to the different metabolic products from Phe **1**/Tyr **2**).

Whatever the case, the different phenylpropanoid and phenylpropanoid-acetate pathway metabolites were necessary for the continuing competitive survival of vascular plants. The purpose of this chapter, therefore, is to attempt to define how evolution of the pathway leading to lignans and neolignans occurred relative to their closely related metabolites. As will become apparent later, this question has an additional complication fairly unique to plant metabolism. The uniqueness is that there are at least *three* distinct enzyme types (oxidase, laccase and peroxidase) in plants capable of catalyzing the initial synthesis of specific lignans/neolignans which can then be further metabolized. Although these enzymes differ

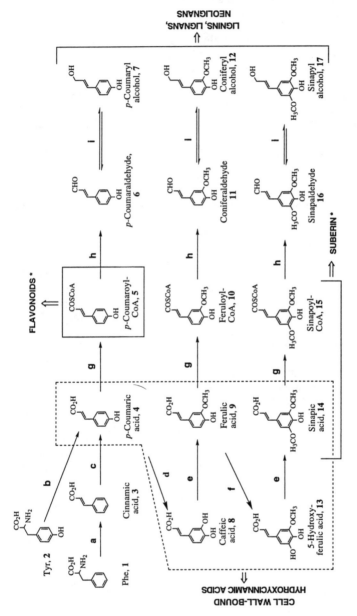

Figure 1. Proposed biosynthetic pathway to monolignols, and related metabolites. **a.** Phenylalanine ammonia-lyase, **b.** Tyrosine ammonia-lyase, **c.** Cinnamate-4-hydroxylase, **d.** p-Coumarate-3-hydroxylase, **e.** O-Methyltransferase, **f.** Ferulate-5-hydroxylase, **g.** Hydroxycinnamate-CoA ligase, **h.** Cinnamoyl-CoA:NADP oxidoreductase, **i.** Cinnamyl alcohol dehydrogenase. (* Note: plus acetate pathway for both flavonoids and suberin)

substantially in mode of action and stereoselectivity, products such as pinoresinol **18** are formed in each case. Clearly, each enzyme type must have been conscripted over different evolutionary time intervals, thereby making the issue of lignan/neolignan and lignin evolution even more challenging and provocative.

Nomenclature

Lignans and neolignans are typically found as dimeric phenylpropanoid [C_6C_3] derivatives *(19)*, although sesqui-, di-, tri-, and higher oligomeric forms also exist *(20-22)*. Typically they are found as single enantiomeric forms (the sign of which can vary from one plant species to another), but racemic products are also encountered [e.g., (±)-syringaresinols **19a/b** *(23)*]. Interestingly, in plants containing them, they are present in all different organs and tissues e.g., roots, shoots, seeds, flowers, needles, leaves, bark, heartwood, etc.

Lignans are usually defined as being constructed through 8.8' carbon-carbon linkages between the phenylpropanoid units *(19)*. They vary substantially in oxidation levels, degree of substitution, and structural complexity *(19, 24)*. Frequently they are subclassified into distinct groups, the most common of which are the dibenzylbutanes (e.g., secoisolariciresinol **20**, dihydroguaiaretic acid **21**), dibenzylbutyrolactones (e.g., arctigenin **22**, arctiin **23**, matairesinol **24**), tetrahydrofurans (e.g., lariciresinol **25**, liovil **26**, sylvone **27**), furanofurans (e.g., pinoresinol **18**, hydroxymedioresinol **28**), aryltetrahydronaphthalenes (e.g., cyclo-lariciresinol **29**), 2,7-cyclolignan-9'-olides (e.g., podophyllotoxin **30**), arylnaphthalenes (e.g., justicidin A **31**, taiwanin H **32**), dibenzocyclooctadienes (e.g., gomisin A **33**, kadsurin **34**), etc.

Neolignans are currently defined as being connected via linkages other than 8.8' *(25-27)*, e.g., 8.3' (e.g., futoquinol **35**, mirandin B **36**), 8.1' (e.g., megaphone **37**, guianin **38**), 8-O-4' (e.g., guaiacylglycerol-8-O-coniferyl alcohol ether **39**, surinamensin **40**, virolin **41**, eusiderin A **42**), 8.5' (e.g., dehydrodiconiferyl alcohol **43**), and 3-O-4' (e.g., obovatol **44**, isomagnolol **45**) linkages. But, this nomenclature has important shortcomings since : (i) Many trimers and higher oligomers contain 8.8' as well as other types of linkages [e.g., the trimer buddlenol D **46** from *Buddleia davidii (19)*] and, (ii) certain lignans and neolignans are formed via the same biochemical steps, e.g., pinoresinol **18** (a lignan) and dehydrodiconiferyl alcohol **43** (a neolignan) are both obtained following treatment of coniferyl alcohol **12** with either laccase/O_2 *(28-31)* or peroxidase/H_2O_2 *(28, 31)*. These limitations may eventually lead to a drop in the use of the term, neolignans, with lignans consequently being retained for all types. [Another concern that also needs to be addressed is the adoption of a systematic (IUPAC-approved) nomenclature for the various lignan skeleta; this is not addressed in this chapter since it is currently under review (O. R. Gottlieb, personal communication).]

Some of the simplest and most abundant lignans/neolignans have structures found as subunits within lignin polymers i.e., dehydrodiconiferyl alcohol **43**, pinoresinol **18**, syringaresinol **19**, and guaiacylglycerol-8-O-coniferyl alcohol ether **39** substructures. Surprisingly there is no clear line of demarcation between what constitutes oligomeric lignans/neolignans and

18a: R = H, (+)-Pinoresinol
19a: R = OCH₃, (+)-Syringaresinol

18b: R = H, (-)-Pinoresinol
19b: R = OCH₃, (-)-Syringaresinol

20a: (+)-Secoisolariciresinol

20b: (-)-Secoisolariciresinol

21: Dihydroguaiaretic acid

22: R = H, Arctigenin
23: R= Glc, Arctiin

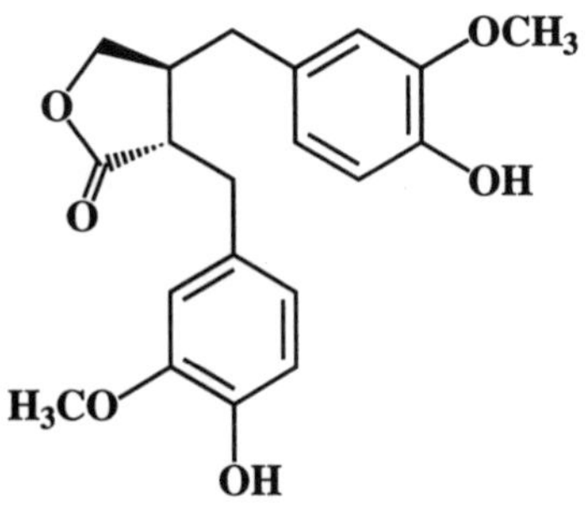

24a: (+)-Matairesinol

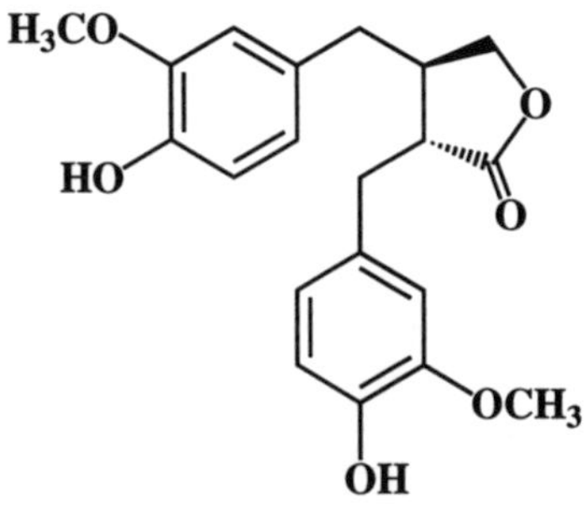

24b: (-)-Matairesinol

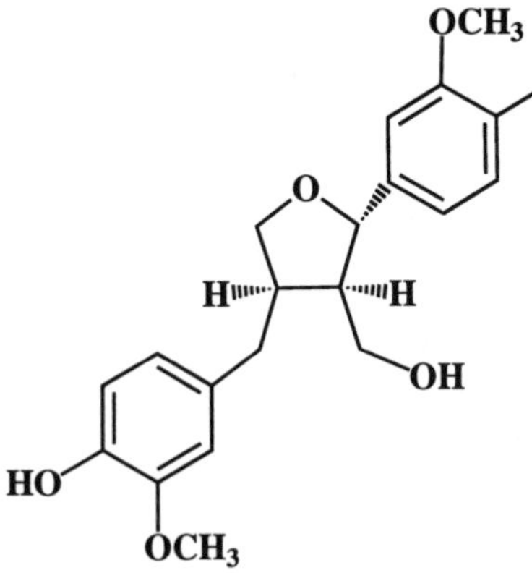

25a: (+)-Lariciresinol

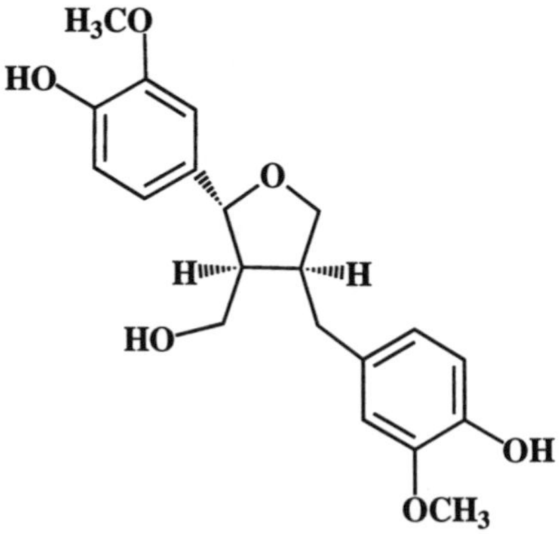

25b: (-)-Lariciresinol

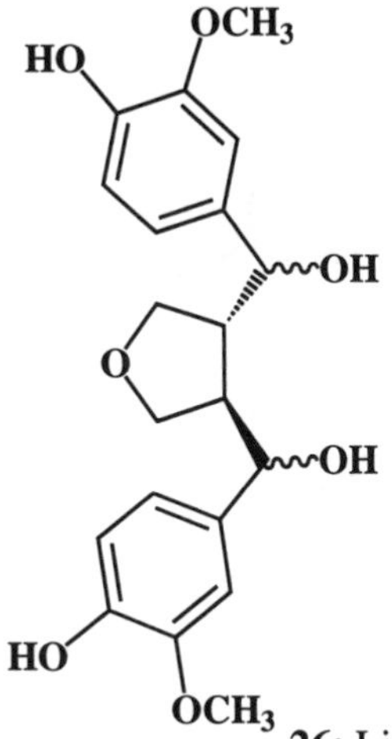

26: Liovil

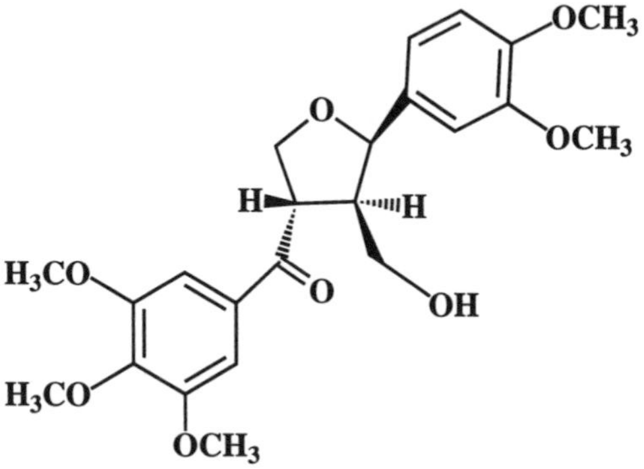

27: Sylvone

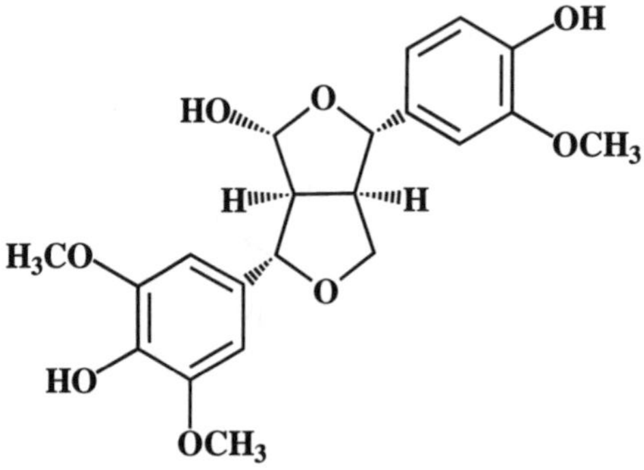

28: 9-Hydroxymedioresinol

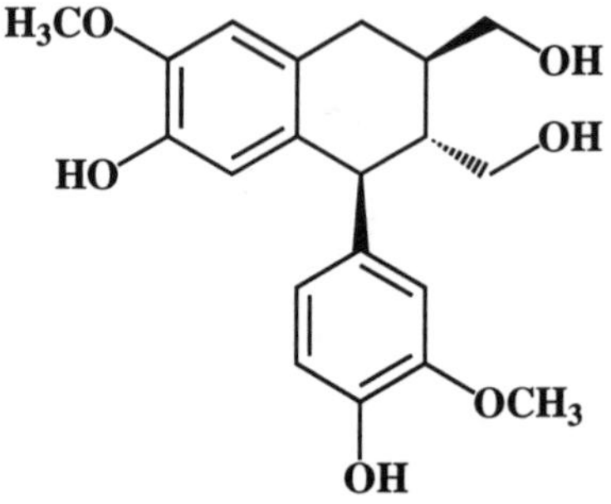

29: Cyclolariciresinol

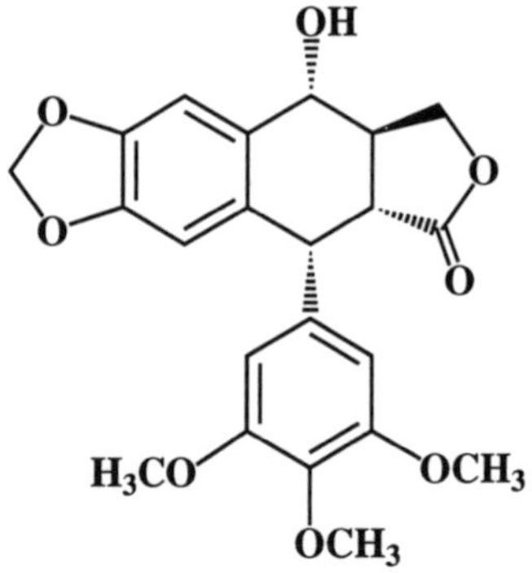

30: Podophyllotoxin

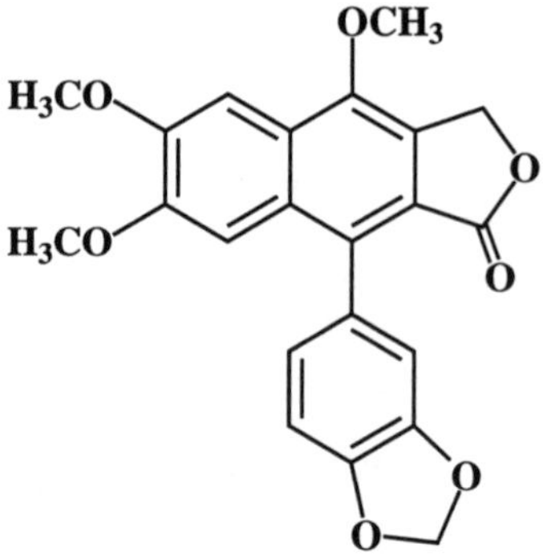

31: Justicidin A

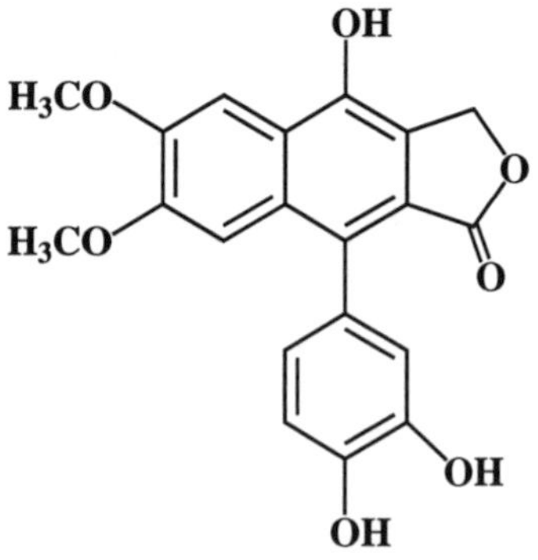

32: Taiwanin H

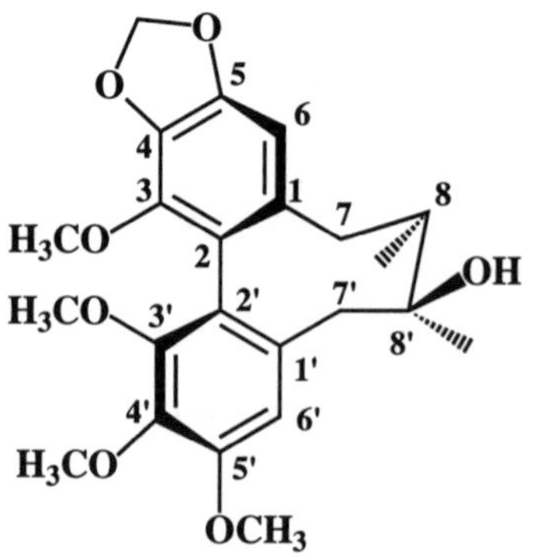

33: Gomisin A

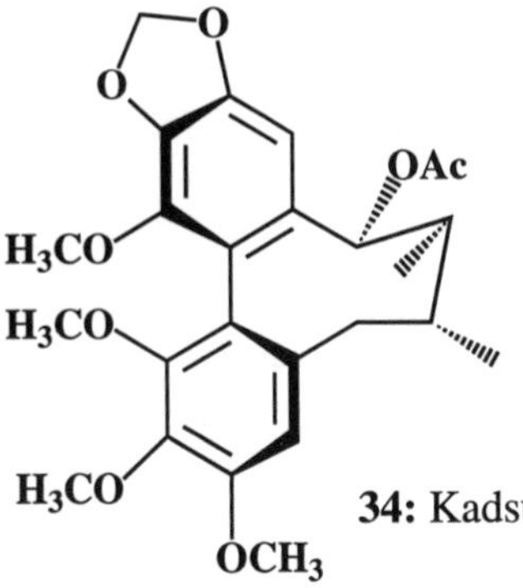

34: Kadsurin

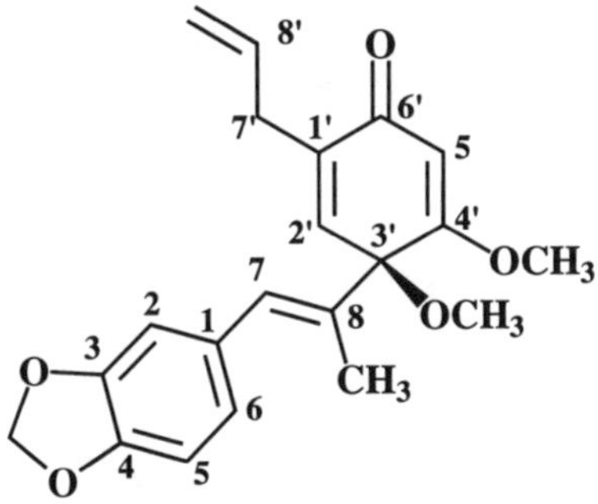

35: Futoquinol

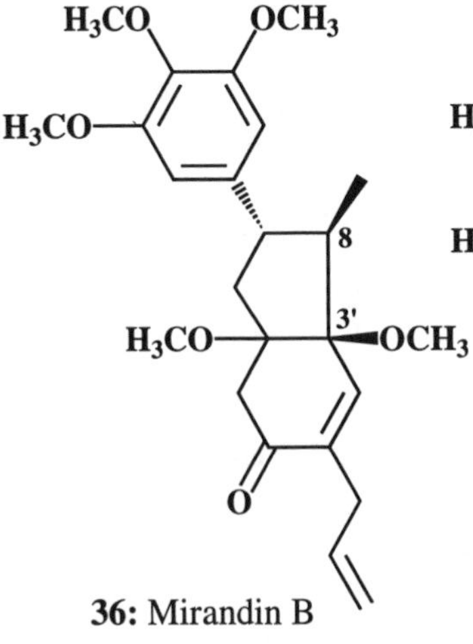

36: Mirandin B

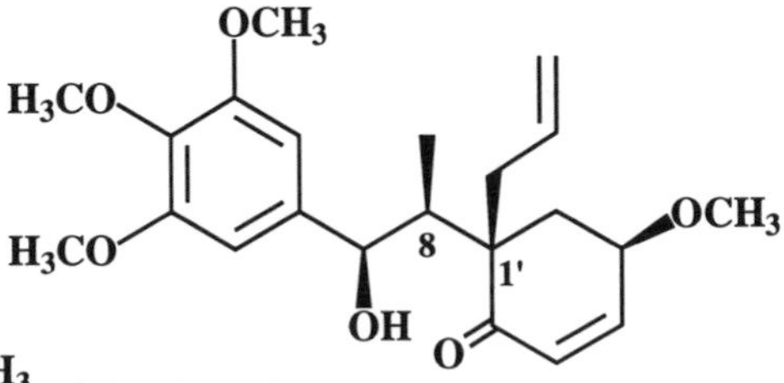

37: Megaphone

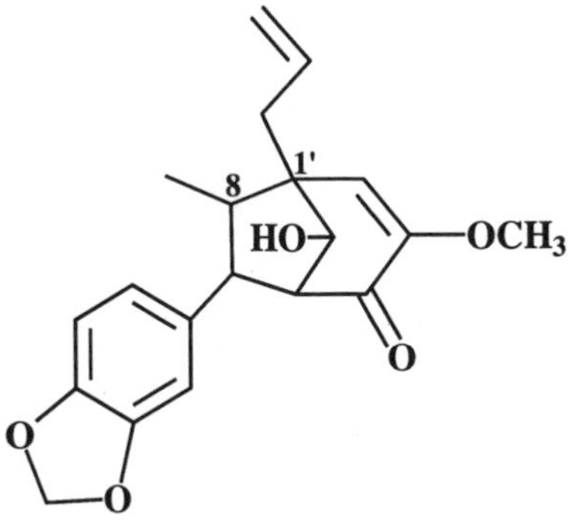

38: Guianin

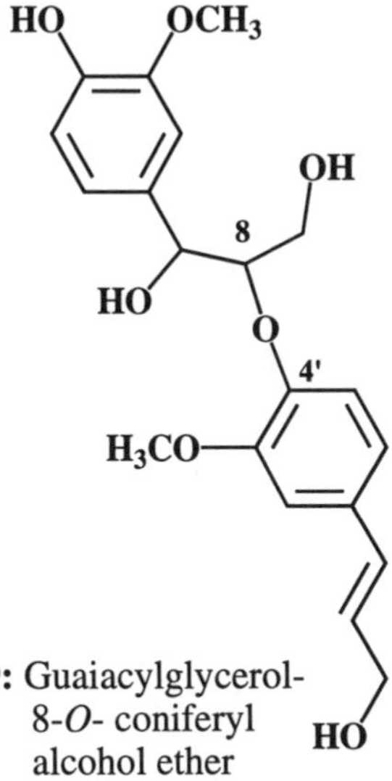

39: Guaiacylglycerol-
8-*O*- coniferyl
alcohol ether

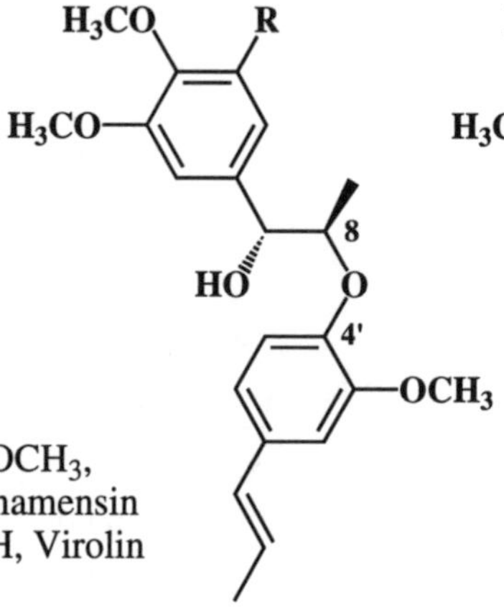

40: R = OCH₃,
Surinamensin
41: R = H, Virolin

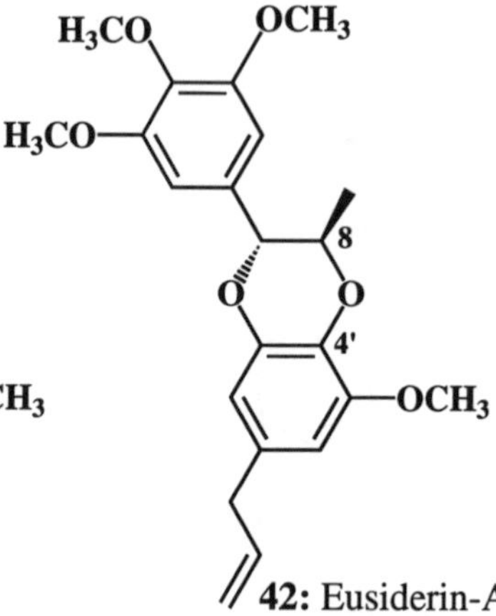

42: Eusiderin-A

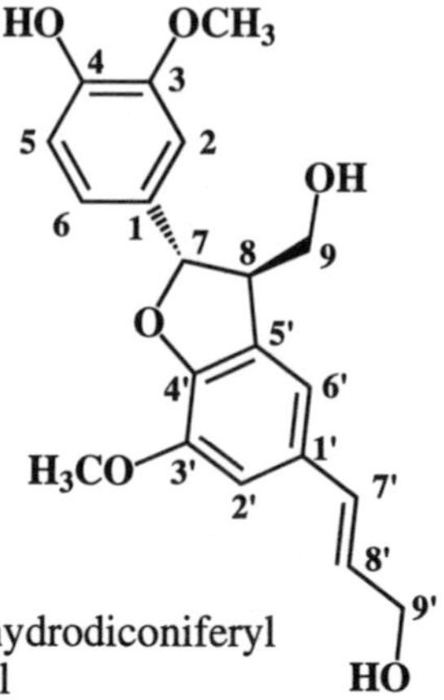

43a: (+)-Dehydrodiconiferyl
alcohol

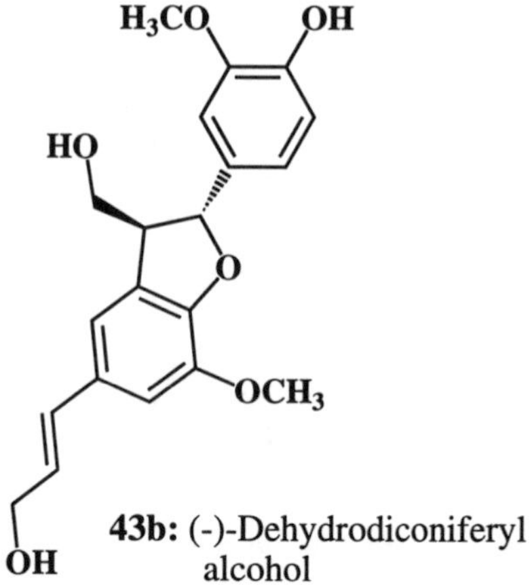

43b: (-)-Dehydrodiconiferyl
alcohol

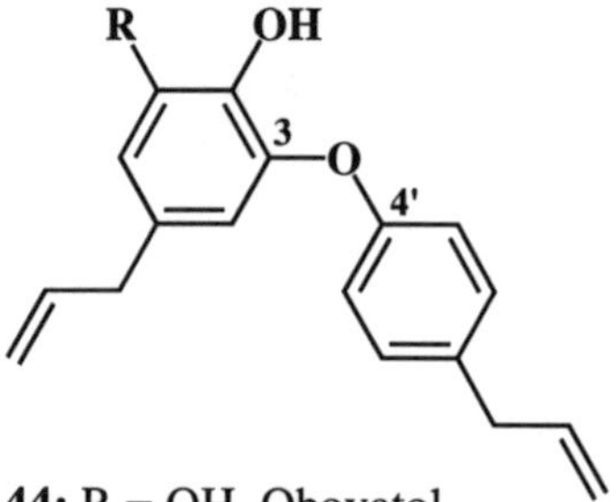

44: R = OH, Obovatol
45: R = H, Isomagnolol

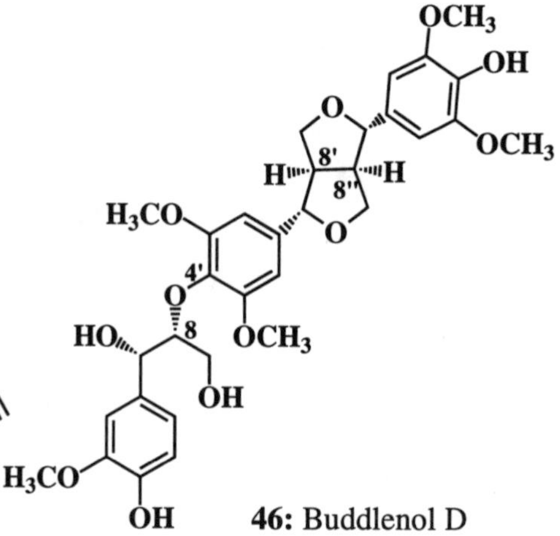

46: Buddlenol D

lignins in terms of either molecular size or structure. Lignins are typically classified as guaiacyl and guaiacyl-syringyl cell wall polymers: Although exceptions occur, gymnosperm lignins are normally built up mainly from coniferyl alcohol **12** together with smaller amounts of *p*-coumaryl alcohol **7** (so-called guaiacyl lignins), whereas with angiosperms, sinapyl alcohol **17** is also a major component (so-called guaiacyl-syringyl lignins). As an added complication, in the grasses, hydroxycinnamic acids can also be incorporated into the lignin polymer.   Lignins are generally considered to be racemic *(32)*, being formed via monolignol polymerization involving either H_2O_2-dependent peroxidases and/or O_2-laccase catalyzed transformations. Depending upon the species, it is proposed that the laccase(s) catalyze the initial formation of lignans such as (±)-pinoresinols **18a/b**, (±)-dehydrodiconiferyl alcohols **43a/b**, (±)-guaiacylglycerol-8-*O*-coniferyl alcohols **39** and (±)-syringaresinols **20a/b** which then undergo polymerization proper via H_2O_2-dependent peroxidase catalyzed transformations. Hence, lignans/neolignans can be viewed as involved in lignin synthesis, even if only as transient species. [The overlap between lignan and lignin biosynthetic pathways is discussed below].

Biosynthesis

Our understanding of lignan biosynthesis is in its infancy. Until the onset of our studies, essentially the only investigations had focused upon formation of 8.8'-linked metabolites, such as podophyllotoxin **30** *(33)* and arctigenin **22**/arctiin **23** *(34)*.   But these studies just extended to preliminary incorporation studies and did not reveal the enzymology involved or intermediates of the pathway. To overcome these limitations, we embarked upon delineating the intermediates/enzymes involved in the biosynthesis of the 8.8'-linked dibenzylbutyrolactone lignan, (-)-matairesinol **24b**, in *Forsythia intermedia* and that of (+)-pinoresinol **18a**, in *F. suspensa*. The first challenge was to identify the nature of the substrate undergoing coupling, and whether the coupling was stereoselective or not. [In this regard, all previously known phenylpropanoid coupling reactions were either O_2-laccase or H_2O_2-peroxidase catalyzed, but which only led to formation of (±)-racemic products.]

Following radiolabeling and time-course experiments using intact plants *(35)* and soluble protein/cell-wall residues *(36)* as needed, it was discovered that the "wall residues" from either *F. suspensa* or *F. intermedia* were capable of converting *E*-[8-^{14}C]coniferyl alcohol **12** into (±)-pinoresinols **18a/b**, with the (+)-antipode **18a** in enantiomeric excess (*ca* 68%). These transformations occurred without addition of exogenously supplied co-factors, and time course studies indicated that both stereoselective and non-stereoselective coupling enzymes were present. [By contrast, the soluble proteins only engendered typical peroxidase-catalyzed coupling of *E*-[8-^{14}C]coniferyl alcohol **12** in reactions requiring H_2O_2 as cofactor *(36)*.]

The next difficulty encountered was solubilization of the stereoselective/non-stereoselective enzyme(s) in the wall residue but this has since been overcome (Paré *et al.*, *Tetrahedron Lett.*, in press). Two enzymes with very distinct coupling modes have been obtained: The first is the stereoselective O_2-requiring protein which converts E-coniferyl alcohol **12** into (+)-pinoresinol **18a** (Figure 2), and hence is the first example of stereoselective coupling in phenylpropanoid metabolism (Paré *et al.*, *Tetrahedron Lett.*, in press) This is an important discovery since it is precisely this form of coupling that explains the preponderance of optical activity within the (neo)lignans. This transformation is presumed to occur via *si-si* coupling of two bound E-coniferyl alcohol **12** moieties as shown in Figure 2. We are currently purifying this enzyme to homogeneity in order to precisely define its mode of *unprecedented* coupling.

The second to be released from the wall residue was a non-stereoselective coupling enzyme, i.e., a laccase-like protein of M.W. ~ 100,000 (Davin *et al.*, *Phytochemistry*, in press). It catalyzes the conversion of E-coniferyl alcohol **12** into (±)-pinoresinols **18a/b** and (±)-dehydrodiconiferyl alcohols **43a/b**, and E-sinapyl alcohol **17** into (±)-syringaresinols **19a/b**; however, it does not utilize E-*p*-coumaryl alcohol **7** efficiently. It is postulated to be involved in the initial stages of lignin synthesis in woody plants, with polymer formation proper occurring via peroxidase-catalyzed transformations *(37)*. It is tempting to speculate that the laccase-catalyzed reaction may be one of the major reasons for woody tissue formation, since it is found in many angiosperm and gymnosperm families and appears to be wall-associated *(38)*. [The substrate specificity differences may also account for the preponderance of E-*p*-coumaryl alcohol **7**-derived moieties in middle lamellae and compression wood, in contrast to secondary walls which are E-coniferyl **12** and/or E-sinapyl **17** alcohols derived.] By contrast, non-woody plants such as monocotyledons *seem* to be generally devoid of laccase, and hence their lignins are considered to be formed via peroxidase-catalyzed transformations. Consequently, the distribution of laccase may be of profound evolutionary significance in forming woody plants.

Returning to the discovery of the stereoselective coupling enzyme (which engenders formation of (+)-pinoresinol **18a** from E-coniferyl alcohol **12**), we next investigated whether this step was of significance for formation of other lignan skeleta. Thus, attention was directed towards the formation of (-)-matairesinol **24b** in *F. intermedia*, where it was established to occur as shown in Figure 3, i.e., via a sequential NAD(P)H dependent reduction of (+)-pinoresinol **18a** to first afford (+)-lariciresinol **25a** and then (-)-secoisolariciresinol **20b**, respectively; subsequent dehydrogenation of the latter affords (-)-matairesinol **24b** *(39, 40)*. Importantly, the conversions were enantiospecific since the corresponding antipodes did not serve as substrates. The reductase catalyzing sequential formation of (+)-pinoresinol **18a** into (+)-lariciresinol **25a** and (-)-secoisolariciresinol **20b** has been purified to homogeneity; it has a M.W. ~40,000 and acts only on (+)-pinoresinol **18a** and (+)-lariciresinol **25a**. It is a type A reductase since only the pro-R hydride of NAD(P)H is abstracted *(41)*. To our knowledge, it is the *first example* of a benzylic ether reductase in plants. In an analogous manner, the NAD(P) conversion of (-)-secoisolariciresinol **20b** into (-)-matairesinol **24b** is also completely stereospecific, and only the (-)-enantiomer **20b** serves as substrate *(42, 43)*. Thus, an overall picture has emerged of a highly unusual stereoselective coupling step, which first affords the furanofuran skeleta (i.e., (+)-pinoresinol **18a**). Enantiospecific reduction can then follow affording entry into the furano (i.e., (+)-lariciresinol **25a**) and dibenzylbutane (i.e.,

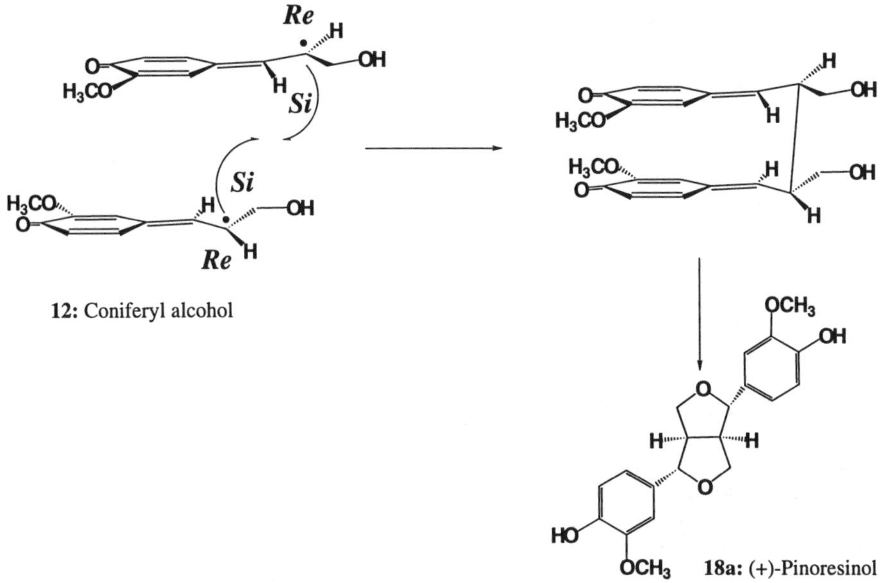

12: Coniferyl alcohol

18a: (+)-Pinoresinol

Figure 2. Proposed mechanism of (+)-pinoresinol **18a** formation in *Forsythia* sp.

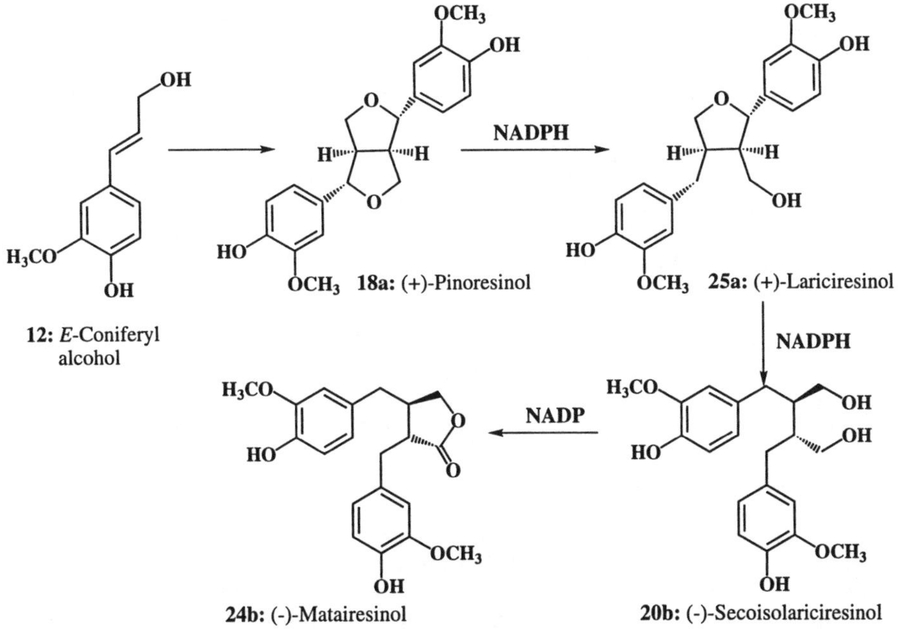

12: *E*-Coniferyl alcohol

18a: (+)-Pinoresinol

25a: (+)-Lariciresinol

24b: (-)-Matairesinol

20b: (-)-Secoisolariciresinol

Figure 3. Biosynthetic pathway from *E*-coniferyl alcohol **7** to (-)-matairesinol **24b** in *Forsythia intermedia*

(-)-secoisolariciresinol **20b**): Subsequent dehydrogenation yields the corresponding dibenzylbutyrolactone class, i.e., (-)-matairesinol **24b**.

Another interesting example of phenylpropanoid coupling was reported from *Bupleurum salicifolium (44)*. In this regard, an H_2O_2-dependent peroxidase was isolated, which converts either caffeic **8** or ferulic **9** acids, into 8.8'-bis (caffeic acid) and 8.8'-bis (ferulic acid), respectively. Although a role was proposed for these compounds in the synthesis of more elaborate arylnaphthalene lignan skeleta, this has not been established.

In summary, our studies have revealed that (+)-pinoresinol **18a** is a central intermediate in the biogenesis of furanofuran, furan, dibenzylbutane and dibenzylbutyrolactone lignans. Thus, an orderly progression of events leading to the various lignan/neolignan skeleta in vascular plants is now emerging.

Distribution in the Plant (and Animal) Kingdom

As implied earlier, the fossil record has not provided definitive information regarding the initiation and elaboration of the phenylpropanoid pathway leading to the lignins, lignans and neolignans. Consequently, it was more instructive, from an evolutionary perspective, to correlate and compare the accumulation of lignans and neolignans relative to that of extant plant forms (including lignin synthesizing types). This approach therefore, relies heavily upon chemotaxonomic findings and correlations with current perceived evolutionary relationships. We thus begin by first examining the evidence for a functional phenylpropanoid pathway in the algae, bryophytes (mosses, liverworts, hornworts and takakiophytes), fungi, and then extend the discussion to the pteridophytes (e.g., ferns), gymnosperms, angiosperms, and mammals.

Algae. In the introductory section, it was pointed out that there is only one known case of a functional biochemical machinery for the deamination of Phe **1** *(8)*. Consequently, the phenylpropanoid pathway (Figure 1) is considered to be generally absent from algae. But curiously there have been numerous reports claiming the existence of polymeric lignins in the brown algae, *Fucus vesiculosus (45)* and *Cystoseira barbata (46, 47)*, thereby implying an intact biochemical pathway to the monolignols and their polymeric product, lignin. But more detailed analyses revealed that these claims were incorrect *(48)*, and that the cell wall polymers were "1,3,5-trihydroxy phenolic-like" **47** and, hence, presumably acetate-derived. It has also been proposed that lignin occurs in *Staurastrum (49)* and *Coleochaete (50)* sp., but again the evidence obtained was neither definitive nor convincing. Hence any claims to date of lignin, lignan, and neolignan occurrence in algae are highly suspect.

Bryophytes. This includes the loosely and, perhaps imprecisely, arranged categories consisting of the Bryophyta (mosses), Hepatophyta (liverworts), Anthocerotophyta (hornworts), and Takakiophyta (takakiophytes) *(6)*. Although there have been protracted discussions regarding the presence of lignins in mosses *(51, 52)* a more careful and demanding examination of *Thamnobryum pandum, Rhizogonium parramatense, Leucobryum candidum, Sphagnum cristatum,* and *Dawsonia superba*, revealed that these claims were incorrect *(53)*. Instead, moss cell-wall constituents contain "1,3,5-trihydroxy phenolic-like" **47** polymers of unknown constitution *(53)*. Thus, there is no definitive basis to propose that either lignins, lignans, or neolignans are present in mosses.

47: Phloroglucinol   **48**: Caffeoyl-CoA   **49**: Malonyl-CoA

Some mosses do, however, synthesize flavone *C*- and *O*-glycosides, biflavonols, dihydroflavones, etc., *(54)* from *p*-coumaroyl-CoA **5** or caffeoyl-CoA **48** esters (and malonyl-CoA **49**). Thus, these organisms have evolved a partially developed phenylpropanoid pathway to afford metabolites presumably helping offset the effects of UV-irradiation, or providing cell wall polymers with properties "akin" to lignin. Related flavonoids are also found in the liverworts and hornworts *(54)*.

The hornworts (*Anthocerotae*) are thought to have originated during the Cretaceous period *(6)* **after** the introduction of mosses, pteridophytes, gymnosperms and some angiosperms. Interestingly, the hornworts, *Dendroceros japonicus, Megaceros flagellaris, Notothylas temperata,* and *Phaeoceros laevis* accumulate an unusual lactone, trivially called megacerotonic acid **50** ([α_D] = +233.0) *(55, 56)*. Its formation can be rationalized to occur via either oxidative coupling of *p*-coumaric **4** and caffeic **8** acids, or alternatively of two *p*-coumaric acid **4** residues with subsequent hydroxylation. According to current definition it is a lignan, since both precursors are C_6C_3 derived and it is 8.8'-carbon-carbon linked. It is tempting to speculate whether the enzyme catalyzing its formation is closely related to that engendering the formation of (+)-pinoresinol **18a** (discussed in Biosynthesis section). Another unusual metabolite found in *Anthoceros punctatus* is anthocerotonic acid **51** ([α_D] = +4.4), presumably formed by photochemical coupling of *p*-coumaric **4** and rosmarinic **52** acids *(55, 56)*. At this point of writing, such a compound would fall into the broad category of neolignans.

Fungi. The evidence for a partially complete phenylpropanoid-acetate pathway leading to cinnamic **3** and *p*-coumaric **4** acids, which then 'condense' with acetate-derived moieties to yield the styrylpyrones, hispidin **53**, *bis*-noryangonin **54**, hymenoquinone **55**, and its leuco derivative **56** has been reported in the genera *Gymnopilus, Polyporus, Pholiota, Phellinus,* and *Hymenochaeta* (Basidiomycetes) *(57)*. Interestingly, fruiting bodies of

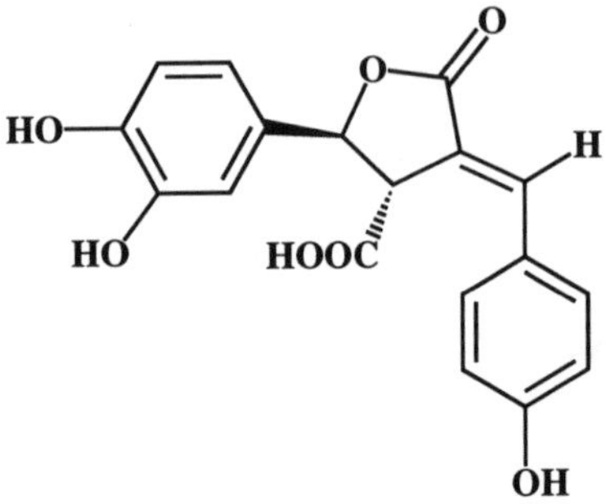

50: Megacerotonic acid

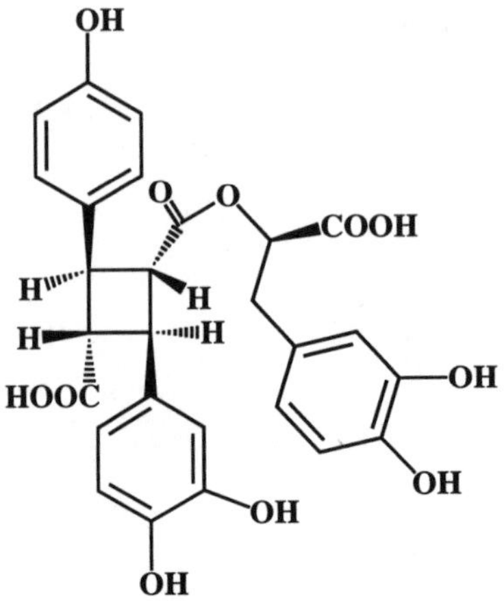

51: Anthocerotonic acid

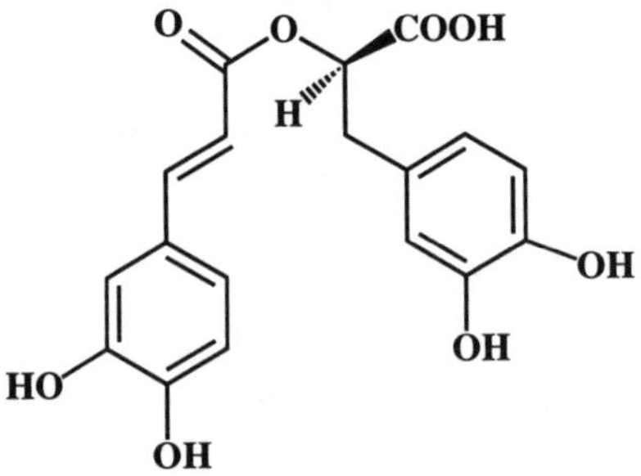

52: Rosmarinic acid

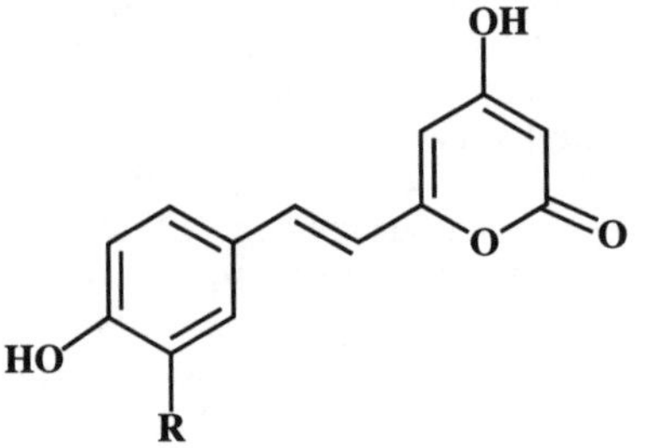

53: R = OH, Hispidin
54: R = H, *Bis*-noryangonin

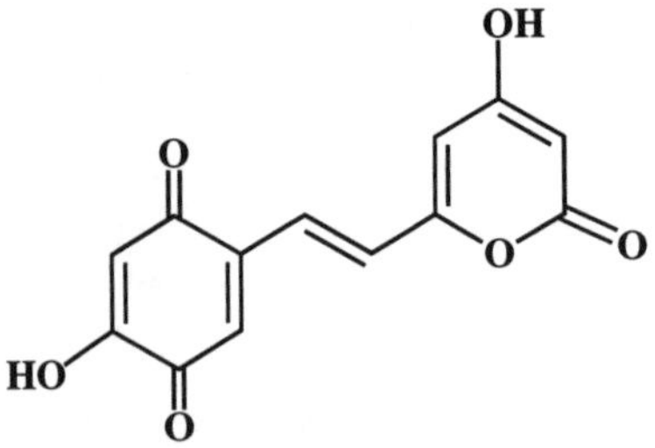

55: Hymenoquinone

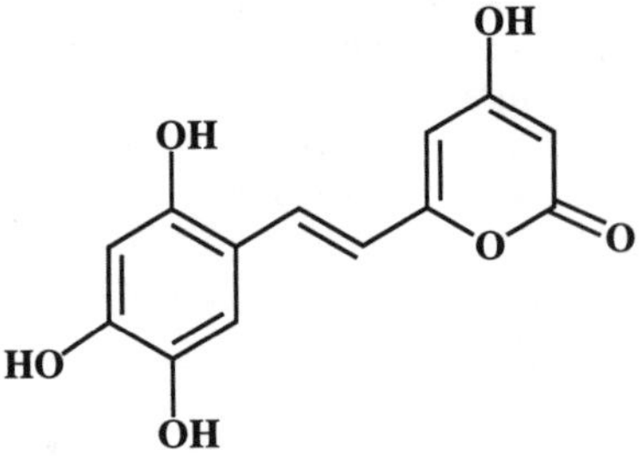

56: *Leuco*-hymenoquinone

Polyporus hispidus catalyze (via a phenol oxidase?) the conversion of hispidin **53** to a cell wall polymeric substance, considered to possess properties akin to lignin *(58, 59)*. Hispidin-like dimers have also been isolated from *Hypholoma fasciculare*, *Pholiota flammans* and *Phellinus pomaceus* *(57)*.

57: Methyl *p*-methoxy cinnamate   **58:** Veratric acid   **59:** Isoferulic acid

The fungi appear to have made an additional important biochemical development when compared to the algae and bryophytes i.e., the ability to *O*-methylate phenols, as in methyl *p*-methoxycinnamate **57** formation in *Lentinus lepideus (57, 60, 61)*, veratric acid **58** synthesis in *Phanerochaete chrysosporium (62)* and isoferulic acid **59** from caffeic acid **8** in *Lentinus lepideus (61)*. But a major evolutionary distinction between the fungi and lignifying higher plants apparently concerns the regiospecificity of *O*-methylation, i.e. although both can methylate caffeic acid **8**, only the higher plants seem able to synthesize ferulic acid **9** (via *O*-methylation of the 3-OH group). This regiospecificity of methylation may have been a crucial advance in the further development of the phenylpropanoid pathway in higher plants.

Pteridophytes. The Pteridophyta can be traced back to the Devonian period; e.g., the Cladoxylales (Devonian-Lower Carboniferous), the Iridopteridales (Middle-Upper Devonian), the Rhacophytales (Middle-Upper Devonian), the Stauropteridales (Devonian-Pennsylvanian), the Zygopteridales (Devonian-Permian) and Coenopteridales (Devonian-Permian). Of these, some members of the Rhacophytales share features with ferns and the progymnosperms e.g., as evidenced by the appearance of secondary xylem. But exactly how the evolution to the gymnosperms precisely occurred still awaits full clarification *(6)*.

In terms of chemotaxonomy, very little work has been carried out on the chemical constituents of the Filicopsida (ferns), Lycopsida (clubmosses), Psilotopsida and Equisetopsida (horsetails), although it is reported that they contain lignin *(32)*. In each case these conclusions were based upon preliminary results from nitrobenzene oxidations, e.g., it was reported that lignins of the Cyatheaceae (i.e., *Dicksonia squarrosa*, *Cibotium barometz* and *Cyathea arborea*) and Adiantaceae (i.e., *Pteris podophylla*) families were mainly derived from *p*-coumaryl **7** and coniferyl **12** alcohols, whereas in the

Dennstaedtiaceae (i.e.*Dennstaedtia bipinnata*), sinapyl alcohol **17** was also apparently involved *(63)*. Evidence for lignins has also been obtained for Lycopodiaceae, Selaginellaceae and Psilopsida *(32)*. Thus, with the appearance of the pteridophytes, three new biochemical features seem to have evolved. These were the position-specific methylation of caffeic acid **8** (or its CoA-derivative **48**) to give ferulic acid **9** (or feruloyl-CoA **10**), the synthesis of sinapic acid **14** (or its CoA-derivative **15**) and the facile stepwise reduction of the corresponding CoA-derivatives **5, 10, 15** to afford monolignols **7, 12** and **17**.

Based upon available evidence, therefore, pteridophytes appear to be the first organisms capable of the important position-specific methylation and reductive steps necessary for lignin synthesis and, hence, the earliest known cases of forming cell wall lignin polymers proper. Consequently, it is not surprising that lignans and neolignans are also found in pteridophytes, although to this point there have only been two reports both in ferns: The first, in 1978, cited the presence of the *optically active* lignans, dihydrodehydrodiconiferyl alcohol glucoside **60** (aglycone $[\alpha]_D$ = -8.5°) and lariciresinol-9-*O*-β-D-glucoside **61** (aglycone $[\alpha]_D$ = +15.7°) in *Pteris vittata (64)*. A more recent report in 1992 established the occurrence of optically active forms of the neolignans, blechnic **62** and 7-epiblechnic **63** acids ($[\alpha]_D$ = -28°, and -145°, respectively) in six Blechnaceous tree-ferns (e.g., *Blechnum orientale*) *(65)*.

In the absence of any biochemical data to date, two important conclusions can be tentatively drawn from these observations: (i) Blechnic acid **62** formation apparently occurs via stereospecific oxidative coupling of two achiral caffeic acid **8** moieties and presumably involves a novel stereoselective coupling enzyme of the kind described for (+)-pinoresinol **18a** synthesis in *Forsythia sp*; the enzymology of this (apparently) stereoselective process needs to be clarified. Blechnic acid **62** synthesis may therefore, represent an early evolutionary step to lignan formation proper in higher plants. [It is noteworthy that hydroxycinnamic acid-derived dimers of this type are not typically encountered in nature] and (ii) dihydrodehydrodiconiferyl alcohol **60** and lariciresinol **61** glucosides are presumably engendered by a different enzyme type, since essentially only the racemates are encountered. They are probably formed via oxidative racemic coupling of two molecules of coniferyl alcohol **12** to first afford (±)-dehydrodiconiferyl alcohols **43a/b** and (±)-pinoresinols **18a/b**, respectively, followed by subsequent reductive modifications and glucosylation, i.e., to afford racemic (±)-dihydrodehydrodiconiferyl alcohol glucosides **60** and (±)-lariciresinol-9-*O*-β-D-glucosides **61**. Based on this limited data, it can be proposed that the pteridophytes developed or acquired two coupling modes which differed substantially in stereoselectivity, as well as in the subsequent reductions to give **60** and **61**, respectively. Their biochemical mechanism of formation, however, awaits experimental verification. Additionally, much more needs to be known regarding the structural variations of lignans/neolignans in this class of plants.

Thus, with the ability of the pteridophytes to form all three monolignols, lignins, lignans and neolignans, the stage was now set for other plant forms to capitalize on these developments.

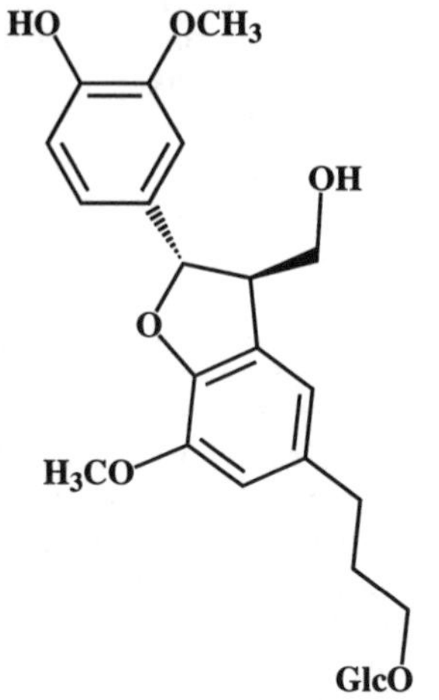

60: Dihydrodehydrodiconiferyl alcohol glucoside

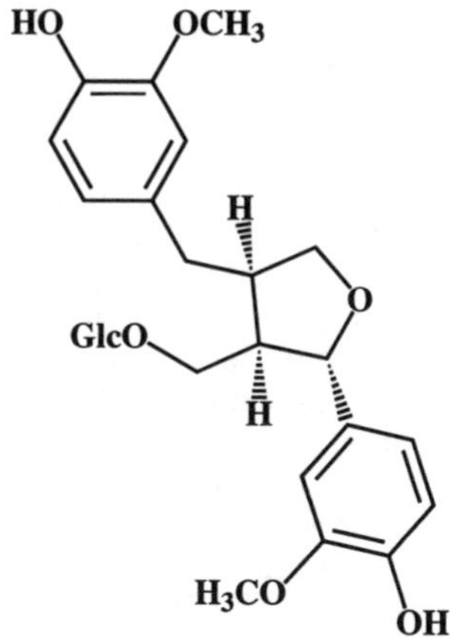

61: Lariciresinol-9-O-β-D-glucoside

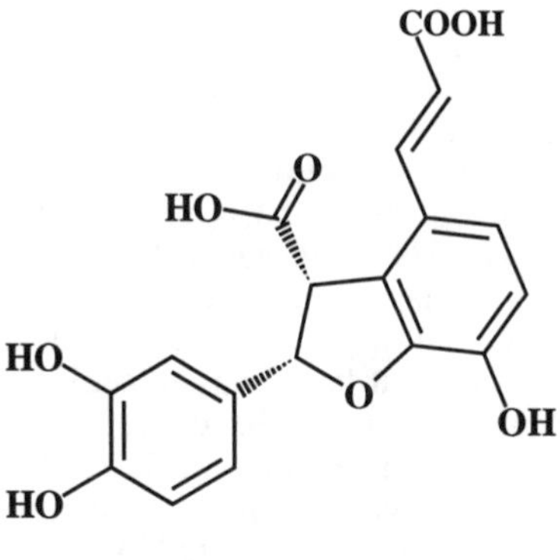

62: Blechnic acid

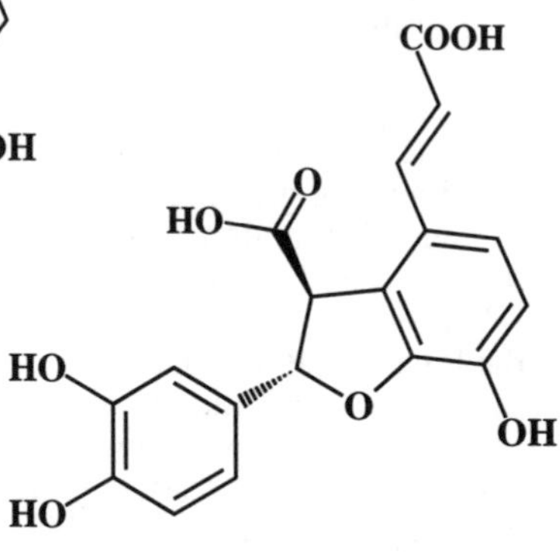

63: 7-Epiblechnic acid

Gymnosperms. Like the pteridophytes, the gymnosperms have a fully functional phenylpropanoid pathway to the monolignols and hence to the lignins and lignans *(32)*. In contrast to the ferns, however, there are numerous examples of lignans and neolignans in the gymnosperms, e.g., in the Araucariaceae, Cupressaceae, Ephedraceae, Podocarpaceae, Pinaceae, Taxaceae and Taxodiaceae families. Almost all are 8.8' linked dimers, with structures which reveal a significant elaboration of the biochemical repertoire from that observed to date in the pteridophytes. Before discussing the possible path of evolution of the (neo)lignan pathway in the gymnosperms, it needs to be emphasized that the 'conclusions' are based upon incomplete data and hence, may be subject to future modification.

As alluded to earlier, the precise origin (ancestors) and sequential progression (evolution) of the present day gymnosperms are quite obscure, although progymnosperms have been implicated *(6)*. This is primarily due to difficulties in finding intact specimens and much of what is known about evolution of conifers comes from analysis of ovulated cones rather than whole plants *(6)*. Nevertheless, based upon current analysis and interpretation of the fossil records to date, the Cycadales and Ginkgophytes apparently appeared in the late Paleozoic era, i.e., in the Carboniferous and Permian periods, respectively. The Podocarpaceae were first discovered in early Mesozoic (lower Triassic period) fossils, based upon the analysis of leafy twig, detached pollen and seed cones from presumed Podocarpaceae fossilized remains found in sediments in South Africa and Australia *(6, 66)*. The Araucariaceae have earliest fossils assigned to the Upper Triassic e.g., the secondary xylem of *Araucarioxylon* found in both the petrified forests of Arizona and the Cerro Cuadrado of Patagonia *(6)*. In the Taxodiaceae, fossiled remains assigned to *Margeriella cretacea* and *Sewardiodendron* have been discovered in the late Triassic and Jurassic sediments, respectively *(6)*. The Cupressaceae have been more difficult to place: old Mesozoic fossils attributed to this family (*Cupressinocladus*) have been reclassified as the extinct Cheirolepidiaceae *(67, 68)*. However, seed cone fossils believed to be of Cupressaceae origin have been found in Jurassic sediments *(69)*. The Pinaceae, which represent one of the largest modern conifer groups, also has obscure beginnings: Some of the oldest fossils, presumed to be Pinaceae in origin, are the ovulate cones and shoots of *Compsostrobus neotericus* found in late Triassic sediments in North Carolina *(70)*, whereas most others have been discovered in Upper Cretaceous sediments e.g., *Pinus haboroensis* from Hokkaido, Japan *(71)*. It is, therefore, generally thought that the Pinaceae were well established by the early Mesozoic. Lastly, the Taxaceae are well represented by ancestral remains of *Palaeotaxus* (vegetative and ovulate parts) which are considered to be from the lower Jurassic time frame *(72)*.

Chemotaxonomical studies of lignans in extant species have revealed interesting biochemical progressions. Surprisingly, they have not yet been found in the Cycadales (cycads or sagopalms) or Ginkgophytes, the fossilized ancestors of which appear in the oldest sediments (paleozoic) examined. However, the Podocarpaceae, *Dacrydium intermedium*, contains the unusual dibenzylfuran lignan, α-intermedianol **64** *(73)*, and (+)-*O*-methyl-α-conidendral **65** *(73, 74)*, the corresponding aryltetrahydronaphthalene derivative, whereas in *Podocarpus spicatus* both (-)-matairesinol **24b** (a dibenzylbutyrolactone) *(75)* and (-)-α-conidendrin **66** (a 2.7 cyclolignan-9'-

olide) *(76)* accumulate. Based upon previous findings in our laboratory, it can be tentatively proposed that these are derived from (+)-pinoresinol **18a**. Thus, in *Podocarpus spicatus* the same overall pathway to (-)-matairesinol **24b** presumably occurs except for the addition of an oxidative ring closure to afford the corresponding aryltetrahydronaphthalene skeleton. It is also tempting to speculate that this presumed enzymatic ring closure may utilize the corresponding 7-*O*-methyl derivative (e.g., as for **64/65**). In the Araucariaceae, e.g., *Araucaria angustifolia* (Parana pine), the biochemical repertoire is further extended as revealed by the accumulation of dimethylpinoresinol **67**, secoisolariciresinol **20**, lariciresinol **25**, cyclolariciresinol **28**, 4-*O*-methyl isolariciresinol **68**, its corresponding 4-*O*-methyl ether and (-)-galbulin **69** *(19, 24, 77)*. [All can be presumed to be derived from (+)-pinoresinol **18a**.] Interesting new biochemical additions have apparently been made which include *O*-methylation at the 4-OH positions (e.g., **64, 67-69**), oxidative ring closures, and an apparent reduction of the hydroxymethyl groups at $C_9/C_{9'}$ as in (-)-galbulin **69**. To our knowledge this is the only case in the gymnosperms where the $C_9/C_{9'}$ carbons are not oxygenated.

Further progressive extension of the biochemical pathway beyond matairesinol **24** seems to be operative in both the Taxodiaceae and the Cupressaceae families, i.e., as evidenced by the accumulation of sventenin **70** *(78)*, taiwanin H **32** *(19)*, taiwanin C **71** *(78)* and diphyllin **72** *(19)* in *Taiwania cryptomerioides* (Taxodiaceae) and, in the Cupressaceae, of dihydroxythujaplicatin **73** [*Thuja plicata (79)*], (-)-yatein **74** [*Libocedrus yateensis (80)*], deoxypodophyllotoxin **75** [*Libocedrus bidwilii (81)* and *Juniperus phoenica (82)*], (-)-β-peltatin **76** [*Juniperus bermudiana (25)*] and austrobailignan-2 **77** [in *Calocedrus formosa* and *Juniperus sabina (27)*]. In these cases, several new features have also been added such as different hydroxylation patterns (e.g., **72-77**), introduction of methylenedioxy groups (e.g., **70**), and formation of the arylnaphthalene skeleta (e.g., **71**).

The Pinaceae, by contrast, seem to have a much more restricted biochemical pathway and only a few variations are known. Thus, dehydrodiconiferyl alcohol **43** has been found in *Pinus taeda* (unpublished results), (+)-pinoresinol **18a** in *Picea abies (83)*, *Picea excelsa (84)* and *Pinus taeda (9)* as well as its 'epimeric' form, **78** in *Picea excelsa (24)*. Others include (+)-lariciresinol **25a**, the dibenzylbutane analog, (-)-secoisolariciresinol **20b** in *Larix leptolepis (85)* and its unusual *meso*-isomer **79** in *Cedrus deodara (86)*, the dibenzylbutyrolactone, (-)-matairesinol **24b** in *Abies nephrolepis, Picea ajanensis, Picea obovata (24)*, and its 7-oxo derivative **80** in *Abies nephrolepis, Picea sibirica, Picea abies, Picea ajanensis* and *Picea jezoensis (24)*, as well as the 2.7-cyclolignan-9'-olide lignans, (-)-α-conidendrin **66**, in *Picea excelsa, Tsuga heterophylla* and *Tsuga sieboldii (24)*, and todolactol **81** in *Abies sachalinensis(87)*. [Once again essentially all can be considered to be (+)-pinoresinol **18a** derived via modifications of the pathway following either (-)-secoisolariciresinol **20b** or (-)-matairesinol **24b** formation.] *Picea abies* also contains a rather rare and unusual lignan, the furan dimer **82**, apparently derived via oxidative coupling of *p*-coumaryl **7** and coniferyl **12** alcohols *(88)*.

In the Taxaceae, taxiresinol **83** *(89)* and cyclotaxiresinol **84** *(90)* have been found in *Taxus baccata*. Interestingly, their structures suggest the

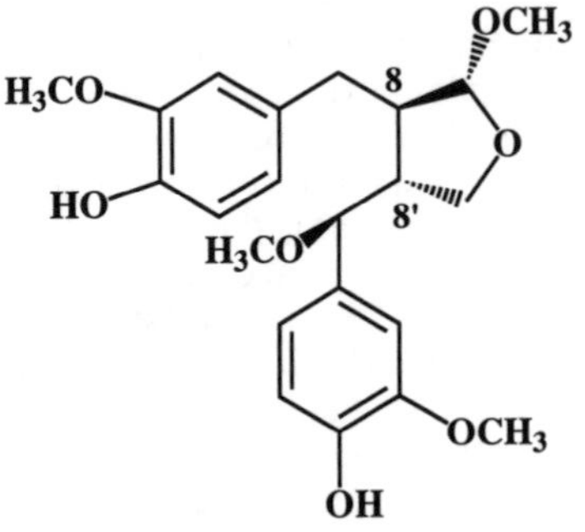

64: α-Intermedianol

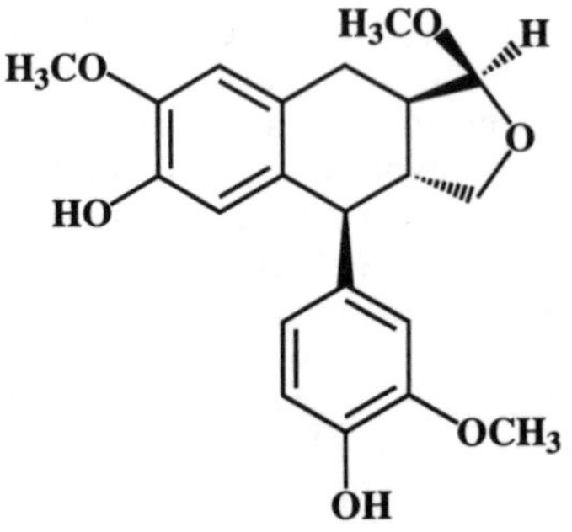

65: (+)-*O*-Methyl-α-conidendral

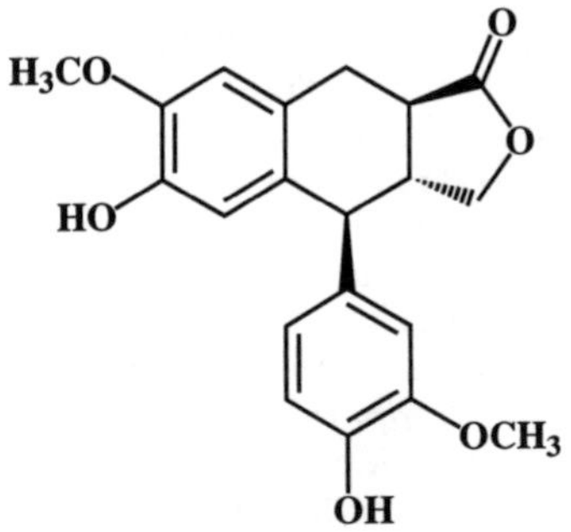

66: (-)-α-Conidendrin

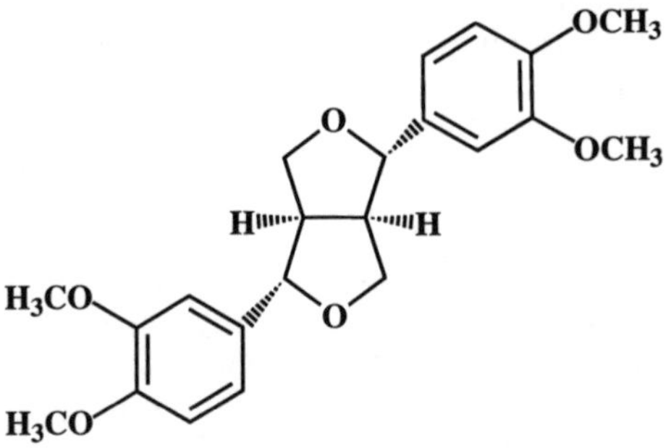

67: Dimethylpinoresinol

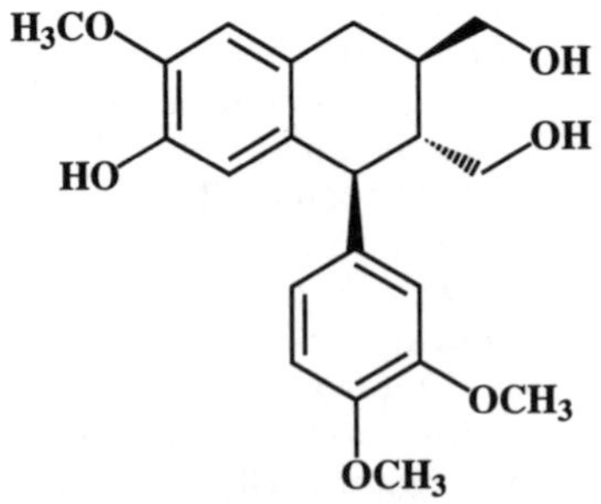

68: 4-*O*-Methyl isolariciresinol
(= 4-*O*-Methyl cyclolariciresinol)

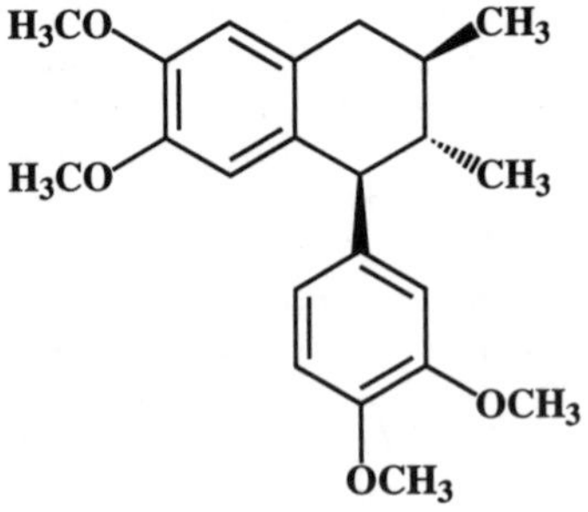

69: (-)-Galbulin

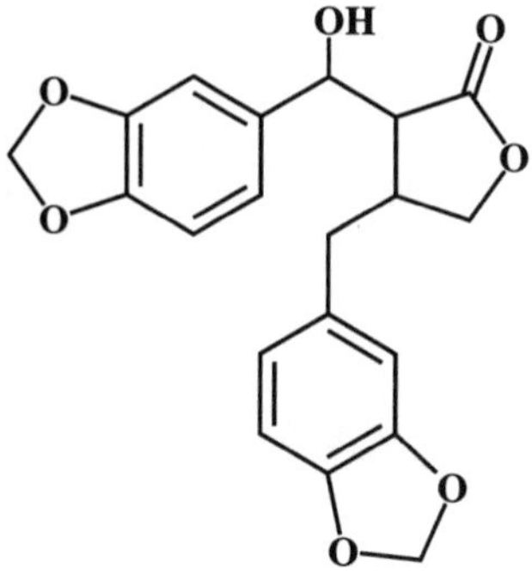

70: Sventenin

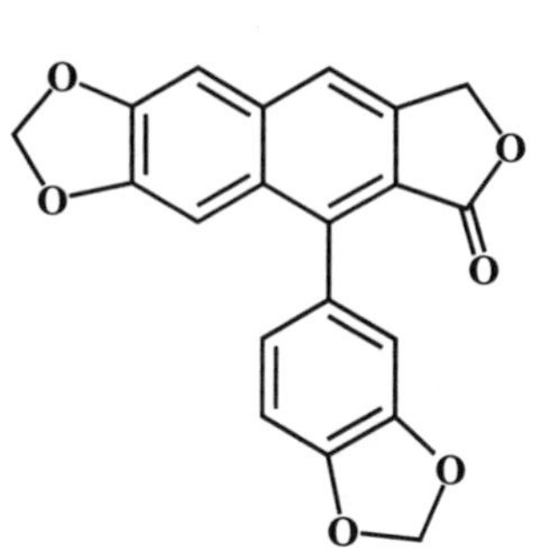

71: Taiwanin C

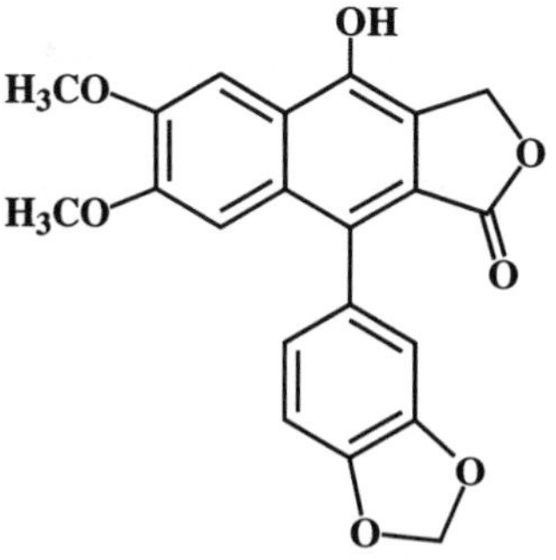

72: Diphyllin

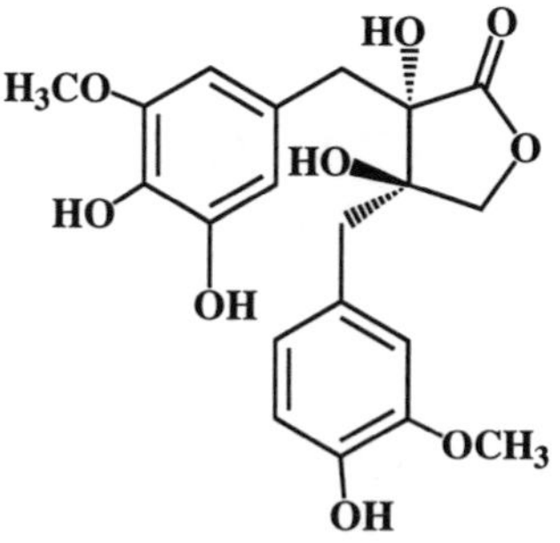

73: Dihydroxythujaplicatin

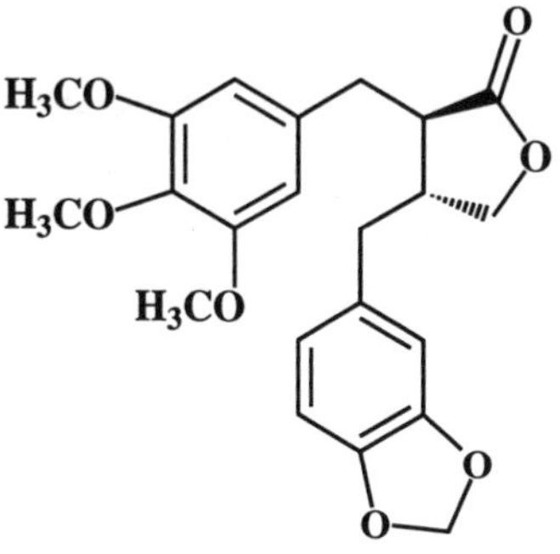

74: (-)-Yatein

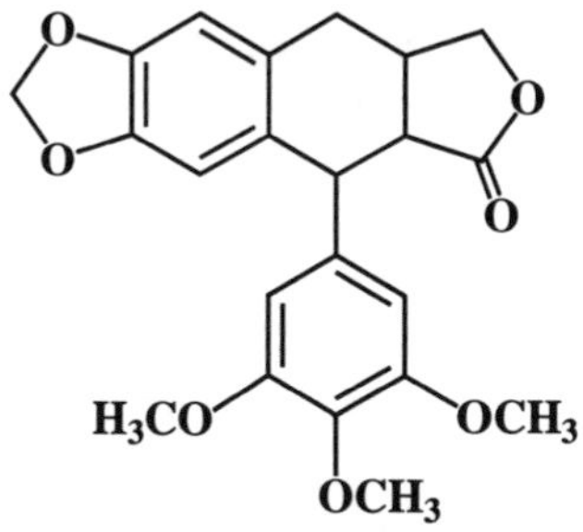

75: Deoxypodophyllotoxin

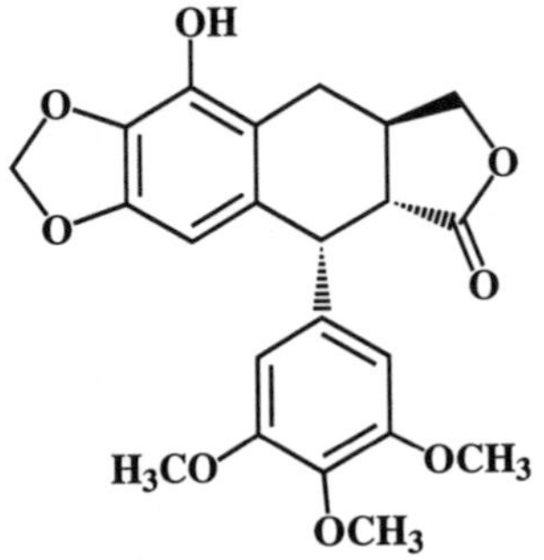

76: β-Peltatin

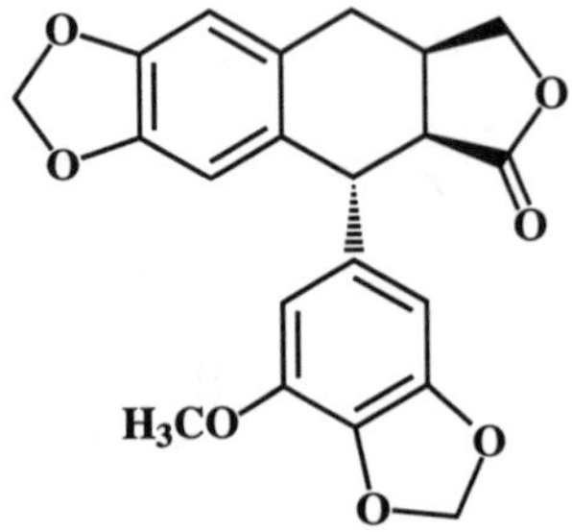

77: Austrobailignan 2

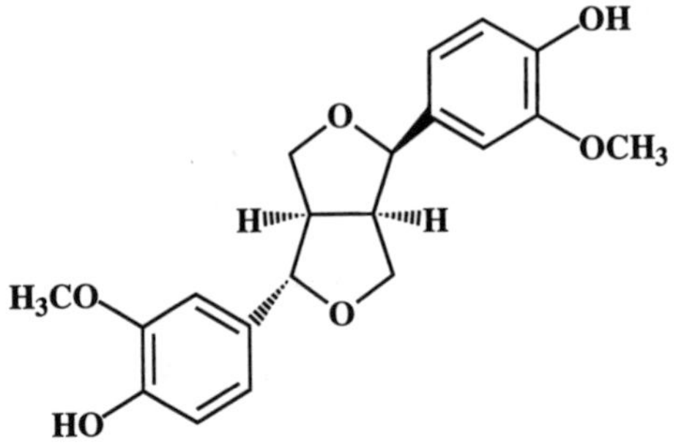

78: *Epi*-pinoresinol

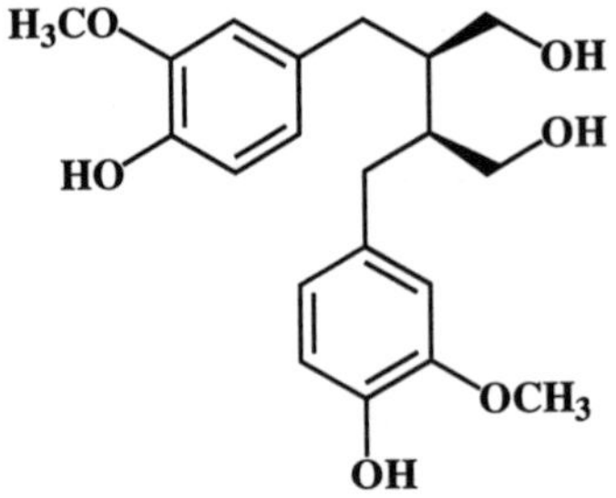

79: *Meso*-secoisolariciresinol

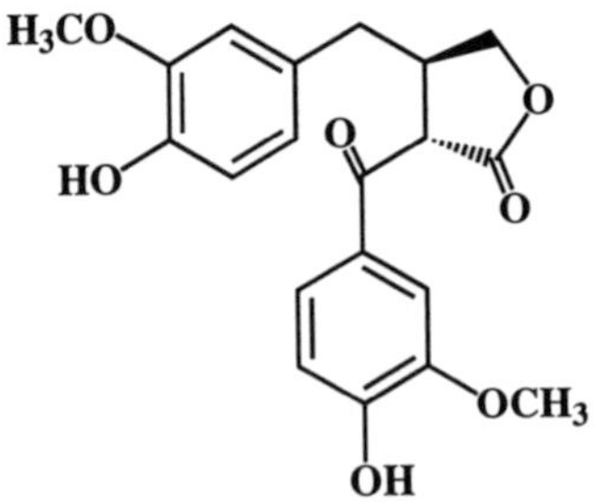

80: 7-Oxo-matairesinol

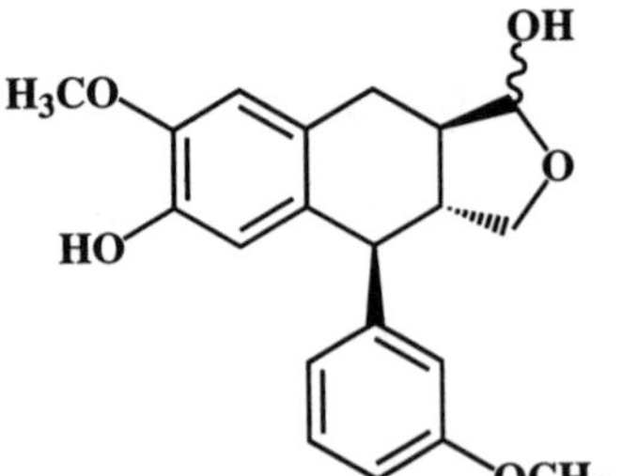

81: Todolactol

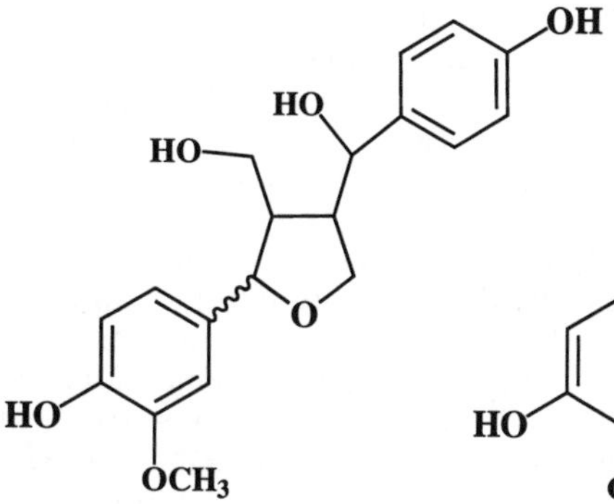

82

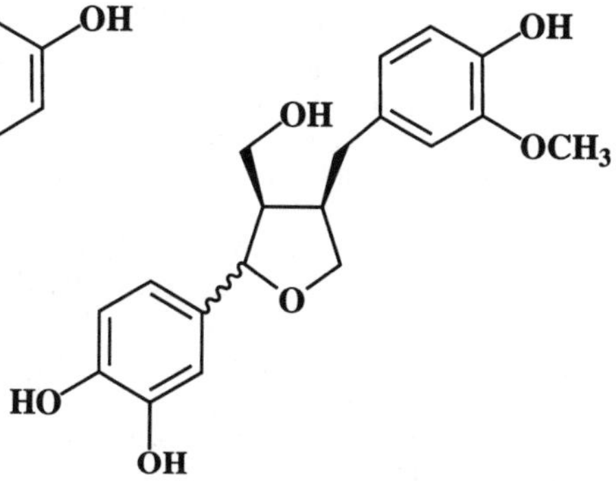

83: Taxiresinol

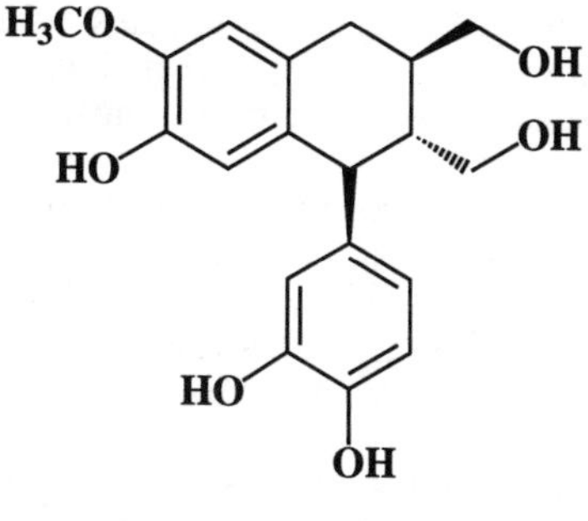

84: Cyclotaxiresinol

involvement of either caffeoyl alcohol in their biogenesis, or demethylation of the dimeric product (if derived from (+)-pinoresinol **18a**). This needs to be clarified.

In summary, the chemotaxonomic findings to date, together with our current knowledge of gymnosperm fossilized remains, seems to suggest a fairly straightforward elaboration of the lignan/neolignan pathway from that observed in the pteridophytes. Essentially all 8.8'-lignans can be proposed to be derived from (+)-pinoresinol **18a** i.e., following stereoselective 8.8'-coupling of *E*-coniferyl alcohol **12**. The major differences from the pteridophytes appear to include specific hydroxylations and formation of aryl(tetrahydro)naphthalene skeleta and methylenedioxy functionalities. Presumably these changes enabled the plants in question to develop more effective mechanisms for defense. This assertion is arguably correct since these substances accumulate in tissues (e.g., heartwood) which are biologically very resistant to herbivore and pathogen attack.

Angiosperms. The angiosperms are the most prevalent forms of plant life (> 200,000 species), with lignins typically derived mainly from both coniferyl **12** and sinapyl **17** alcohols *(32)*. Hence, angiosperms possess a fully functional phenylpropanoid pathway. They are also a rich and varied source of lignans, with representatives found in several species from many superorders *(27)* (Figure 4), following the classification of Dahlgren *(91, 92)*. In the dicotyledons, this includes the Magnoliiflorae, Nymphaeiflorae, Rosiflorae, Malviflorae, Myrtiflorae, Rutiflorae, Ranunculiflorae, Violiflorae, Primuliflorae, Santaliflorae, Araliiflorae, Asteriflorae, Corniflorae, Gentianiflorae and Lamiiflorae, whereas in the monocots, they have only been found in the Ariflorae and Commeliniflorae. As described below, structural trends have emerged which will be useful in predicting the type of species within a particular superorder which may harbor a specific biochemical pathway. But before describing these trends, a brief discussion of our current knowledge of angiosperm evolution is given.

Although both the origins and the progressive evolution of angiosperms are obscure *(6)*, most paleobiologists agree that they were well-entrenched by the Cretaceous period *(6, 93)*. These include the Dicotyledonae (i.e., the Magnoliiflorae, Nymphaeiflorae, Rosiflorae, Malviflorae, Myrtiflorae and Rutiflorae) and the Monocotyledonae (Areciflorae and Ariflorae,) respectively.

As for the gymnosperms and pteridophytes, this is based upon the results of analyses of fossilized tissues, such as fruit, flower, leaf and pollen specimens. For the dicots, the earliest examples in the Magnoliiflorae include the Magnoliales, [e.g., *Lesqueria* fruiting axis found in mid-Cretaceous specimens from Kansas *(94)*], the Laurales [e.g., *Crassidenticulum* leaf from the Cenomanian of southeastern Nebraska *(95)*], and the Chloranthales [e.g., from the upper Cretaceous androecia, *Chloranthistemon*, of Sweden *(96)*]. *Barclayopsis* seeds *(97)* and *Zonosulcites* pollen *(98)* from the Maastrichtian are representatives from the Nymphaeiflorae. In the Rosiflorae, these include the Trochodendrales [e.g., *Lhassoxylon* in the upper Aptian of Lhasa in Tibet, *(99)*], the Cercidiphydales [e.g., *Cercidiphyllites* from the Campanian of Alberta *(100)*], the Saxifragales [e.g., *Scandianthus*, (*Saxifragaceae* from the Upper Cretaceous of southern Sweden *(101)*] and the Fagales [e.g.,

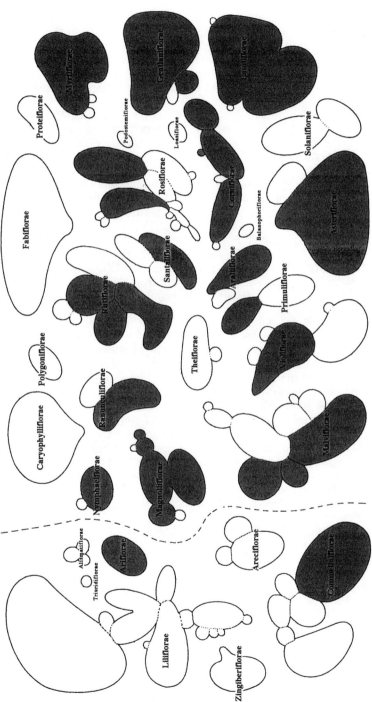

Figure 4. Superorders of the Dicotyledonae and Monocotyledonae (adapted from Ref. 91, 92). Shaded areas represent documented occurrence of (neo)lignans (adapted from Ref. 27).

Tricolporopollenites in the Santonian/Campanian of British Columbia *(102)]*. Examples in the Malviflorae have been found in sediments from the mid- to upper-Cretaceous period as evidenced by *Ulmipollenites* (Ulmaceae, Malviflorae) from California *(103)*. Other mid- to upper-Cretaceous fossils are of the Myrtiflorae and Rutiflorae, as exemplified by the *Myrtaceidites* (Myrtaceae) from Antarctica *(104)* and *Cupanieidites* (Sapindaceae) from Wyoming, respectively *(105)*.

Other dicotlyledon superorders (Malviflorae, Corniflorae, Violiflorae, Asteriflorae, Araliiflorae, Primidiiflorae, Santaliflorae) have been observed in later fossils, e.g., in the Tertiary subera, as exemplified by *"Crotonoidipollis"* (Thymelaeaceae, Malviflorae) pollen from the Middle Eocene of Tennessee *(93)*, *Swida discimontana* (Cornaceae, Corniflorae) a fruit from the Middle Miocene of Germany *(106)*, *Paraphyllanthoxylon abbottii* (Euphorbiaceae, Malviflorae) wood from the Paleocene of Big Bend National Park, Texas *(107)* and *Berhamniphyllum* (Rhamnaceae, Malviflorae) leaves from the Eocene of North America*(108)*. In the Salicaceae (Violiflorae), both *Populus wilmattae* leaves from the Middle Eocene of Utah *(109)* and *Salix* wood from the Upper Miocene of Colorado *(110)* have been described. In the Asteraceae (Asteriflorae) pollen grains were found in Oligocene sediments *(111)*, *Oreopanax* (Araliaceae, Araliiflorae) a fruit from the Paleocene of North Dakota *(112)*, *Austrodiospyros* (Ebenaceae, Primuliflorae) leaves from the Upper Eocene of Australia *(113)* and compressed seeds of the genus *Vitis* (Vitaceae, Santaliflorae) from the Lower Eocene have been reported *(114)*.

But as for the pteridophytes and gymnosperms, most of our knowledge of evolution of the lignan pathway in the angiosperm dicotyledons comes from chemotaxonomical studies rather than from the existing fossil record. In this context, analysis of the 500 or so known lignan structures in angiosperms reveals that the most widely observed mode of monomer coupling is again that leading to the 8.8' coupled products *(27)*. These are present in every superorder where lignans have been isolated (see Figure 4), with more than two-thirds containing these linkages. Moreover, with one exception in the Fagales (Rosiflorae) where the aryltetrahydronaphthalene lignan, lyoniresinol **85**, occurs, pinoresinol **18** (or a closely related furanofuran derivative thereof) has been found in every instance. Since lyoniresinol **85** is presumably also (+)-pinoresinol **18** derived, this again underscores the importance of this stereoselective coupling step, and of its very broad significance in lignan synthesis/evolution.

Of the remaining 8.8'-linked lignans (all presumed to be pinoresinol **18** derived) reported to date, these include the oxydiarylbutanes, dibenzylbutanes, dibenzylbutyrolactones, dibenzylfurans, arylnaphthalenes, 2,7'-cyclolignan-9'-olides and the aryltetrahydronaphthalenes (as previously described for the gymnosperms). Additionally, dibenzylcyclooctadienes now appear in the Araliiflorae e.g. steganacin **86** from *Steganotaenia araliaceae (115)* which involves an unusual 8.8',2.2'-coupling. The occurrence of each lignan class (or family) in specific plant superorders is shown in Table I; for specific examples the reader is referred to previous compilations *(19, 27)*. Of these known to date, the dibenzylbutyrolactone and 2,7'-cyclolignan-9'-olide lignans are most widespread, with fewer representatives of other types being found e.g., the arylnaphthalenes apparently only occur in the Euphorbiales, Polygalales, Ranunculales, Rutales, and Scrophulariales orders, etc. This study, therefore, reveals two major findings: The first is that stereoselective coupling of coniferyl alcohol **12** has been extended to a wide range of superorders in the woody dicots (see Figure 4), and secondly that subsequent conversions appear to be superorder specific. Clearly, this is not only of evolutionary significance, but also allows us to be able to begin to systematically address questions of biodiversity in terms of prediction of molecular composition.

There is, however, a very significant departure in lignan synthesis in the woody dicots from that encountered previously in the gymnosperms/ angiosperms, i.e., formation of lignans containing alkylphenol or propenylphenol-derived moieties, such as (+)-guaiacin **87** from *Virola carinata (116)*. At this point, it is unknown whether they are derived via coupling of coniferyl alcohol **12** moieties, or result from eugenol **88** or isoeugenol **89**. [Nor are the biochemical pathways to **88/89** known.] These dimers are mainly distributed in the Magnoliiflorae and Nymphaeiflorae *(27)*, with two apparent exceptions in the Myrtiflorae *(117)* and Rutiflorae *(118)*, respectively.

Formation of such lignans is also accompanied by substantial changes in the range of coupling modes encountered. For example, several 8.8', 2.2'- linked dibenzylcyclooctadiene lignans have been found in the Schisandraceae (Magnoliiflorae) and Verbenaceae (Lamiiflorae) e.g., gomisin A **33** in *Schizandra chinensis (119)*. Other modified 8.8'-coupling modes include 8.8',7-*O*-7'-epoxylignan synthesis [e.g., (+)-aristolignin **90** from *Aristolochia chilensis (120)*], the formation of the 8.8', 2.7', 8:8→7'-linked lignan **91**, the 8.8', 2.7', 7→8.8'-linked lignan **92** and the 8.8', 6.7', seco 1.7- linked lignan **93** in *Virola sebifera (121, 122)* and 8.8', 2.7', 3-*O*-6'-linked lignan e.g., carpanone **94** in *Cinnamomum* species *(117)* . Interestingly, all of these are restricted to the Magnoliiflorae superorder. Coupling processes other than 8.8' linkages are also almost exclusively found in the Magnoliiflorae and Nymphaeiflorae. These include: 8.1' [e.g., megaphone **37** in *Aniba megaphylla (123)*], 8.3' [e.g., mirandin A **95** in *Nectandra miranda (117)*], 8.5 [e.g., licarin A **96** in *Magnolia kachirachirai (124)*], 8-*O*-4' [e.g., surinamensin **41** in *Virola surinamensis (125)*] and 5.5' [e.g., dehydrodieugenol **97** in *Litsea turfosa (126)*]. Less common are the 3-*O*-4'- [e.g., isomagnolol **45** in *Sassafras randaiense (127)*], 2-*O*-3'- [e.g., lancilin **98** in *Aniba lancifolia (25, 128)*], 7,1'- [e.g., chrysophyllon IA **99** in *Licaria chrysophylla (129)*], 8,7'- [e.g., magnosalicin **100** in *Magnolia salicifolia (130)*] and 1.5'- [e.g., (-)-isoasatone **101** in *Asarum taitonense (117)*], these dimers being restricted to the Myristicaceae, Magnoliaceae, Lauraceae and Aristolochiaceae families. The only examples of alkylphenol-derived lignans in the Myrtales (Myrtiflorae) and Polygalales (Rutiflorae), to date, are the 8.5'- linked dimer e.g., conocarpin **102** isolated from *Conocarpus erectus (131)* and ratanhiaphenol I **103** from *Krameria cystisoides (118)*. These different coupling modes now raise the tantalizing prospect that there exists in nature a series of stereoselective coupling enzymes with properties analogous to the unprecedented (+)-pinoresinol **18a** synthase. This urgently needs to be established.

In summary, there is a further extrapolation of the biochemical repertoire in the woody dicotyledons from that previously observed in the

Table I. Distribution of 8.8'-linked lignans, presumably derived from pinoresinol 18, and classified according to structural types and plant superorders (Lignan occurrence was compiled from ref. 27, and plant classification is from Dahlgren 91, 92).

Superorders	Oxydiaryl-butanes	Dibenzyl-butanes	Dibenzyl-butyro-lactones	Dibenzyl-furans	Aryl-naphthalenes	2.7'-Cyclolignan-9'-olides	Aryltetra-hydro-naphthalenes	Dibenzyl-cyclo-octadienes
Araliiflorae			Araliales			Araliales		Araliales
Asteriflorae	Asterales		Asterales			Asterales		
Balanophori-florae								
Caryophylli-florae								
Celastriflorae								
Corniflorae			Cornales			Eucommiales	Eucommiales Ericales	
Fabiflorae								
Gentiani-florae		Gentianales	Gentianales					
Lamiiflorae			Scrophulariales		Scrophulariales	Scrophulariales	Oleales	
Loasiflorae								
Magnolii-florae		Aristolachiales	Aristolachiales Annonales Laurales	Laurales		Annonales Laurales	Annonales Laurales	

Malviflorae	Urticales				Thymelaeales		
Myrtiflorae			Euphorbiales		Piperales		Euphorbiales
Nymphaei-florae						Piperales	
Podostemoni-florae							
Polygoni-florae							
Primuliflorae				Ebenales			
Proteiflorae							
Ranunculi-florae		Ranunculales	Ranunculales	Ranunculales			
Rosiflorae	Rosales Fagales						
Rutiflorae		Rutales Polygalales	Rutales Polygalales		Rutales		
Solaniflorae							
Theiflorae							
Violiflorae							

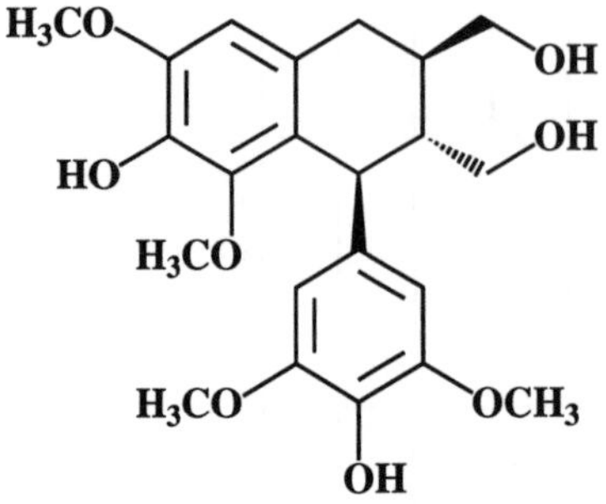

85: (+)-Lyoniresinol

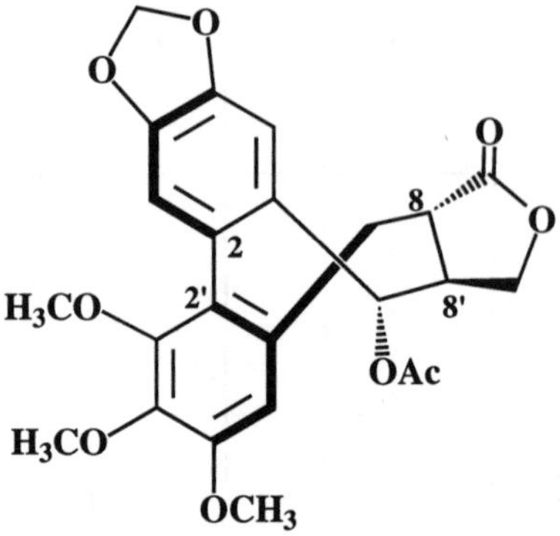

86: (-)-Steganacin

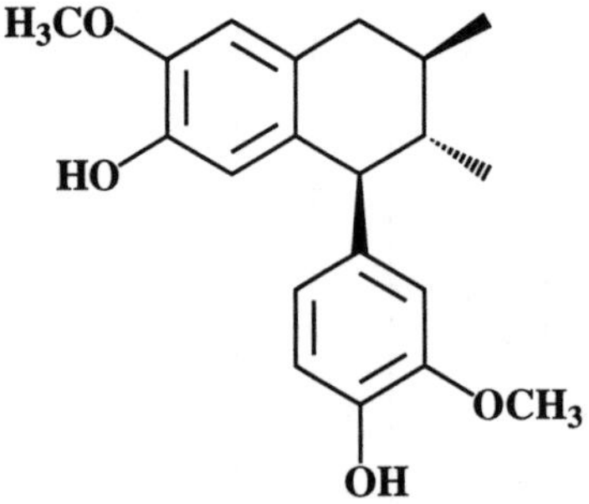

87: (+)-Guaiacin

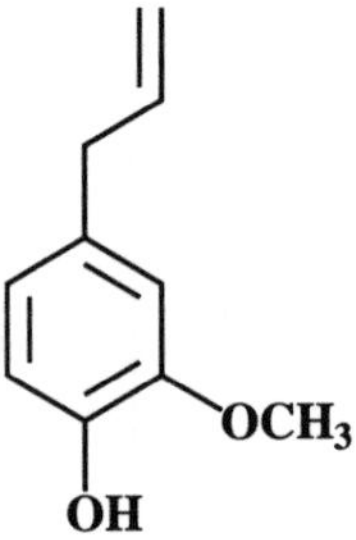

88: Eugenol

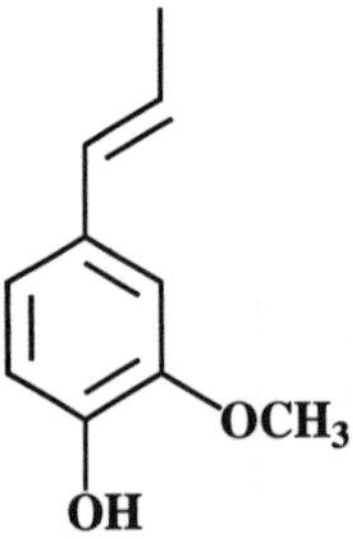

89: Isoeugenol

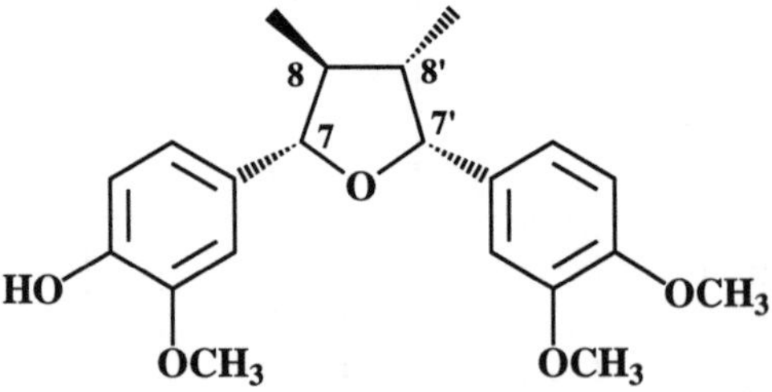

90: (+)-Aristolignin

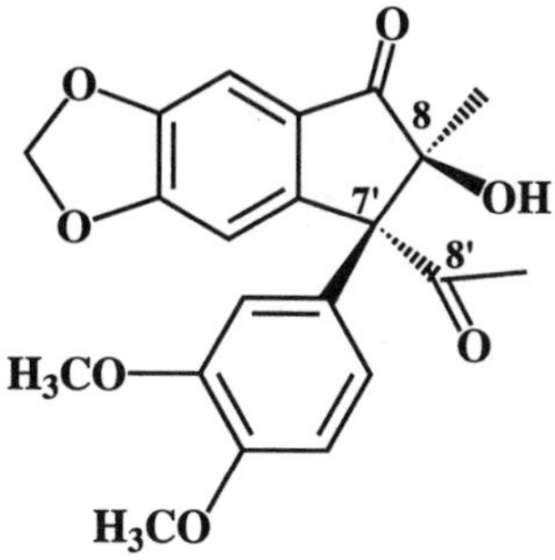

91

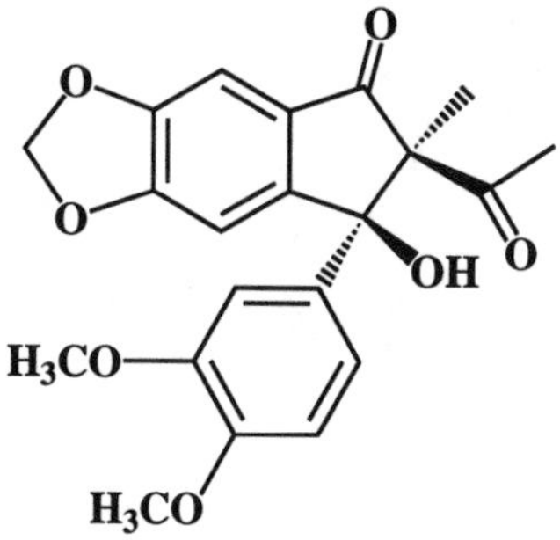

92

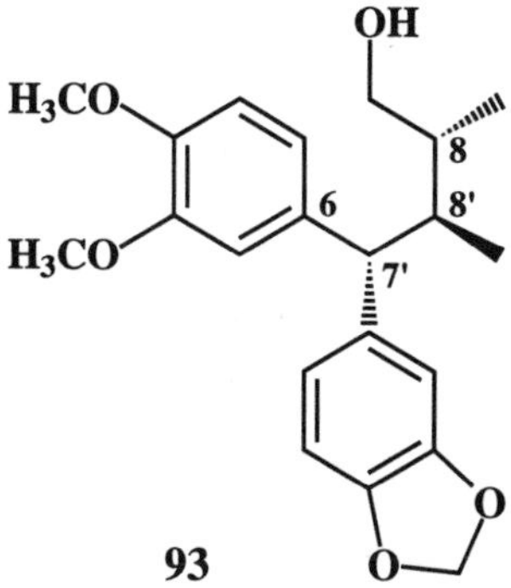

93

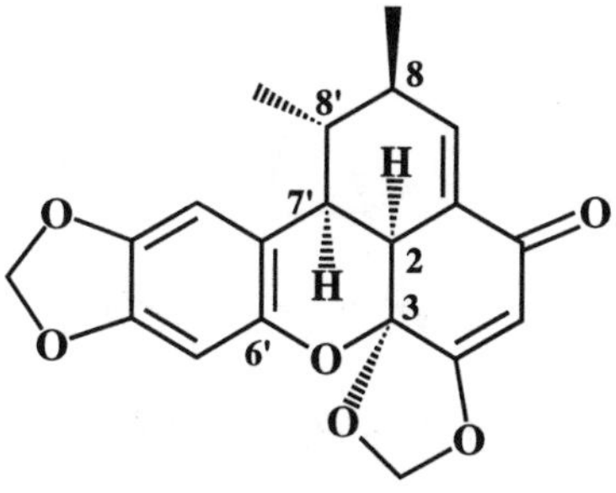

94: Carpanone

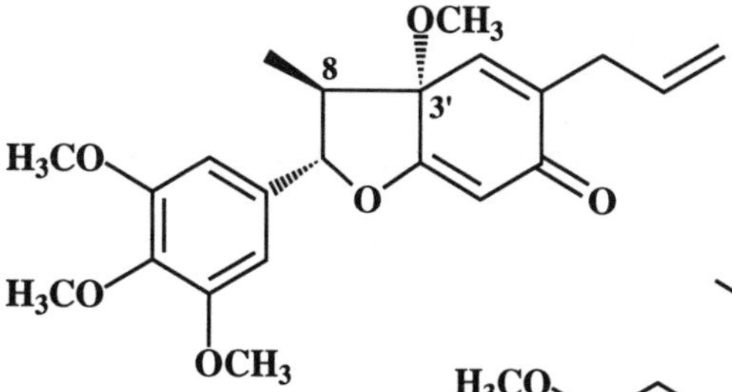

95: Mirandin A

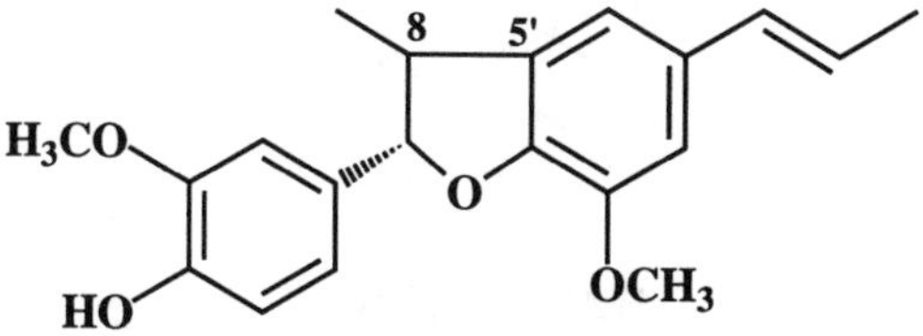

96: Licarin A

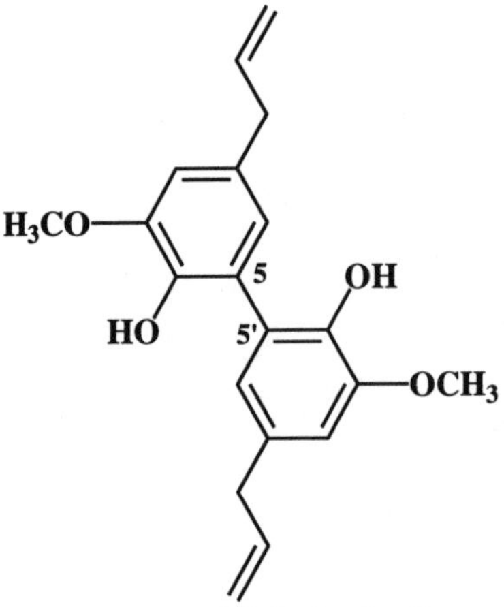

97: Dehydrodieugenol

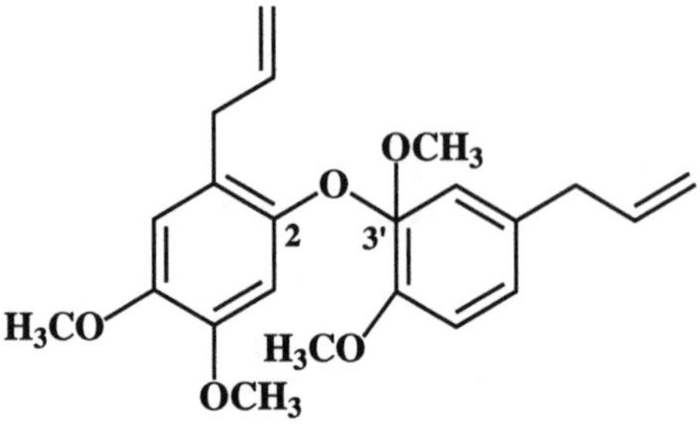

98: Lancilin

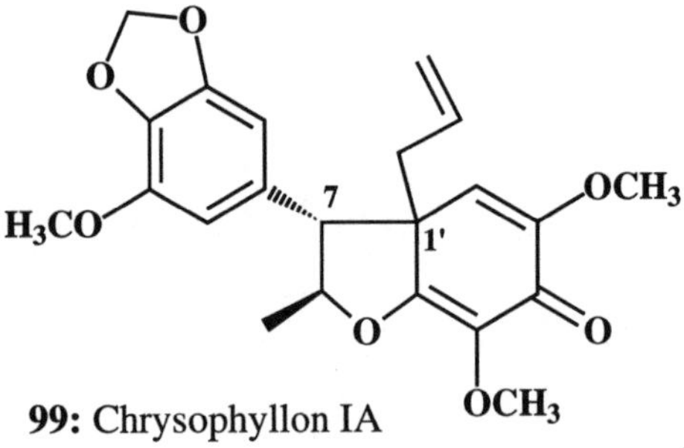

99: Chrysophyllon IA

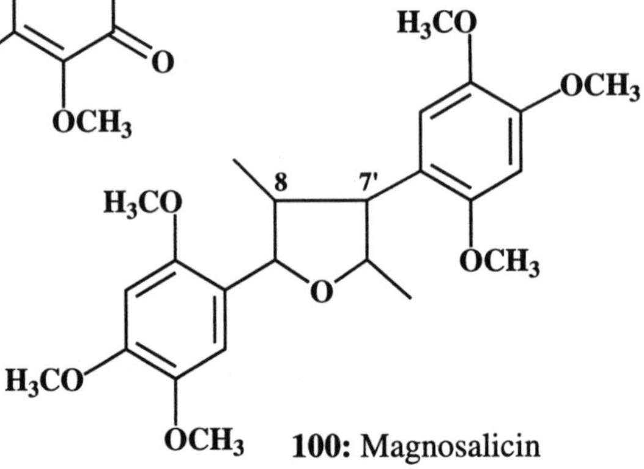

100: Magnosalicin

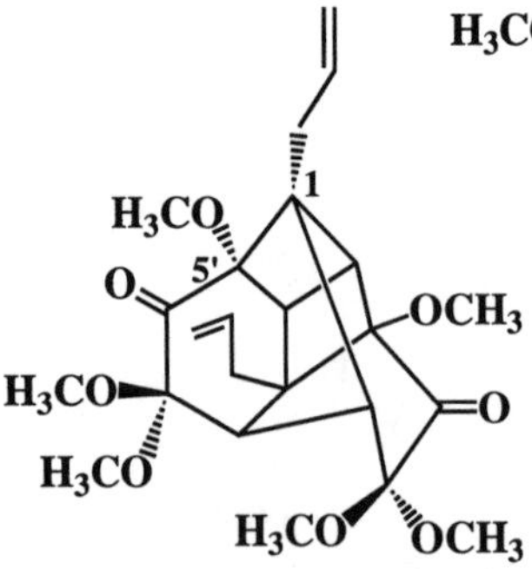

101: Isoasatone

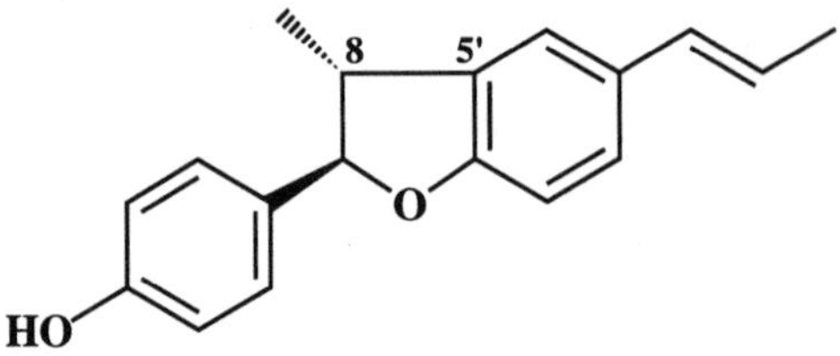

102: Conocarpin

pteridophytes and gymnosperms. Of these, the most unusual variations are mainly restricted to the Magnoliiflorae and Nymphaeiflorae superorders.

Turning our attention next to the monocotyledons, it needs to be emphasized that their origins are very obscure. However, like the dicots they were also well established in the Cretaceous period, as indicated from the results of analyses of pollen and leaves of the Pandanaceae (Areciflorae) and Potamogetonaceae (Alismaliflorae) *(132)*. Others fossilized remains have been found in the Tertiary subera e.g., *Acorites* (Araceae, Ariflorae) flowers from the middle Eocene *(133)*, *Pollia tugenensis* (Commelinaceae, Commeliniflorae) leaves from Miocene deposits in Kenya *(134)* and a permineralized rhizome fossil of the grass *Tomlinsonia* (Poaceae, Commeliniflorae) in the upper Miocene of California *(135)*.

But lignans (whether monolignol or alkylphenol-derived) are, for the most part, apparently absent in the monocotyledons based on chemotaxonomical reports to date. The only exceptions to this trend are in the occurrence of unusual cyclobutane dimers. One example is the lignan acoradin **104** (stereochemistry unknown), found in *Acorus calamus* (Ariflorae) *(136)*. The only other reports of the occurrence of metabolites such as this are in the putatively closely aligned Magnoliiflorae [i.e., magnosalin **105** from *Magnolia salicifolia*, Magnoliaceae *(137)* and heterotropin **106** from *Heterotropa takahoi*, Aristolochiaceae *(138, 139)*], suggesting some evolutionary overlap (Figure 4). Their biosynthetic pathways are unknown. Based on melting point and spectroscopic data, acoradin **104** and heterotropan **106** may be the same compound.

The remaining examples of cyclobutane lignans known to date are in the Poaceae (Commeliniflorae). These are primarily found in the cell wall fractions (both leaf and stem) of various grasses and grains, and are putatively involved in cross-linking adjacent hemicellulosic polymers containing ester-linked, p-coumaric **4** or ferulic **9** acids e.g., in tropical grasses such as *Setaria anceps* cv. *Nandi (140, 141)*. The dimers isolated are mainly (substituted) dihydroxytruxillic acids **107** formed by head-to-tail coupling, but there is also evidence for head-to-head coupling (e.g., with p-coumaric acid **4**) to give the corresponding dihydroxytruxinic acid **108** derivatives. Further, in the stem cell walls of these grasses, there is also preliminary evidence for mixed cyclobutane dimers derived from both ferulic acid **9** and coniferyl alcohol **12** *(140, 141)*. [A role for structural support can thus also be envisaged.] The current working hypothesis is that these lignans are derived via photochemical coupling, which may also explain the formation of acoradin **104**. To our knowledge, there are no reports of the presence of these cyclodimers in other monocotyledon superorders to this point.

Mammalian Lignans. Lignans are not only found in plants, but have also been isolated from serum, urine, bile and seminal fluid of human and several animals *(142-144)*. The major mammalian lignans are enterolactone **109** and enterodiol **110**, with each possessing only one phenolic hydroxyl group in the *meta* position of their aromatic rings as shown. They are excreted in large amounts in urine by persons on diets rich in whole-grain products, berries or fruits *(145)*. Both are presumed to be produced by the metabolic action of intestinal bacteria on matairesinol **24** and secoisolariciresinol **20**, respectively, these being released from foodstuffs during digestion *(146)*. Lariciresinol **25**

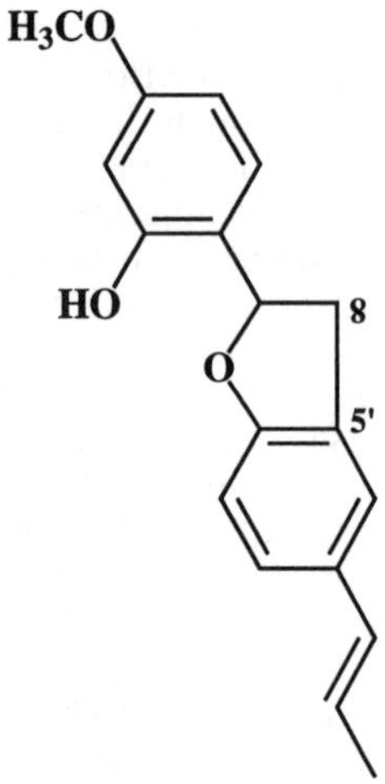

103: Ratanhiaphenol I

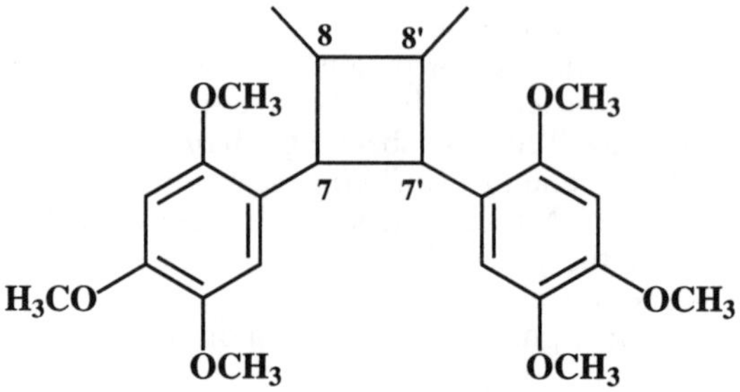

104: Acoradin

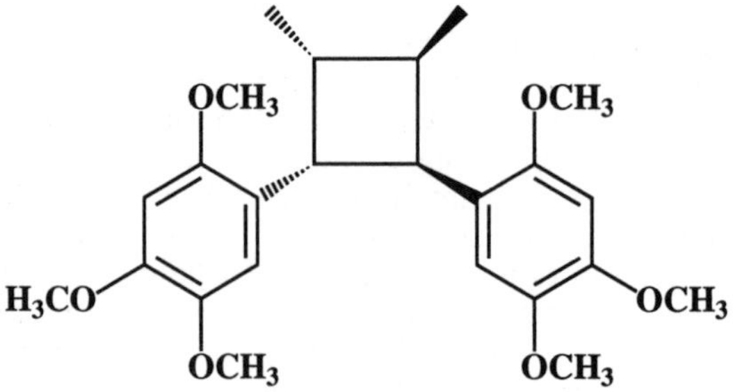

105: Magnosalin

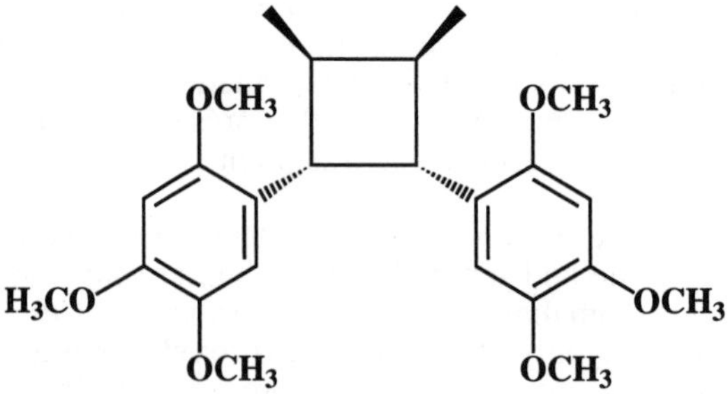

106: Heterotropan

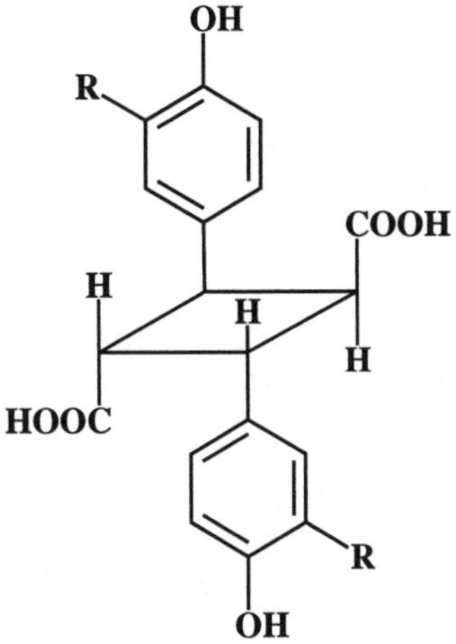

107: R = H or OCH$_3$
Dihydroxytruxillic acid

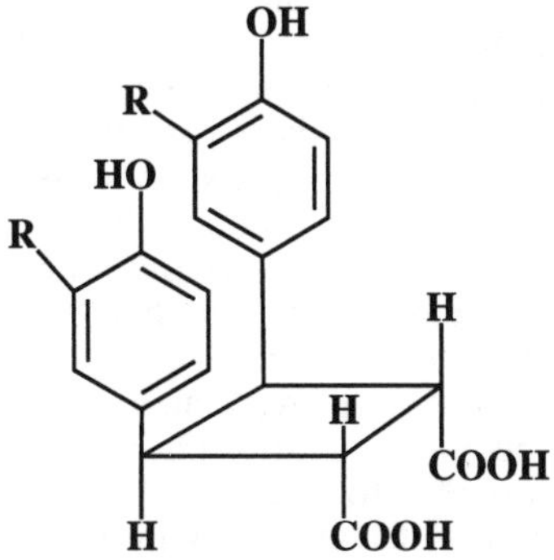

108: R = H or OCH$_3$
Dihydroxytruxinic acid

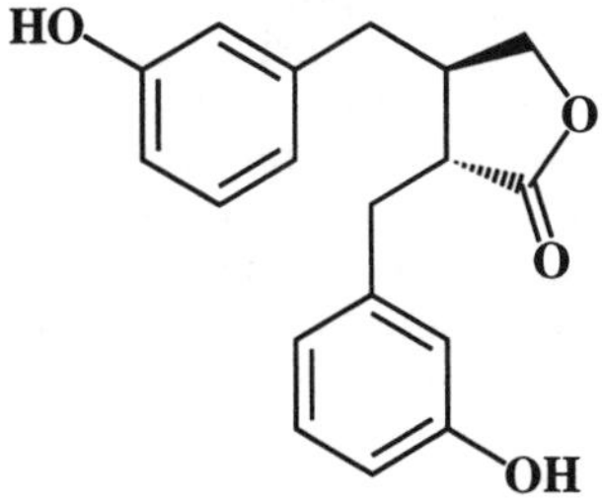

109 : Enterolactone

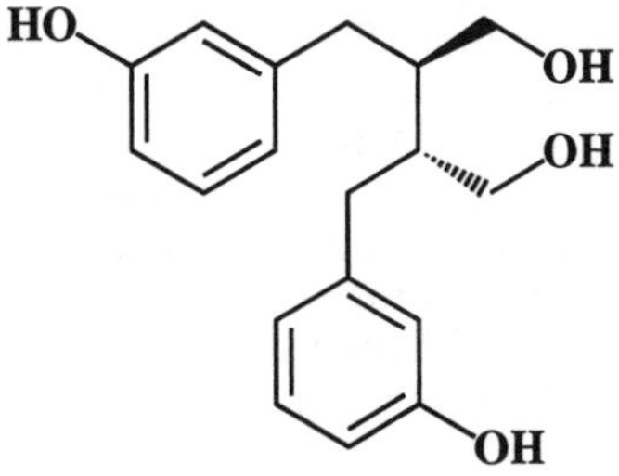

110 : Enterodiol

and isolariciresinol **29** have also been detected in small amounts in human urine *(147)*. It is considered that these mammalian lignans may be of profound importance in reducing incidence rates of breast and prostatic cancers via modulation of hormone synthesis *(145)*.

Concluding Remarks

Although the existing fossil record does not provide an effective means to determine how evolution of the lignan/neolignan pathway (or related phenylpropanoid metabolites) occurred, the analysis of chemotaxonomical data from extant plants has revealed important insights. First, the pathways to the monolignols, and hence lignans/neolignans, are apparently absent in the algae, fungi and mosses. The reason for this absence seems to be strongly tied to the inability of these organisms to synthesize ferulic **9** and sinapic **12** acids. This, in turn, suggests an important evolutionary role for regiospecific *O*-methylation in vascular plants.

The pathway to the monolignols apparently first evolved in the pteridophytes. But whether elaboration of the pathway was initially for structural support (i.e., lignin synthesis), defense (e.g., against UV irradiation or oxidation) or to provide an ability to survive in a desiccating environment can only be guessed at. There are only a few examples of lignans in the pteridophytes to date, i.e., blechnic acid **62** and its epimer **63**, together with lariciresinol **61** and dehydrodiconiferyl alcohol **60** glucosides. Importantly, their occurrence signifies the introduction (evolution) of novel biochemical pathways, i.e., stereoselective coupling yielding blechnic acid **62**, and non-stereoselective processes generating racemic (±)-pinoresinol **18a/b** and dehydrodiconiferyl alcohols **43a/b**; the latter enzymes are presumably laccase-like. It is our current working hypothesis that the appearance of these oxidases is of profound importance in woody tissue formation.

The gymnosperms further capitalized on these advances, with particular reliance upon the pathway affording (+)-pinoresinol **18a** and (+)-lariciresinol **25a**. Further extension beyond (+)-lariciresinol **25a** provided entry into all of the major subclasses, i.e., dibenzylfuran, dibenzylbutyrolactone, 2,7' cyclolignan-9'-olides, arylnaphthalene lignans, etc. Interestingly, this was also accompanied by a whole range of oxidative modifications, albeit in a species dependent manner. It can be postulated that these modifications were necessary to enhance antioxidant and biocidal capabilities, although the role of certain representatives in lignin synthesis was also essential for competitive survival.

Evolution of the pathway continued unabated in the woody dicots, where two major new advances occurred. These were introduction of novel (i.e., non 8.8') coupling modes, with a greater emphasis upon the formation of lignans lacking oxygen functionalities at C9/C9'. These can be considered to further expand the biochemical repertoire of these plants, again probably for defense purposes such as enhanced biocidal/antioxidant properties. It now appears that there exists in nature a new class of stereoselective coupling enzymes whose mode of action needs to be systematically determined.

By contrast, the non-woody monocots significantly de-emphasised exploitation of the pathways to the lignans and neolignans (i.e., regardless of whether monolignol or alkyl phenol derived). This de-emphasis can be speculated as due to either a reduction or loss of both stereoselective and laccase-like coupling processes, although this needs to be unambiguously determined. This apparent loss may also help explain why the monocots do not form woody tissue. The grasses do, however, engender a new form of coupling which gives truxillic and, to a lesser extent, truxinic acids. These are viewed to be formed via photochemical coupling, and are putatively involved in cell wall reinforcement. The lack of either lignans or formation of woody tissue in these plants, presumably accounts for their ease of biodegradation, in contrast to their woody counterparts.

Finally, the occurrence of specific lignan families (or classes) in the few representatives of each superorder examined to date has yielded some very important trends. These will help guide future chemotaxonomic studies via predictive analyses. This has obvious ramifications in drug discovery and searches for pharmacologically active compounds, as well as in the other biotechnological targets described earlier.

Acknowledgments. Financial support from the National Science Foundation (MCB 9219586), the U.S. Department of Agriculture (91371036638) and the U.S. Department of Energy (DEFG0691ER20022) is gratefully acknowledged.

Literature Cited

1.   Stebbins, G. J.; Hill, G. J. C. *Am. Nat.* **1980**, *115*, 342-353.
2.   Manhart, J. R.; Palmer, J. D. *Nature.* **1990**, *345*, 268-270.
3.   Taylor, T. N. *Taxon.* **1988**, *37*, 805-833.
4.   Niklas, K. J. In *Biochemical Aspects of Evolutionary Biology*; Nitecki, M. H. Ed.; The University of Chicago Press: Chicago, IL, 1982; pp 29-91.
5.   Niklas, K. J.; Pratt, L. M. *Science.* **1980**, *209*, 396-397.
6.   Taylor, T. N.; Taylor, E. L. *The Biology and Evolution of Fossil Plants*; Prentice Hall, Inc,: Englewood Cliffs, NJ, 1993; pp 982.
7.   Davin, L. B.; Lewis, N. G. In *Rec. Adv. Phytochemistry*; Stafford, H. A., Ibrahim, R. K. Eds.; Plenum Press: New York, NY, 1992; Vol. 26; pp 325-375.
8.   Löffelhardt, W.; Ludwig, B.; Kindl, H. *Hoppe-Seyler's Z. Physiol. Chem.* **1973**, *354*, 1006-1012.
9.   Eberhardt, T. L.; Bernards, M. A.; He, L.; Davin, L. B.; Wooten, J. B.; Lewis, N. G. *J. Biol. Chem.* **1993**, *268*, 21088-21096.
10. Fry, S. C. *Ann. Rev. Plant Physiol.* **1986**, *37*, 165-186.
11. El-Basyouni, S. Z.; Neish, A. C.; Towers, G. H. N. *Phytochemistry.* **1964**, *3*, 627-639.
12. Rahman, M. M. A.; Dewick, P. M.; Jackson, D. E.; Lucas, J. A. *Phytochemistry.* **1990**, *29*, 1971-1980.
13. Markkanen, T.; Makinen, M. L.; Maunuksela, E.; Himanen, P. *Drugs Exptl. Clin. Res.* **1981**, *7*, 711-718.
14. Nitao, J. K.; Nair, M. G.; Thorogood, D. L.; Johnson, K. S.; Scriber, J. M. *Phytochemistry.* **1991**, *30*, 2193-2195.
15. Hattori, M.; Hada, S.; Watahiki, A.; Ihara, H.; Shu, Y.-Z.; Kakiuchi, N.; Mizuno, T.; Namba, T. *Chem. Pharm. Bull.* **1986**, *34*, 3885-3893.

16. Taniguchi, E.; Imamura, K.; Ishibashi, F.; Matsui, T.; Nishio, A. *Agric. Biol. Chem.* **1989**, *53*, 631-643.
17. Lynn, D. G.; Chen, R. H.; Manning, K. S.; Wood, H. N. *Proc. Natl. Acad. Sci. USA.* **1987**, *84*, 615-619.
18. Binns, A. N.; Chen, R. H.; Wood, H. N.; Lynn, D. G. *Proc. Natl. Acad. Sci. USA.* **1987**, *84*, 980-984.
19. Ayres, D. C.; Loike, J. D. *Chemistry and Pharmacology of Natural Products. Lignans. Chemical, Biological and Clinical Properties*; Cambridge University Press: Cambridge, England, 1990; pp 402.
20. Abe, F.; Yamauchi, T.; Wan, A. S. C. *Phytochemistry.* **1988**, *27*, 3627-3631.
21. Abe, F.; Yamauchi, T.; Wan, A. S. C. *Phytochemistry.* **1989**, *28*, 3473-3476.
22. Sakakibara, A.; Sasaya, T.; Miki, K.; Takahashi, H. *Holzforschung.* **1987**, *41*, 1-11.
23. Yamaguchi, H.; Nakatsubo, F.; Katsura, Y.; Murakami, K. *Holzforschung.* **1990**, *44*, 381-385.
24. Rao, C. B. S. *Chemistry of Lignans*; Andhra University Press: Andhra Pradesh, India, 1978; pp 377.
25. Whiting, D. A. *Nat. Prod. Rep.* **1985**, *2*, 191-211.
26. Gottlieb, O. R. *Phytochemistry.* **1972**, *11*, 1537-1570.
27. Gottlieb, O. R.; Yoshida, M. In *Natural Products of Woody Plants. Chemicals Extraneous to the Lignocellulosic Cell Wall*; Rowe, J. W., Kirk, C. H. Eds.; Springer Verlag: Berlin, 1989; pp 439-511.
28. Freudenberg, K.; Harkin, J. M.; Reichert, M.; Fukuzumi, T. *Chem. Ber.* **1958**, *91*, 581-590.
29. Freudenberg, K. *Nature.* **1959**, *183*, 1152-1155.
30. Higuchi, T. *Physiol. Plant.* **1957**, *10*, 356-372.
31. Higuchi, T.; Ito, Y. *J. Biochem.* **1958**, *45*, 575-579.
32. Lewis, N. G.; Yamamoto, E. *Annu. Rev. Plant Physiol. Plant Mol. Biol.* **1990**, *41*, 455-496.
33. Kamil, W. M.; Dewick, P. M. *Phytochemistry.* **1986**, *25*, 2093-2102.
34. Stöckigt, J.; Klischies, M. *Holzforschung.* **1977**, *31*, 41-44.
35. Umezawa, T.; Davin, L. B.; Yamamoto, E.; Kingston, D. G. I.; Lewis, N. G. *J. Chem. Soc., Chem. Commun.* **1990**, 1405-1408.
36. Davin, L. B.; Bedgar, D. L.; Katayama, T.; Lewis, N. G. *Phytochemistry.* **1992**, *31*, 3869-3874.
37. Sterjiades, R.; Dean, J. F. D.; Gamble, G.; Himmelsbach, D. S.; Eriksson, K.-E. L. *Planta.* **1993**, *190*, 75-87.
38. Driouich, A.; Lainé, A.-C.; Vian, B.; Faye, L. *Plant J.* **1992**, *2*, 13-24.
39. Lewis, N. G.; Davin, L. B.; Katayama, T.; Bedgar, D. L. *Bull. Soc. Groupe Polyphénols.* **1992**, *16*, 98-103.
40. Katayama, T.; Davin, L. B.; Lewis, N. G. *Phytochemistry.* **1992**, *31*, 3875-3881.
41. Chu, A.; Dinkova, A.; Davin, L. B.; Bedgar, D. L.; Lewis, N. G. *J. Biol. Chem.* **1993**, *268*, 27026-27033.
42. Umezawa, T.; Davin, L. B.; Lewis, N. G. *Biochem. Biophys. Res. Commun.* **1990**, *171*, 1008-1014.
43. Umezawa, T.; Davin, L. B.; Lewis, N. G. *J. Biol. Chem.* **1991**, *266*, 10210-10217.

44. Frias, I.; Siverio, J. M.; González, C.; Trujillo, J. M.; Pérez, J. A. *Biochem. J.* **1991**, *273*, 109-113.
45. Reznikov, V. M.; Mikhaseva, M. F.; Zil'bergleit, M. A. *Chem. Nat. Comp.* **1978**, *14*, 554-556.
46. Dovgan, I. V.; Medvedeva, E. I. *Chem. Nat. Comp.* **1983**, *19*, 81-84.
47. Dovgan, I. V.; Medvedeva, E. I. *Chem. Nat. Comp.* **1983**, *19*, 84-87.
48. Ragan, M. A. *Phytochemistry.* **1984**, *23*, 2029-2032.
49. Gunnison, D.; Alexander, M. *Applied Microbiol.* **1975**, *29*, 729-738.
50. Delwiche, C. F.; Graham, L. E.; Thomson, N. *Science.* **1989**, *245*, 399-401.
51. Siegel, S. M. *Amer. J. Bot.* **1969**, *56*, 175-179.
52. Bland, D. E.; Logan, A.; Menshun, M.; Sternhell, S. *Phytochemistry.* **1968**, *6*, 1373.
53. Wilson, M. A.; Sawyer, J.; Hatcher, P. G.; Lerch III, H. E. *Phytochemistry.* **1989**, *28*, 1395-1400.
54. Markham, K. R. In *Bryophytes. Their Chemistry and Chemical Taxonomy*; Zinsmeister, H. D., Mues, R. Eds.; Oxford University Press: New York, NY, 1990; Vol. 29; pp 143-159.
55. Takeda, R.; Hasegawa, J.; Sinozaki, K. In *Bryophytes. Their Chemistry and Chemical Taxonomy*; Zinsmeister, H. D., Mues, R. Eds.; Oxford University Press: New York, NY, 1990; Vol. 29; pp 201-207.
56. Takeda, R.; Hasegawa, J.; Shinozaki, M. *Tetrahedron Lett.* **1990**, *31*, 4159-4162.
57. Wat, C.-K.; Towers, G. H. N. In *Rec. Adv. Phytochem.*; Conn, E. E. Ed.; Plenum Press: New York, NY, 1979; Vol. 12; pp 371-432.
58. Bu'Lock, J. D.; Smith, H. G. *Experientia.* **1961**, *17*, 553-556.
59. Bu'Lock, J. D.; Leeming, P. R.; Smith, H. G. *J. Chem. Soc.* **1962**, 2085-2089.
60. Shimazono, H.; Schubert, W. J.; Nord, F. F. *J. Amer. Chem. Soc.* **1958**, *80*, 1992-1994.
61. Shimada, M.; Ohta, A.; Kurosaka, H.; Hattori, T.; Higuchi, T.; Takahashi, M. In *Plant Cell Wall Polymers. Biogenesis and Biodegradation*; Lewis, N. G., Paice, M. G. Eds.; ACS Symposium Series: Washington, DC, 1989; Vol. 399; pp 412-425.
62. Shimada, M.; Nakatsubo, F.; Kirk, T. K.; Higuchi, T. *Arch. Microbiol.* **1981**, *129*, 321-324.
63. Logan, K. J.; Thomas, B. A. *New Phytol.* **1985**, *99*, 571-585.
64. Satake, T.; Murakami, T.; Saiki, Y.; Chen, C.-M. *Chem. Pharm. Bull.* **1978**, *26*, 1619-1622.
65. Wada, H.; Kido, T.; Tanaka, N.; Murakami, T.; Saiki, Y.; Chen, C.-M. *Chem. Pharm. Bull.* **1992**, *40*, 2099-2101.
66. Townrow, J. A. *Proc. Roy. Soc. Tasmania.* **1967**, *101*, 103-136.
67. Alvin, K. L. *Rev. Palaeobot. Palynol.* **1982**, *37*, 71-98.
68. Watson, J. *Phyta, Studies on Living and Fossil Plants.* **1982**, 265-273.
69. Vaudois, N.; Privé, C. *Palaeontographica.* **1971**, *134B*, 61-86.
70. Delevoryas, T.; Hope, R. C. *Rev. Palaeobot. Palynol.* **1987**, *51*, 59-64.
71. Stockey, R. A.; Nishida, M. *Can. J. Bot.* **1986**, *64*, 1856-1866.
72. Florin, R. *Acta Horti Bergiani.* **1951**, *15*, 285-388.

73. Cambie, R. C.; Clark, G. R.; Craw, P. A.; Jones, T. C.; Rutledge, P. S.; Woodgate, P. D. *Aust. J. Chem.* **1985**, *38*, 1631-1645.
74. Cambie, R. C.; Parnell, J. C.; Rodrigo, R. *Tetrahedron Lett.* **1979**, *12*, 1085-1088.
75. Easterfield, T. H.; Bee, J. *J. Chem. Soc.* **1910**, 1028-1032.
76. Briggs, L. H.; Cambie, R. C.; Hoare, J. L. *Tetrahedron.* **1959**, *7*, 262-269.
77. Fonseca, S. F.; Nielsen, L. T.; Rúveda, E. A. *Phytochemistry.* **1979**, *18*, 1703-1708.
78. Lee, C. L.; Hirose, Y.; Nakatsuka, T. *Mokuzai Gakkaishi.* **1974**, *20*, 558-563.
79. MacLean, H.; MacDonald, B. F. *Can. J. Chem.* **1967**, *45*, 739-740.
80. Erdtman, H.; Harmatha, J. *Phytochemistry.* **1979**, *18*, 1495-1500.
81. Russel, G. B. *Phytochemistry.* **1975**, *14*, 2708.
82. Cairnes, D. A.; Ekundayo, O.; Kingston, D. G. I. *J. Nat. Prod.* **1980**, *43*, 495-497.
83. Popoff, T.; Theander, O.; Johansson, M. *Physiol. Plant.* **1975**, *34*, 347-356.
84. Weinges, K. *Tetrahedron Lett.* **1960**, *20*, 1-2.
85. Miki, K.; Ito, K.; Sasaya, T. *Mokuzai Gakkaishi.* **1979**, *25*, 665-670.
86. Agrawal, P. K.; Rastogi, R. P. *Phytochemistry.* **1982**, *21*, 1459-1461.
87. Ozawa, S.; Sasaya, T. *Mokuzai Gakkashi.* **1988**, *34*, 851-857.
88. Ekman, R. *Holzforschung.* **1976**, *30*, 79-85.
89. Mujumdar, R. B.; Srinivasan, R.; Venkataraman, K. *Indian J. Chem.* **1972**, *10*, 677-680.
90. King, F. E.; Jurd, L.; King, T. J. *J. Chem. Soc.* **1952**, 17-24.
91. Dahlgren, R. *Nord. J. Bot.* **1983**, *3*, 119-149.
92. Dahlgren, R. M. T. *Bot. J. Linn. Soc.* **1980**, *80*, 91-124.
93. Taylor, D. W. *Bot. Rev.* **1990**, *56*, 279-417.
94. Crane, P. R.; Dilcher, D. L. *Ann. Missouri Bot. Gard.* **1984**, *71*, 384-402.
95. Upchurch, G. R., Jr.; Dilcher, D. L. *U.S. Geol. Surv. Bull.* **1990**, *1915*, 1-55.
96. Crane, P. R.; Friis, E. M.; Pedersen, K. R. *Pl. Syst. Evol.* **1989**, *165*, 211-226.
97. Knoblock, E.; Mai, D. H. *Feddes Repertorium.* **1984**, *95*, 3-41.
98. Jerzykiewicz, T.; Sweet, A. R. *Canad. J. Earth Sci.* **1986**, *23*, 1356-1374.
99. Vozenin-Serra, C.; Pons, D. *Palaeontographica.* **1990**, *216B*, 107-127.
100. Jarzen, D. M.; Norris, G. *GeoScience and Man.* **1975**, *11*, 47-60.
101. Friis, E. M.; Skarby, A. *Ann. Bot.* **1982**, *50*, 569-583.
102. Rouse, G. E.; Hopkins, W. S., Jr.; Piel, K. M. In *Symposium on Palynology of the Late Cretaceous and Early Tertiary*; Kosanke, R. M., Cross, A. T. Eds.; Geol. Soc. Amer. Spec. Pap.: 1970; Vol. 127; pp 213-246.
103. Chmura, C. A. *Palaeontographica.* **1973**, *141B*, 89-171.
104. Krutzsch, W. *Pollen et Spores.* **1969**, *11*, 397-424.
105. Stone, J. F. *Bull. Amer. Paleontol.* **1973**, *64*, 1-135.
106. Mai, D. H.; Gregor, H.-J. *Feddes Repertorium.* **1982**, *93*, 405-435.
107. Wheeler, E. A. *Amer. J. Bot.* **1991**, *78*, 658-671.

108. Jones, J. H.; Dilcher, D. L. *Amer. J. Bot.* **1980**, *67*, 959-967.
109. Manchester, S. R.; Dilcher, D. L.; Tidwell, W. D. *Amer. J. Bot.* **1986**, *73*, 156-160.
110. Wheeler, E. A.; Matten, L. C. *Bot. Gaz.* **1977**, *138*, 112-118.
111. Muller, J. *Bot. Rev.* **1981**, *47*, 1-142.
112. Melchior, R. C. *Sci. Publ. Sci. Mus. Minnesota.* **1976**, *3*, 1-18.
113. Christophel, D. C.; Basinger, J. F. *Nature.* **1982**, *296*, 439-441.
114. Tiffney, B. H.; Barghoorn, E. S. *Rev. Palaeobot. Palynol.* **1976**, *22*, 169-191.
115. Kupchan, S. M.; Britton, R. W.; Ziegler, M. F.; Gilmore, C. J.; Restivo, R. J.; Bryan, R. F. *J. Amer. Chem. Soc.* **1973**, *95*, 1335-1336.
116. Gottlieb, O. R.; Maia, J. G. S.; De S. Ribeiro, M. N. *Phytochemistry.* **1976**, *15*, 773-774.
117. Gottlieb, O. R. *Progr. Chem. Org. Nat. Prod.* **1978**, *35*, 1-72.
118. Achenbach, H.; Groß, J.; Dominguez, X. A.; Cano, G.; Star, J. V.; Brussolo, L. D. C.; Muñoz, G.; Salgado, F.; López, L. *Phytochemistry.* **1987**, *26*, 1159-1166.
119. Nakajima, K.; Taguchi, H.; Ikeya, Y.; Endo, T.; Yosioka, I. *Yakugaku Zasshi.* **1983**, *103*, 743-749.
120. Urzúa, A.; Freyer, A. J.; Shamma, M. *Phytochemistry.* **1987**, *26*, 1509-1511.
121. Lopes, L. M. X.; Yoshida, M.; Gottlieb, O. R. *Phytochemistry.* **1984**, *23*, 2647-2652.
122. Lopes, L. M. X.; Yoshida, M.; Gottlieb, O. R. *Phytochemistry.* **1984**, *23*, 2021-2024.
123. Kupchan, S. M.; Stevens, K. L.; Rohlfing, E. A.; Sickles, B. R.; Sneden, A. T.; Miller, R. W.; Bryan, R. F. *J. Org. Chem.* **1978**, *43*, 586-590.
124. Ito, K.; Ichino, K.; Iida, T.; Lai, J. *Phytochemistry.* **1984**, *23*, 2643-2645.
125. Barata, L. E. S.; Baker, P. M.; Gottlieb, O. R.; Rùveda, E. A. *Phytochemistry.* **1978**, *17*, 783-786.
126. Holloway, D. M.; Scheinmann, F. *Phytochemistry.* **1973**, *12*, 1503-1505.
127. EI-Feraly, F. S.; Cheatham, S. F.; Breedlove, R. L. *J. Nat. Prod.* **1983**, *46*, 493-499.
128. Diaz, D. P. P.; Yoshida, M.; Gottlieb, O. R. *Phytochemistry.* **1980**, *19*, 285-288.
129. Lopes, M. N.; da Silva, M. S.; Barbosa-Filho, J. M.; Ferreira, Z. S.; Yoshida, M.; Gottlieb, O. R. *Phytochemistry.* **1986**, *25*, 2609-2612.
130. Tsuruga, T.; Ebizuka, Y.; Nakajima, J.; Chun, Y.-T.; Noguchi, H.; Iitaka, Y.; Sankawa, U. *Tetrahedron Lett.* **1984**, *25*, 4129-4132.
131. Hayashi, T.; Thomson, R. H. *Phytochemistry.* **1975**, *14*, 1085-1087.
132. Daghlian, C. P. *Bot. Rev.* **1981**, *47*, 517-555.
133. Crepet, W. L. *Rev. Palaeobot. Palynol.* **1978**, *25*, 241-252.
134. Jacobs, B. F.; Kabuye, C. H. S. *Rev. Palaeobot. Palynol.* **1989**, *59*, 67-76.
135. Nambudiri, E. M. V.; Tidwell, W. D.; Smith, B. N.; Hebbert, N. P. *Nature.* **1978**, *276*, 816-817.
136. Patra, A.; Mitra, A. K. *Indian J. Chem.* **1979**, *17B*, 412-414.
137. Whiting, D. A. *Nat. Prod. Rep.* **1987**, *4*, 499-525.

138. Yamamura, S.; Niwa, M.; Nonoyama, M.; Terada, Y. *Tetrahedron Lett.* **1978**, *49*, 4891-4894.
139. Yamamura, S.; Niwa, M.; Terada, Y.; Nonoyama, M. *Bull. Chem. Soc. Jpn.* **1982**, *55*, 3573-3579.
140. Ford, C. W.; Hartley, R. D. *J. Sci. Food Agric.* **1990**, *50*, 29-43.
141. Hartley, R. D.; Ford, C. W. In *Plant Cell Wall Polymers. Biogenesis and Biodegradation*; Lewis, N. G., Paice, M. G. Eds.; ACS Symposium Series: Washington, DC, 1989; Vol. 399; pp 137-145.
142. Setchell, K. D. R.; Lawson, A. M.; Mitchell, F. L.; Adlercreutz, H.; Kirk, D. N.; Axelson, M. *Nature.* **1980**, *287*, 740-742.
143. Setchell, K. D. R.; Lawson, A. M.; Conway, E.; Taylor, N. F.; Kirk, D. N.; Cooley, G.; Farrant, R. D.; Wynn, S.; Axelson, M. *Biochem. J.* **1981**, *197*, 447-458.
144. Axelson, M.; Setchell, K. D. R. *FEBS Let.* **1981**, *123*, 337-342.
145. Adlercreutz, H. In *Nutrition, Toxicity, and Cancer*; Rowland, I. R. Ed.; CRC Press: Boca Raton, Fl, 1991; pp 137-195.
146. Borriello, S. P.; Setchell, K. D. R.; Axelson, M.; Lawson, A. M. *J. Appl. Bacteriol.* **1985**, *58*, 37-43.
147. Bannwart, C.; Aldercreutz, H.; Wähälä, K.; Brunow, G.; Hase, T. *Clinica Chimica Acta.* **1989**, *180*, 293-302.

RECEIVED May 2, 1994

Author Index

Bisseret, Philippe, 31
Brassell, Simon C., 2
Buntel, Christopher J., 44
Davin, Laurence B., 202
Feldlaufer, Mark F., 126
Griffin, John H., 44
Kerwin, James L., 163
Lewis, Norman G., 202

Morré, D. James, 142
Nes, W. David, 55
Parish, Edward J., 109
Patterson, Glenn W., 90
Rohmer, Michel, 31
Svoboda, James A., 126
Venkatramesh, M., 55
Weirich, Gunter F., 126

Affiliation Index

Agricultural Research Service, 126
Auburn University, 109
Ecole Nationale Supérieure de Chimie
  de Mulhouse, 31
Indiana University, 2
Purdue University, 142

Stanford University, 44
Texas Tech University, 55
U.S. Department of Agriculture, 126
University of Maryland, 90
University of Washington, 163
Washington State University, 202

Subject Index

A

Absolute stereochemistry, determination,
  57,59
Acid-catalyzed cyclopropyl ring opening
  of cycloartenol, hypothetical
  mechanism, 78f,79
S-Adenosyl-L-methionine, binding
  model, 81,83f
S-Adenosyl-L-methionine C-24 methyl
  transferases, 76–78
Algae, lignan–neolignan distribution, 216
Algal phyla, sterols, 91,93f,94t,95
Analytical approaches in isopentenoid
  geochemistry
  carbon isotope composition, 11,12f
  compound identification, 10–11
Angiosperms, lignan–neolignan
  distribution, 228–239

Animals, sterol identification,
  101,103,104f
Ants, sterol metabolism, 131,132f,133
Archaebacteria, fatty acid occurrence,
  174f,175–176
Asymmetric carbon atom, description, 57
Asymmetric molecule, enantiomorphic
  forms, 57
Asymmetry, 56–57,58f

B

Bacteria, synthesis, 65
Bacteriohopane skeleton formation,
  pentacyclic C_{35}, See Pentacyclic C_{35}
  bacteriohopane skeleton formation
Bees, sterol metabolism, 131,132f,133
Beetles, sterol metabolism, 133–134

247

Production: Meg Marshall
Indexing: Deborah H. Steiner
Acquisition: Rhonda Bitterli
Cover design: Amy Hayes

Printed and bound by Maple Press, York, PA

Bestsellers from ACS Books

The ACS Style Guide: A Manual for Authors and Editors
Edited by Janet S. Dodd
264 pp; clothbound ISBN 0–8412–0917–0; paperback ISBN 0–8412–0943–X

The Basics of Technical Communicating
By B. Edward Cain
ACS Professional Reference Book; 198 pp;
clothbound ISBN 0–8412–1451–4; paperback ISBN 0–8412–1452–2

Chemical Activities (student and teacher editions)
By Christie L. Borgford and Lee R. Summerlin
330 pp; spiralbound ISBN 0–8412–1417–4; teacher ed. ISBN 0–8412–1416–6

Chemical Demonstrations: A Sourcebook for Teachers,
Volumes 1 and 2, Second Edition
Volume 1 by Lee R. Summerlin and James L. Ealy, Jr.;
Vol. 1, 198 pp; spiralbound ISBN 0–8412–1481–6;
Volume 2 by Lee R. Summerlin, Christie L. Borgford, and Julie B. Ealy
Vol. 2, 234 pp; spiralbound ISBN 0–8412–1535–9

Chemistry and Crime: From Sherlock Holmes to Today's Courtroom
Edited by Samuel M. Gerber
135 pp; clothbound ISBN 0–8412–0784–4; paperback ISBN 0–8412–0785–2

Writing the Laboratory Notebook
By Howard M. Kanare
145 pp; clothbound ISBN 0–8412–0906–5; paperback ISBN 0–8412–0933–2

Developing a Chemical Hygiene Plan
By Jay A. Young, Warren K. Kingsley, and George H. Wahl, Jr.
paperback ISBN 0–8412–1876–5

Introduction to Microwave Sample Preparation: Theory and Practice
Edited by H. M. Kingston and Lois B. Jassie
263 pp; clothbound ISBN 0–8412–1450–6

Principles of Environmental Sampling
Edited by Lawrence H. Keith
ACS Professional Reference Book; 458 pp;
clothbound ISBN 0–8412–1173–6; paperback ISBN 0–8412–1437–9

Biotechnology and Materials Science: Chemistry for the Future
Edited by Mary L. Good (Jacqueline K. Barton, Associate Editor)
135 pp; clothbound ISBN 0–8412–1472–7; paperback ISBN 0–8412–1473–5

For further information and a free catalog of ACS books, contact:
American Chemical Society
Distribution Office, Department 225
1155 16th Street, NW, Washington, DC 20036
Telephone 800–227–5558

DATE DUE

GAYLORD			PRINTED IN U.S.A.